WITHDRAWN

Well-Posedness for General
2 × 2 Systems of
Conservation Laws

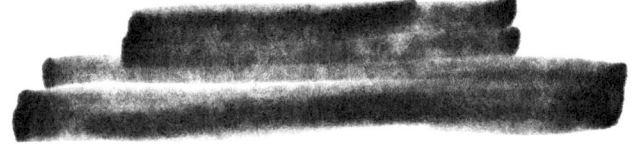

Memoirs
of the
American Mathematical Society

Number 801

Well-Posedness for General
2 × 2 Systems of
Conservation Laws

Fabio Ancona
Andrea Marson

May 2004 • Volume 169 • Number 801 (second of 4 numbers) • ISSN 0065-9266

American Mathematical Society
Providence, Rhode Island

2000 *Mathematics Subject Classification.* Primary 35L65.

Library of Congress Cataloging-in-Publication Data

Well-posedness for general 2 × 2 systems of conservation laws / Fabio Ancona, Andrea Marson.
    p. cm. — (Memoirs of the American Mathematical Society, ISSN 0065-9266 ; no. 801)
  "Volume 169, number 801 (second of 4 numbers)."
  Includes bibliographical references.
  ISBN 0-8218-3435-5 (alk. paper)
  1. Conservation laws (Mathematics).   I. Ancona, Fabio, 1964– .   II. Marson, Andrea, 1968– . III. Series.

QA3.A57   no. 801
[QA377]
510 s—dc22
[515′.353]   2003070915

Memoirs of the American Mathematical Society

This journal is devoted entirely to research in pure and applied mathematics.

Subscription information. The 2004 subscription begins with volume 167 and consists of six mailings, each containing one or more numbers. Subscription prices for 2004 are $583 list, $466 institutional member. A late charge of 10% of the subscription price will be imposed on orders received from nonmembers after January 1 of the subscription year. Subscribers outside the United States and India must pay a postage surcharge of $31; subscribers in India must pay a postage surcharge of $43. Expedited delivery to destinations in North America $35; elsewhere $130. Each number may be ordered separately; *please specify number* when ordering an individual number. For prices and titles of recently released numbers, see the New Publications sections of the *Notices of the American Mathematical Society*.

Back number information. For back issues see the *AMS Catalog of Publications*.

Subscriptions and orders should be addressed to the American Mathematical Society, P. O. Box 845904, Boston, MA 02284-5904, USA. *All orders must be accompanied by payment.* Other correspondence should be addressed to 201 Charles Street, Providence, RI 02904-2294, USA.

Copying and reprinting. Individual readers of this publication, and nonprofit libraries acting for them, are permitted to make fair use of the material, such as to copy a chapter for use in teaching or research. Permission is granted to quote brief passages from this publication in reviews, provided the customary acknowledgment of the source is given.

Republication, systematic copying, or multiple reproduction of any material in this publication is permitted only under license from the American Mathematical Society. Requests for such permission should be addressed to the Acquisitions Department, American Mathematical Society, 201 Charles Street, Providence, Rhode Island 02904-2294, USA. Requests can also be made by e-mail to reprint-permission@ams.org.

Memoirs of the American Mathematical Society is published bimonthly (each volume consisting usually of more than one number) by the American Mathematical Society at 201 Charles Street, Providence, RI 02904-2294, USA. Periodicals postage paid at Providence, RI. Postmaster: Send address changes to Memoirs, American Mathematical Society, 201 Charles Street, Providence, RI 02904-2294, USA.

© 2004 by the American Mathematical Society. All rights reserved.
This publication is indexed in *Science Citation Index*®, *SciSearch*®, *Research Alert*®, *CompuMath Citation Index*®, *Current Contents*®/*Physical, Chemical & Earth Sciences*.
Printed in the United States of America.

∞ The paper used in this book is acid-free and falls within the guidelines established to ensure permanence and durability.
Visit the AMS home page at http://www.ams.org/

10 9 8 7 6 5 4 3 2 1   09 08 07 06 05 04

Acknowledgments

The authors wish to thank Prof. Tommaso Ruggeri for suggesting the problem and Prof. Alberto Bressan for useful discussions.

Fabio Ancona and Andrea Marson

Contents

1. Introduction	1
2. Preliminaries	10
3. Outline of the proof	18
4. The algorithm	37
4.1. Approximate flux function	38
4.2. Approximate Hugoniot curves	40
4.3. Interpolations	41
4.4. Approximate mixed-curves	46
4.5. Approximate elementary curves	48
5. Basic interaction estimates	51
5.1. Strength estimates for interaction between wave-fronts of the same family	52
5.2. Speed estimates for interaction between wave-fronts of the same family	53
5.3. Interaction between waves of different families	55
5.4. Strength estimates for two coupled Riemann problems	56
6. Bounds on the total variation and on the interaction potential	60
6.1. Wave-front basic notations	63
6.2. Main estimates	72
6.3. Estimates on the increase of the interaction potential	86
6.4. Proof of Proposition 3.1	95
7. Estimates on the number of discontinuities	97
7.1. Proof of Proposition 3.2	98
7.2. Proof of Proposition 3.3	100
7.3. Proof of Proposition 3.4	108
8. Estimates on shift differentials	112
8.1. Estimates on shifting interactions	114
8.2. Basic notations	119
8.3. Main estimates	125
8.4. Estimates on the increase of the shift-interaction functional	135
8.5. Proof of Propositions 3.5-3.6	151

9. Completion of the proof   157

10. Conclusion   162

Bibliography   169

Abstract

We consider the Cauchy problem for a strictly hyperbolic 2×2 system of conservation laws in one space dimension

$$(1) \qquad u_t + [F(u)]_x = 0, \qquad u(0,x) = \bar{u}(x),$$

which is neither linearly degenerate nor genuinely non-linear. We make the following assumption on the characteristic fields. If $r_i(u)$, $i = 1, 2$, denotes the i-th right eigenvector of $DF(u)$ and $\lambda_i(u)$ the corresponding eigenvalue, then the set $\{u : \nabla \lambda_i \cdot r_i(u) = 0\}$ is a smooth curve in the u-plane that is transversal to the vector field $r_i(u)$.

Systems of conservation laws that fulfill such assumptions arise in studying elastodynamics or rigid heat conductors at low temperature.

For such systems we prove the existence of a closed domain $\mathcal{D} \subset L^1$, containing all functions with sufficiently small total variation, and of a uniformly Lipschitz continuous semigroup $S : \mathcal{D} \times [0, +\infty) \to \mathcal{D}$ with the following properties. Each trajectory $t \mapsto S_t \bar{u}$ of S is a weak solution of (1). Viceversa, if a piecewise Lipschitz, entropic solution $u = u(t, x)$ of (1) exists for $t \in [0, T]$, then it coincides with the trajectory of S, i.e. $u(t, \cdot) = S_t \bar{u}$.

This result yields the uniqueness and continuous dependence of weak, entropy-admissible solutions of the Cauchy problem (1) with small initial data, for systems satysfying the above assumption.

Received by the editor June 8, 2000, and in revised form November 7, 2002.

2000 *Mathematics Subject Classification.* 35 L 65.

Key words and phrases. Conservation Laws, Non Genuine non-Linearity, Lipschitz Semigroup.

This work has been partially supported by TMR project HCL ERBFMRXCT960033.

1. Introduction

Consider the Cauchy problem for a strictly hyperbolic 2×2 system of conservation laws in one space dimension

$$u_t + F(u)_x = 0, \tag{1.1}$$

$$u(0, x) = \overline{u}(x). \tag{1.2}$$

Let the flux function $F : \Omega \mapsto \mathbb{R}^2$ be a smooth vector field defined on a neighborhood of the origin $\Omega \subset \mathbb{R}^2$. Denote by $\lambda_1(u) < \lambda_2(u)$ the eigenvalues of the Jacobian matrix $A(u) \doteq DF(u)$, and let $\{r_1(u), r_2(u)\}$ be a basis of right eigenvectors.

We assume that system (1.1) may admit non genuinely nonlinear (NGNL) characteristic fields, i.e. characteristic fields that are neither genuinely nonlinear (GNL) nor linearly degenerate (LD) in the sense of Lax [**La, Sm**]. Instead, letting

$$r_k \bullet \phi(u) \doteq \lim_{h \to 0} \frac{\phi(u + h\, r_k(u)) - \phi(u)}{h} \tag{1.3}$$

denote the directional derivative of a function $\phi = \phi(u)$ in the direction of the characteristic field $r_k(u)$, we make the following assumption.

(A) If the k-th characteristic family is NGNL, then the set

$$\Gamma_k \doteq \{\, u \in \Omega;\ r_k \bullet \lambda_k(u) = 0 \,\} \tag{1.4}$$

is a smooth curve in the u-space that is transversal to the vector field r_k (see figure 1.1), and there holds

$$r_k \bullet (r_k \bullet \lambda_k)(u) \neq 0 \qquad \forall\, u \in \Gamma_k. \tag{1.5}$$

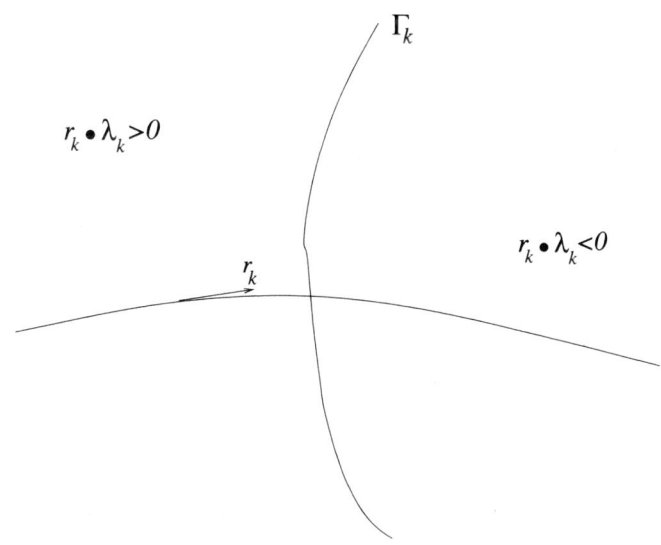

FIGURE 1.1. The set Γ_k

Systems of conservations laws that fulfill this type of assumptions physically arise in several contexts, in particular in studying elastodynamic (e.g. see [**DP3**]) or rigid heat conductors at low temperature. An example is given by the generalized Cattaneo model proposed by T. Ruggeri and co-workers [**RMS1, RMS2**] to describe the heat propagation in high-purity crystals (IIe, NaF, Bi):

$$(1.6) \qquad \big[\rho\, e\big]_t + q_x = 0, \qquad \big[\alpha\, q\big]_t + \nu_x = -\frac{\nu'}{\kappa}q.$$

Here ρ denotes the constant mass density, $q = q(t,x)$ the one-dimensional heat flux, $e = e(\theta)$ the internal energy depending on the absolute temperature $\theta = \theta(t,x)$, while $\alpha = \alpha(\theta)$ and $\nu = \nu(\theta)$ are constitutive scalar functions (depending on the material and prescribed by the second law of thermodynamic), ν' is the derivative of ν with respect to θ, and $\kappa = \kappa(\theta)$ is the heat conductivity. Equation $(1.6)_1$ represents the energy balance, while $(1.6)_2$ is a generalization of the Maxwell-Cattaneo equation. In the range of temperature in which can be observed the so called "second sound" effect the heat conductivity becomes very high, and thus the system (1.6) can be assumed to be homogeneous. A direct computation of the corresponding characteristic speeds shows that such a system shares the property (**A**).

We recall that, for hyperbolic systems of conservation laws, to achieve uniqueness of solutions it is necessary to consider weak solutions that satisfy some additional admissibility condition, possibly motivated by physical considerations. In [**Li1, Li2**] T.P. Liu proposed an admissibility criterion valid for general $n \times n$ systems of conservation laws having non genuinely nonlinear characteristic fields r_i with the property that the directional derivative $r_i \bullet \lambda_i$ vanishes on a finite, disjoint union of $n-1$-dimensional manifolds. This criterion, in the case of a system where $r_i \bullet \lambda_i = 0$ on a single connected hypersurface in the u-space, is equivalent to the classical stability condition introduced by Lax:

(**L**) A shock connecting the left state u^L and the right state u^R, travelling with speed s is an admissible discontinuity of the i-th family if
$$\lambda_i(u^L) \geq s \geq \lambda_i(u^R).$$

Thus, for a system satisfying the assumption (**A**), with an *entropy weak solution* of the Cauchy problem (1.1)-(1.2) we shall always mean a weak solution of (1.1)-(1.2) that has admissible discontinuities in the sense of Lax.

For systems with NGNL characteristic fields of the type considered in [**Li1, Li2**], T.P. Liu established the existence of a unique, self-similar, entropy weak solution of the Riemann problem with initial data

$$u(0,x) = \begin{cases} u^L & \text{if} \quad x < 0, \\ u^R & \text{if} \quad x > 0, \end{cases}$$

with $|u^R - u^L|$ sufficiently small. The existence of global weak solutions of the Cauchy problem (1.1)-(1.2) with small BV data was first obtained in [**Li3**], using the Glimm scheme. An alternative method for constructing solutions of the Cauchy problem, as limit of piecewise constant approximations defined by a front tracking algorithm, is developed in [**AM4**] for $n \times n$ systems with NGNL characteristic fields whose directional derivative $r_i \bullet \lambda_i$ vanishes on a single hypersurface in the u-space. In the present paper, following the approach introduced by A. Bressan, R.M. Colombo in [**BC1**], we construct a new front tracking algorithm, valid for 2×2 systems with NGNL characteristic fields satisfying the assumption (**A**), which yields a Cauchy sequence of flow maps S_n whose trajectories are approximate solutions of (1.1) that depend on the initial conditions in a uniformly Lipschitz continuous way. Therefore, the entire sequence S_n converges, on a set $\mathcal{D}$ of integrable functions with small total variation, to a globally Lipschitz flow S whose trajectories are weak solutions of (1.1). More precisely, the following holds.

THEOREM 1. *Let F be a smooth map from a neighborhood of the origin $\Omega \subset \mathbb{R}^2$ into $\mathbb{R}^2$. Assume that the system (1.1) is strictly hyperbolic and that every characteristic family which is neither GNL nor LD satisfy the assumption* (**A**). *Then, there exist a closed domain $\mathcal{D} \subset \mathbb{L}^1(\mathbb{R}; \mathbb{R}^2)$, constants $L, \overline{\delta} > 0$, and a continuous semigroup $S : [0, \infty[\times \mathcal{D} \mapsto \mathcal{D}$ with the properties:*

(i) *Every function $\overline{u} \in \mathbb{L}^1$ with $Tot.Var.(\overline{u}) \leq \overline{\delta}$ lies in $\mathcal{D}$.*
(ii) *For all $\overline{u} \in \mathcal{D}$, one has $S_0 \overline{u} = \overline{u}$, $S_t S_s \overline{u} = S_{t+s} \overline{u}$, $t, s \geq 0$.*
(iii) *For all $\overline{u}, \overline{v} \in \mathcal{D}$, $t, s \geq 0$, one has*

$$\left\| S_t \overline{u} - S_s \overline{v} \right\|_{\mathbb{L}^1} \leq L \left(|t - s| + \left\| \overline{u} - \overline{v} \right\|_{\mathbb{L}^1} \right).$$

(iv) *If $\overline{u} \in \mathcal{D}$ is piecewise constant, then for $t > 0$ sufficiently small the function $u(t, \cdot) \doteq S_t \overline{u}$ coincides with the solution of (1.1)-(1.2) obtained by piecing together the entropy-admissible, self-similar solutions of the corresponding Riemann problems.*
(v) *Each trajectory $t \mapsto u(t, \cdot) \doteq S_t \overline{u}$ yields a weak solution to the Cauchy problem (1.1)-(1.2).*

Following [**B4**], we say that a map with the properties (i)-(iv) is a *Standard Riemann Semigroup* (SRS) generated by (1.1). The existence of such a map plays a key role in proving the uniqueness of entropy weak solutions for a given Cauchy-problem with small BV data. Indeed, with the same technique in [**B4**] one can show that a map S with the properties (i)-(iv) is necessarily unique (up to the domain). Besides, by similar arguments as in [**BG, BL**], one can prove that any entropy weak solution to (1.1)-(1.2), satisfying some mild assumptions either on the oscillation or on the local boundedness of the total variation, must coincide with the trajectory of S. Results in this direction can be found in [**BLP**]. A further consequence of the existence of a SRS that is worth mentioning is the derivation of sharp error estimates on approximate solutions constructed by the Glimm scheme as in the case of genuinely nonlinear systems (cfr. [**BM2**]).

The existence of a map S with the properties (i)-(v), under the assumption that the characteristic fields were GNL or LD, was first proved by A. Bressan, R.M. Colombo in [**BC1**], for systems of two equations, and by A. Bressan, G. Crasta, B. Piccoli in [**BCP**] and, more recently, by A. Bressan, T.P. Liu, T. Yang, in [**BLY**], for $n \times n$ systems. Our main theorem here, is a first step towards the extensions of these results to the more general class of NGNL systems considered by T.P. Liu in [**Li1, Li2, Li3**].

We recall that the standard procedure to construct a front tracking approximate solution (see [**B2, BJ, B5**]) starts at time $t = 0$ by taking a piecewise constant approximation of the initial data. The resulting Riemann problems are then solved within the class of piecewise constant functions by using an approximate Riemann solver. Next, one tracks the outgoing fronts until the first time two waves interact. The corresponding Riemann problems are then solved applying again the approximate Riemann solver, etc ... The main source of technical difficulty in this construction stems from the fact that the number of lines of discontinuity may approach infinity in finite time. To overcome such a difficulty, a simplified procedure for solving the Riemann problem is usually adopted by introducing "non-physical" fronts that propagate with a constant speed greater than all characteristic speeds. In particular, in the case of NGNL systems considered in [**AM4**], two different simplified procedures to approximately solving the Riemann problem were introduced in order to control the number of wave-fronts of the NGNL characteristic families. Clearly, the approximate solution generated by a construction of this type will not depend continuously on the initial data.

Instead, the algorithm developed in the present paper, as the one proposed in [**BC1**], allows to keep finite the total number of fronts without introducing any non-physical wave. This is possible thanks to the existence of a coordinate system of Riemann invariants, but it is achieved at the price of a quite technical construction of suitable approximations of the Hugoniot curves.

The main features of our wave-front tracking algorithm are:

1. Every Riemann Problem is solved by the same procedure. In particular, we never introduce any non-physical wave-front.

2. Centered rarefaction waves are cut along a fixed grid of step size ε and approximated by a corresponding piecewise constant rarefaction fan.

3. Shock waves of small amplitude coincide with rarefactions.

A Riemann Solver with the above properties, valid for GNL or LD systems, is obtained in [**BC1**] by constructing ε-approximate solutions in which

- shocks of strength $|\sigma| \geq 2\sqrt{\varepsilon}$ satisfy the exact Rankine-Hugoniot equations;
- shocks of strength $|\sigma| \leq \sqrt{\varepsilon}$ are rarefactions;
- shocks of strength $|\sigma| \in \left[\sqrt{\varepsilon}, 2\sqrt{\varepsilon}\right]$ connect a left and right state lying on a smooth interpolation between the shock and the rarefaction curve.

Unfortunately, in the case of NGNL systems, approximate solutions of this type fail to produce a Riemann Solver with the properties **1-3**, depending continuously on the initial data. This is the consequence of the different structure of the elementary waves contained in the general solution of the Riemann Problem. In fact, for such systems, the exact solution to the Riemann Problem may consist of composed waves made of one-sided contact discontinuities adjacent to rarefaction waves. Here (and throughout the paper), following [**Li1, Li2**] we use with a slight abuse of notation the term "contact-discontinuity" to indicate a (truly compressive) Lax-admissible shock that travels with the characteristic speed of its left or right state, although in gas dynamics it is customary to reserve this term to not compressive shocks (that do not change the entropy). Then, to guarantee the stability of the composite wave pattern of a piecewise constant function that approximates a composed wave of this type, we need to modify the Rankine-Hugoniot equations for shocks of any amplitude. This fact, in turn, forces us to introduce three different types of interpolations between the rarefaction and the approximate shock curves defined in connection with the modified Rankine-Hugoniot conditions.

A basic step in the proof of the global existence of approximate solutions generated by wave-front tracking consists in deriving a-priori bounds on the total variation. As customary, these a-priori estimates are obtained using a functional Q that measures the *potential interaction* of waves in the solution.

In particular, for NGNL systems, T.P. Liu [**Li3**] introduced a functional Q in which the amount of potential interaction between waves of the same family is proportional to the product of the strength of the waves and of their angle. This choice is motivated by the fact that no interaction between two waves is expected when their angle is zero. On the other hand, this corresponds to consider an interaction functional that is fourth order w.r.t. the strength of the waves of the same family. In fact, the angle between two waves of strength $|\sigma'|$, $|\sigma''|$, belonging to the same family, is proportional either to $|\sigma'\sigma''|$ or to $\max\{|\sigma'|^2, |\sigma''|^2\}$.

In our analysis, however, to establish the Lipschitz continuous dependence of the approximate solutions on the initial data, it is essential to have at disposal an interaction-potential that is second order w.r.t. the strength of all waves. For this reason we define here the amount of potential interaction between *any* couple of wave-fronts in the approximate solution as proportional to the product of their strengths, no matter if they belong to the same family or not. More precisely, for any given approximate solution u generated by wave-front tracking, with several wave-fronts, say located at $x_\alpha = x_\alpha(t)$, $\alpha = 1, \ldots, N$, calling $|\sigma_i^\alpha|$ the strength of the wave at x_α belonging to the i-th family, we define the interaction potential $Q(t) \doteq Q(u(t))$ by setting

$$(1.7) \quad Q(t) = K_1 \left[\sum_{x_\alpha < x_\beta} |\sigma_2^\alpha \cdot \sigma_1^\beta| + \sum_{i=1,2} \sum_{\sigma_i^\alpha \cdot \sigma_i^\beta < 0} |\sigma_i^\alpha \cdot \sigma_i^\beta| \right] + \sum_{i=1,2} \sum_{\sigma_i^\alpha \cdot \sigma_i^\beta > 0} |\sigma_i^\alpha \cdot \sigma_i^\beta|$$

where K_1 is a suitable constant > 1 in the case where at least one of the family is NGNL (while $K_1 = 1$ in the classical case where both families are GNL or LD). Notice that, since we are assuming by this definition that the amount of interaction-potential between two wave-fronts of the same family is non zero no matter which angle they make, one would expect that the global interaction-potential may increase after an interaction that produces a (possibly composed) wave containing a piecewise constant rarefaction fan. However, interactions of this type always involve waves of different families or waves of the same family having an opposite sign (and belonging to a NGNL family). Therefore, observing that the total strength of i-rarefaction waves produced by such an interaction is dominated by the total strength of j-waves (or of i-waves of opposite sign w.r.t. the outgoing waves) involved in the interaction, one can always control the increase of Q due to the new rarefaction waves by assigning a sufficiently large weight K_1 to the amount of interaction-potential of the interacting waves.

We point out that a major technical difficulty that arises here in establishing an a-priori bound on the total variation of an approximate solutions u by using the Glimm-type functional Tot.Var.$(u) + Q(u)$, stems from the fact that our front-tracking algorithm yields a class of "bad interactions" involving shocks of an i-th NGNL family, which produce an increase of $Q(u)$ of a magnitude which is not comparable with *any* order of the strength of the incoming fronts. To overcome this problem we will perform an accurate analysis of the future history of the wave-fronts originated by such bad interactions which allows to derive a uniform a-priori bound on the growth of $Q(u)$ and of $\text{Tot.Var.}(u) + Q(u)$ caused by the occurrence of these interactions.

The proof of the Lipschitz continuous dependence on the initial data of the approximate solutions constructed by our algorithm, is based on the same technique developed in [**BC1**]. The basic idea consists in "differentiating" a family of approximate solutions w.r.t. a parameter which determines the location of the jumps. More precisely, consider a piecewise constant approximate solution u having

$$(1.8) \qquad u(0,x) = \sum_{\alpha=1}^{N_0} \overline{\omega}_\alpha \cdot \chi_{]\overline{x}_{\alpha-1}, \overline{x}_\alpha[}(x),$$

as initial data. Here χ_I denotes the characteristic function of the interval I. Then one can construct a one-parameter family of initial data $u^\theta(0,\cdot)$ obtained from $u(0,\cdot)$ by shifting the positions of the jumps $\overline{x}_0, \overline{x}_1, \ldots, \overline{x}_{N_0}$, at the constant rates $\overline{\xi}_0, \overline{\xi}_1, \ldots, \overline{\xi}_{N_0}$, i.e. letting

$$u^\theta(0,x) = \sum_{\alpha=1}^{N_0} \overline{\omega}_\alpha \cdot \chi_{]x^\theta_{\alpha-1}, x^\theta_\alpha[}(x), \qquad x^\theta_\alpha = \overline{x}_\alpha + \theta \overline{\xi}_\alpha, \qquad \theta \in [-a, a].$$

Here the states $\overline{\omega}_\alpha$ remain fixed for all values of the parameter θ. Let $u^\theta = u^\theta(t,x)$ be the corresponding approximate solution. If a is sufficiently small, at any time t where no interaction occurs, the functions $u^\theta(t,\cdot)$ are still obtained from $u(t,\cdot)$ by shifting the positions of its jumps:

$$u^\theta(t,x) = \sum_{\beta=1}^{N_t} \omega_\beta \cdot \chi_{]x^\theta_{\beta-1}(t), x^\theta_\beta[}(x), \qquad x^\theta_\beta = x_\beta + \theta \xi_\beta,$$

for some shift rates $\xi_0, \xi_1, \ldots, \xi_{N_t}$. The $\mathbb{L}^1$-length of the path $\gamma_t : \theta \mapsto u^\theta(t,\cdot)$, is then computed by

$$(1.9) \qquad \|\gamma_t\|_{\mathbb{L}^1} = \int_{-a}^a \sum_\beta \left|\Delta u^\theta(t, x^\theta_\beta)\right| |\xi_\beta| \, d\theta, \qquad \xi_\beta = \frac{dx^\theta_\beta}{d\theta},$$

where $\Delta u^\theta = \omega_{\beta+1} - \omega_\beta$ is the jump of u^θ at x^θ_β.

Thus, to estimate the $\mathbb{L}^1$-distance between two nearby approximate solutions u^{-a}, u^a, one has to provide an a-priori bound on the integrand in (1.9). As soon as an estimate of this type is obtained, a standard argument yields the stability of the approximate solutions. Clearly the integrand in (1.9) can change (as a function of time) only at times where some interaction takes place between wave-fronts in u^θ. Therefore the key point is to study how the summation in (1.9) is changed across each interaction time. As one would expect, the major difference in this analysis, between the standard case of GNL systems and the case of systems which are not GNL, regards the behaviour of the summation in (1.9) at any time where two waves of the same NGNL family interact. We consider the following basic case.

EXAMPLE 1.1. Let $u = u(t,x)$ be a piecewise constant solution, consisting of two shocks of the first family which interact at some time τ. Let $x'(t) < x''(t)$ be the locations of the shocks for $t < \tau$, and assume that the interaction produces an outgoing shock for each i-th family, at $x_i^+(t)$ (see figure 1.2). Call σ', σ'' the sizes of the two incoming waves, and σ_i^+ the size of the outgoing i-shock. Assume that σ', σ'' are both positive.

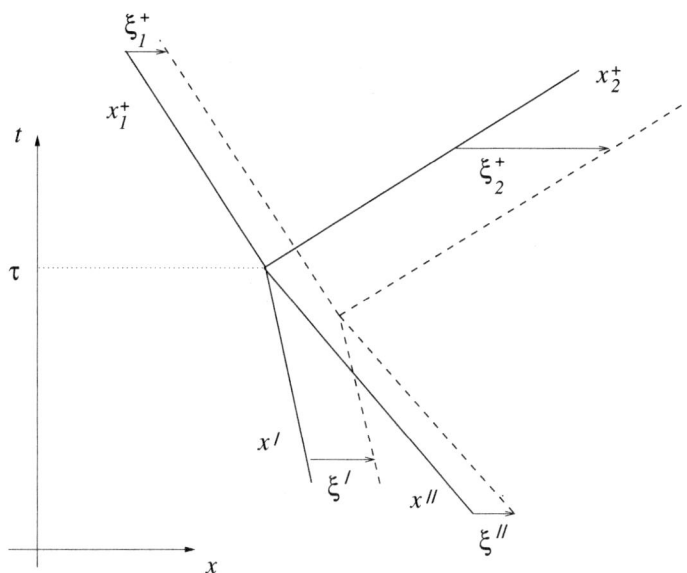

FIGURE 1.2. The waves and the shift rates involved in Example 1.1

Next, consider a family of perturbed solution u^θ, $\theta \in [0,1]$, obtained from u by shifting the position of x', x'' at the constant rates ξ', ξ''. The incoming shocks of u^θ thus occur on the lines $x = x'_\theta(t)$, $x = x''_\theta(t)$, with

$$x'_\theta(t) = x'(t) + \xi' \theta, \qquad x''_\theta(t) = x''(t) + \xi'' \theta, \qquad \theta \in [0,1].$$

For any fixed $\theta \in [0,1]$, after the interaction is taken place (both in u and in u^θ) the perturbed solution u^θ will still contain wave-fronts of exactly the same strength as u. The locations $x_{1,\theta}^+$, $x_{2,\theta}^+$, of these wave-fronts, however, will be shifted by certain amounts, say ξ_1^+, ξ_2^+. An elementary calculation shows that the shift rates of the outgoing i-th shock is

$$(1.10) \qquad \xi_i^+ = \frac{\partial x_{i,\theta}^+}{\partial \theta} = \frac{\xi''(\dot{x}' - \dot{x}_i^+) - \xi'(\dot{x}'' - \dot{x}_i^+)}{\dot{x}' - \dot{x}''}, \qquad i = 1, 2,$$

where $\dot{x}$ denotes the speed of the wave-front at x. By the above arguments, the quantities

$$(1.11) \qquad E^- \doteq |\sigma' \xi'| + |\sigma'' \xi''|, \qquad E^+ \doteq |\sigma_1^+ \xi_1^+| + |\sigma_2^+ \xi_2^+|,$$

provide a measure of the $\mathbb{L}^1$-distance between the two solutions u, u^1 before and after the interaction time τ. Standard interaction estimates on the size and the speed of the outgoing waves imply that

(1.12) $$|\sigma_2^+| + |\sigma_1^+ - (\sigma' + \sigma'')| = \mathcal{O}(1) \cdot |\sigma' \cdot \sigma''| \cdot \max\{|\sigma'|, |\sigma''|\},$$

(1.13) $$\left|\dot{x}_1^+ - \frac{\sigma'\dot{x}' + \sigma''\dot{x}''}{\sigma' + \sigma''}\right| = \mathcal{O}(1) \cdot |\sigma' \cdot \sigma''|,$$

where $\mathcal{O}(1)$ denotes a quantity uniformly bounded by a constant that depends only on the system (1.1). Observe that the shifts ξ_i^+ in (1.10) may be very large compared with ξ', ξ'', because of the small denominator. However, in the case the system (1.1) is GNL, a supplementary bound holds

(1.14) $$|\dot{x}' - \dot{x}''| \geq c(|\sigma'| + |\sigma''|) \qquad \text{for some} \quad c > 0.$$

Thus, from (1.12)-(1.14) we obtain the basic estimate

(1.15) $$E^+ - E^- \leq |\sigma_1^+ \xi_1^+ - \sigma'\xi' - \sigma''\xi''| + |\sigma_2^+ \xi_2^+| = \mathcal{O}(1) \cdot |\sigma' \cdot \sigma''|(|\xi'| + |\xi''|).$$

This estimates indicates that, although the quantity in (1.9) may well increase in time, one can derive a uniform bound on a weighted sum of the form

(1.16) $$\sum_\beta \left|\Delta u^\theta(t, x_\beta^\theta)\right| |\xi_\beta| W^\beta$$

provided that the weights W^β are suitably chosen, depending on the amount of waves in u which approach the wave located at x_β. Clearly this arguments fails for NGNL systems since an estimate as (1.14) no more holds in the case the incoming fronts belong to a NGNL family.

For this reason, it is necessary to provide sharper interaction estimates than the ones in (1.12)-(1.13). Indeed, if the first family (of the incoming front σ', σ'') is NGNL and satisfies the assumption **(A)**, the following holds

(1.17) $$|\sigma_2^+| + |\sigma_1^+ - (\sigma' + \sigma'')| = \mathcal{O}(1) \cdot |\sigma' \cdot \sigma''| \cdot \Big[|\sigma'| + |\sigma''|\Big] \cdot \Big[|\sigma'| + |\sigma''| + |\delta_1|\Big],$$

(1.18) $$\left|\dot{x}_1^+ - \frac{\sigma'\dot{x}' + \sigma''\dot{x}''}{\sigma' + \sigma''}\right| = \mathcal{O}(1) \cdot |\sigma' \cdot \sigma''| \cdot \Big[|\sigma'| + |\sigma''| + |\delta_1|\Big],$$

(1.19) $$|\dot{x}' - \dot{x}''| \geq c' \cdot \Big[|\sigma'| + |\sigma''|\Big] \cdot \Big[|\sigma'| + |\sigma''| + |\delta_1|\Big] \quad \text{for some } c' > 0,$$

where $\delta_1 \doteq \delta_1(u^L)$ denotes the distance between the state u^L on the left of the wave at x', and the curve Γ_1 defined in (1.4). Estimates of this type were first established in [**Li3**] in the case of interacting waves of the same family and of the same sign. A derivation of such estimate in the general case of interacting waves of the same family and of any sign can be found in [**AM2**]. Thanks to (1.17)-(1.19), one can then recover the basic estimate (1.15) also in the case of a system with NGNL fields satisfying the assumption **(A)**.

Relying on (1.15), we can now derive a uniform a-priori bound on a weighted sum as in (1.16), choosing weights of the form

(1.20)
$$W_i^\alpha \doteq \left[1 + K_3\left[\sum_{\substack{j\neq i \\ (x_\alpha - x_\beta)\cdot(i-j)<0}} |\sigma_j^\beta| + \sum_{\sigma_i^\beta \cdot \sigma_i^\alpha < 0} |\sigma_i^\beta|\right] + K_2 \sum_{\sigma_i^\beta \cdot \sigma_i^\alpha > 0} |\sigma_i^\beta|\right] e^{K_4 Q},$$

where K_2, K_3, K_4, ($K_3 > K_2$ in the case where at least one of the family is NGNL) are suitable positive constants, and Q is the interaction potential at (1.7). Here, with σ_i^β (σ_j^β) we denote the size of a wave-front of the i-th (j-th) characteristic family, and W_i^α is the weight assigned to an i-th wave located at x_α. Of course, the weighted sum in (1.16) may well increase after the same type of bad interactions that produce a growth of the interaction potential. Therefore, in order to establish an a-priori bound on the sum in (1.16), we will perform for the integrand in (1.9) the same type of analysis of the wave-fronts originated by such bad interactions that we have developed to control the growth of the interaction potential.

For sake of simplicity, we shall often refer in our analysis throughout the paper to a system in which both characteristic families are NGNL and satisfy the assumption (**A**). Namely, we will consider a 2×2 NGNL system satisfying the following assumption.

(**A'**) For each $k = 1, 2$, the set $\Gamma_k \subset \Omega$ in (1.4) is a smooth curve transversal to the vector filed r_k, and there holds

(1.21) $\quad r_1 \bullet (r_1 \bullet \lambda_1)(u) < 0, \quad \forall\, u \in \Gamma_1$

(1.22) $\quad r_2 \bullet (r_2 \bullet \lambda_2)(u) > 0, \quad \forall\, u \in \Gamma_2.$

The general case in which one of the families is GNL or LD, or both families are NGNL but the second derivatives $r_1 \bullet (r_1 \bullet \lambda_1)$, $r_2 \bullet (r_2 \bullet \lambda_2)$ have the same sign, can be easily covered with an entirely similar analysis and relying on the construction developed in [**BC1**] for 2×2 systems with GNL or LD fields.

The paper is organized as follows. In section 2 we introduce the main notations and we define the elementary curves, associated to each characteristic field, used to define the exact solution of the Riemann Problem. Section 3 contains an outline of the proof of Theorem 1, given as a sequence of Propositions. The technical details involved in the proof of these Propositions are then worked out in Sections 6 to 9, relying on the basic interaction estimates collected in Section 5. The construction of the approximate elementary curves that determine the approximate solution of the Riemann Problem is carried out in Section 4.

Note added. After the completion of the present paper, S. Bianchini and A. Bressan [**BB**] have provided an entirely different proof of the existence of a Lipschitz continuous semigroup of solutions (enjoying the same properties stated in Theorem 1) generated by a general $n \times n$ strictly hyperbolic system of conservation laws, without any assumptions on the nonlinearity of the characteristic fields (and hence including as a particular case the class of 2×2 NGNL systems considered here). Their proof is obtained by studying the evolution of the first order perturbation to a solution of the (artificial) viscous parabolic approximation

$u_t + F(u)_x = \varepsilon\, u_{xx}$ of (1.1), and by establishing uniform (w.r.t. the viscosity coefficient ε) a-priori bounds on its $\mathbb{L}^1$ norm. Although this new approach yields definite $\mathbb{L}^1$ well-posedness theory for the whole class of strictly hyperbolic systems, our construction remains of interest in itself providing an alternative, true hyperbolic, proof of this result for 2×2 systems with NGNL characteristic fields.

2. Preliminaries

Let $\Omega \subseteq \mathbb{R}^2$ be an open neighborhood of the origin and $F : \Omega \to \mathbb{R}^2$ a smooth map. Denote with $\lambda_1(u) < \lambda_2(u)$ the eigenvalues of the Jacobian matrix $A(u) \doteq DF(u)$. Performing a linear change of coordinates in the (t,x)-plane, we can assume that

$$\lambda_1 < 0 < \lambda_2, \qquad 0 < \lambda^{\min} < |\lambda_i|, \tag{2.1}$$

for some constant $\lambda^{\min}$. The directional derivative of a function $\phi = \phi(u)$ in the direction of a vector field $r = r(u)$ is written

$$r \bullet \phi(u) \doteq D_r \phi(u) \doteq \lim_{h \to 0} \frac{\phi\big(u + h\, r(u)\big) - \phi(u)}{h}. \tag{2.2}$$

Since we deal with a 2×2 system, it is convenient to work with a set of Riemann coordinates $v = (v_1, v_2)$ associated with a local diffeomorphism $v \mapsto u(v)$, defined on a neighborhood of zero $\mathcal{V} =]a_1, b_1[\times]a_2, b_2[$. We recall that (v_1, v_2) form a coordinate system of Riemann invariants for (1.1) if and only if each $Dv_i(u)$, $i=1,2$ is a left eigenvector of the Jacobian matrix $A(u) \doteq DF(u)$. We normalize right and left eigenvectors $r_i(u)$, $l_i(u)$, $i=1,2$, of $A(u)$ so that (cfr. [**D**] Section 7.3)

$$\begin{aligned} Dv_i(u) &= l_i(u) \qquad i,j = 1,2, \\ \langle l_i(u), r_j(u) \rangle &= \begin{cases} 1 & \text{if } i=j \\ 0 & \text{if } i \neq j, \end{cases} \qquad \forall\, u, \end{aligned} \tag{2.3}$$

which, in turn, implies

$$\begin{aligned} [r_i, r_j](u) &\doteq r_j \bullet r_i(u) - r_i \bullet r_j(u) \equiv 0, \qquad i,j=1,2, \\ \frac{\partial u}{\partial v_i}(v) &= r_i(u(v)) \qquad \forall\, v \in \mathcal{V}, \quad i=1,2. \end{aligned} \tag{2.4}$$

Because of (2.4), the choice of the parameterization in (2.3) offers, in particular, some technical advantages to prove the convergence of the approximate flux functions $F^{i,\varepsilon}(u)$ that will be introduced in Section 4 (see Proposition 3.1 in [**AM2**]). In the Riemann coordinates, the rarefaction curves $R_i(v)$, $i=1,2$, through a point $v = (v_1, v_2) \in \mathcal{V}$ are naturally parameterized by

$$R_{i,j}(v, \sigma) = \begin{cases} v_i + \sigma & \text{if } j=i, \\ v_j & \text{if } j \neq i, \end{cases} \tag{2.5}$$

where $R_{i,j}(v)$, $j=1,2$, denote the components of $R_i(v)$. On the other hand, since shock and rarefaction curves have second order contact, the Hugoniot curves

$S_i(v)$, $i = 1, 2$, through v can be expressed in the form

$$(2.6) \qquad S_{i,j}(v, \sigma) = \begin{cases} v_i + \sigma & \text{if } j = i, \\ v_j + \widetilde{S}_i(v, \sigma) \cdot \sigma^3 & \text{if } j \neq i, \end{cases}$$

for a suitable smooth function $\widetilde{S}_i(v)$. We shall often use the notations $\lambda_i(v) \doteq \lambda_i(u(v))$, $r_i \bullet \lambda_i(v) \doteq r_i \bullet \lambda_i(u(v))$, for the eigenvalues of $A(u(v))$ and for their derivatives, and let $\lambda_i(v^L, v^R)$ denote the i-th eigenvalue of the averaged matrix

$$(2.7) \qquad A\big(u(v^L), u(v^R)\big) = \int_0^1 A\big(u(v^L) + \xi(u(v^R) - u(v^L))\big) \, d\xi.$$

We shall also adopt the notation $\lambda_i^s(v, \sigma) \doteq \lambda_i\big(v, S_i(v, \sigma)\big)$ for the speed of the shock wave connecting the state v with the state $S_i(v, \sigma)$. Often we will refer to σ as the *size* of the shock wave joining $u(v)$ to $u\big(S_i(v, \sigma)\big)$.

Throughout the paper, for any function $\phi = \phi(v, \sigma)$ depending on the variables $v = (v_1, v_2)$ and σ, we shall denote by $\dot\phi(v, \sigma)$ its derivative with respect to σ, and by $\partial_i \phi(v, \sigma)$ the derivative of ϕ with respect to v_i. Moreover, we will use the Landau symbol $\mathcal{O}(1)$ to denote quantities uniformly bounded by a constant which depends only on the system (1.1), and will let $\{e_1, e_2\}$ denote the canonical basis of $\mathbb{R}^2$.

We briefly review next the standard procedure adopted to solve a Riemann problem for a system with NGNL characteristic fields satisfying the assumption **(A)**. In particular, we will describe the construction of the Riemann solver in the case where both characteristic families are NGNL and satisfy the assumptions (1.21)-(1.22), while we will only sketch this construction in the general case referring to [**Li1, Li2, AM1**] for a more detailed discussion.

General solutions of the Riemann Problem. In contrast with the standard GNL or LD systems, the entropy-admissible solution of the Cauchy Problem (1.1) with initial data

$$u(0, x) = \begin{cases} u^L \doteq u(v^L) & \text{if } x < 0 \\ u^R \doteq u(v^R) & \text{if } x > 0 \end{cases} \qquad v^L, v^R \in \mathcal{V}$$

for a NGNL system satisfying the assumption **(A)** consists of rarefaction waves, compressive shocks and *composed waves* made of one side contact-discontinuity adjacent to a rarefaction wave. In particular, in the case where the k-th family is NGNL and there holds (1.5), a composed wave of the k-th family connecting a left state u^L with a right state u^R may appear only if u^L and u^R lie on opposite sides with respect to the curve Γ_k. Such a wave consists of

CASE 1: a left contact-discontinuity on the right of a rarefaction wave, both of the k-th characteristic family, if

$$(2.8) \qquad r_k \bullet (r_k \bullet \lambda_k)(u^0) < 0, \qquad \forall \, u^0 \in \Gamma_k,$$

i.e. if, for each $u = u(v) \in \Omega$, the characteristic speed λ_k attains its maximum value along the k-th rarefaction curve $R_k(v)$ at the intersection point with Γ_k;

CASE 2: a right contact-discontinuity on the left of a rarefaction wave, both of the k-th characteristic family, if

$$(2.9) \qquad r_k \bullet (r_k \bullet \lambda_k)(u^0) > 0, \qquad \forall \, u^0 \in \Gamma_k,$$

i.e. if, for each $u = u(v) \in \Omega$, the characteristic speed λ_k attains its minimum value along the k-th rarefaction curve $R_k(v)$ at the intersection point with Γ_k.

We will say that a k-th NGNL characteristic family is NGNL1 if (2.8) holds, while it is NGNL2 in the case where (2.9) is verified.

In order to simplify the general expression of the elementary curve associated to every characteristic family (describing the states connected by a simple wave of that family), we shall define such a curve as a curve of *right states* for a NGNL1 family, and as a curve of *left states* for a NGNL2 family. Namely, if the k-th family is NGNL, for any given state $u = u(v)$, $v \in \mathcal{V}$, we will define

1. the *elementary curve through u of right states of the k-th family*, that consists of all right states that are connected with the left state u by a (possibly composed) wave of the k-th characteristic family, if Case 1 holds (i.e. if a composed wave of the k-th family has always a contact discontinuity on the right);

2. the *elementary curve through u of left states of the k-th family*, that consists of all left states that are connected with the right state u by a (possibly composed) wave of the k-th characteristic family, if Case 2 holds (i.e. if a composed wave of the k-th family has always a contact discontinuity on the left).

Of course, in the case the k-th characteristic family is GNL or LD, the k-th elementary curve through a given state $u = u(v)$, $v \in \mathcal{V}$, will be defined as usual as the curve of all right states connected with the left state u by a k-th shock or a k-th rarefaction wave.

The k-th elementary curve is in general a concatenation of a rarefaction curve, a shock curve and, in the case the k-th family is NGNL, a *mixed curve* describing the states connected by a k-th wave, which lie on opposite sides w.r.t. the curve Γ_k at (1.4). Therefore, if the k-th family is NGNL1, for any fixed $u^0 = u(v^0) \in \Gamma_k$ we will define a *mixed curve of right states of the k-th family*, parametrized by $\sigma \mapsto T_k(v^0, \sigma)$ and denoted $T_k(v^0)$, that represents all the right states v^R that are connected on the left to a state $v^L \in R_k(v^0)$ by a left-sided contact discontinuity of the k-th family, and thus has the property:

M1. For every point $v^R \in T_k(v^0)$, there exists a unique point $v^L \in R_k(v^0)$ such that $v^R \in S_k(v^L)$ and $\lambda_k(v^L, v^R) = \lambda_k(v^L)$.

Instead, if the k-th family is NGNL2, for any fixed $u^0 = u(v^0) \in \Gamma_k$ we will define a *mixed curve of left states of the k-th family*, parametrized by $\sigma \mapsto T_k(v^0, \sigma)$ and denoted $T_k(v^0)$, that represents all the left states v^L that are connected on the right to a state $v^R \in R_k(v^0)$ by a right-sided contact discontinuity of the k-th family, and thus has the property:

M2. For every point $v^L \in T_k(v^0)$, there exists a unique point $v^R \in R_k(v^0)$ such that $v^R \in S_k(v^L)$ and $\lambda_k(v^L, v^R) = \lambda_k(v^R)$.

Observe that the mixed curve $T_k(v^0)$ has second order contact with the rarefaction curve $R_k(v^0)$, and hence it admits a representation (in v-coordinates) of the same form as the Hugoniot curve $S_k(v^0)$ i.e., letting $T_{k,i}(v^0)$, $i = 1, 2$ denote the components of $T_k(v^0)$, there holds

(2.10) $$T_{k,i}(v^0, \sigma) = \begin{cases} v_k^0 + \sigma & \text{if } i = k, \\ v_i^0 + \widetilde{T}_k(v^0, \sigma) \cdot \sigma^3 & \text{if } i \neq k, \end{cases}$$

for a suitable smooth function $\widetilde{T}_k(v^0)$. On the other hand, since the curve Γ_k at (1.4) is transversal to the characteristic fields r_k, we can express Γ_k (in the Riemann coordinates) in the form

(2.11) $$\mathcal{V}_k^0 \doteq \{v \in \mathcal{V} : u(v) \in \Gamma_k\} = \{v \in \mathcal{V} : v_k = \widetilde{\gamma}_k(v_i) \ i \neq k\},$$

where $\widetilde{\gamma}_k$ denotes some smooth function which we may assume to be defined on the whole interval $]a_i, b_i[$, $i \neq k$ (by possibly considering a sufficiently small restriction of the domain $\mathcal{V} =]a_1, b_1[\times]a_2, b_2[$). Then, define the maps $\delta_k : \mathcal{V} \to \mathbb{R}$, $\pi_k : \mathcal{V} \to \mathcal{V}_k^0$, by setting

(2.12) $$\delta_k(v) \doteq v_k - \widetilde{\gamma}_k(v_i) \qquad (i \neq k),$$

(2.13) $$\pi_{k,i}(v) \doteq \begin{cases} v_i & \text{if } i \neq k, \\ \widetilde{\gamma}_k(v_j) \quad (j \neq k) & \text{if } i = k, \end{cases}$$

where $\pi_{k,i}$, $i = 1, 2$, are the components of π_k. The quantity $\delta_k(v)$ represents the signed distance of v from $\mathcal{V}_k^0$, measured along the rarefaction curve $R_k(v)$, while $\pi_k(v)$ is the projection on $\mathcal{V}_k^0$ along the rarefaction curve $R_k(v)$. The following basic properties hold.

i) For any fixed $v^0 \in \mathcal{V}_k^0$, there exists a smooth map $\sigma \mapsto \mu_k(v^0, \sigma)$ defined on a neighborhood of zero, such that

(2.14) $$\mu_k(v^0, 0) = 0,$$

(2.15)
$$\lambda_k^s\Big(R_k\big(v^0, \sigma - \mu_k(v^0, \sigma)\big), \mu_k(v^0, \sigma)\Big) = \lambda_k\Big(R_k\big(v^0, \sigma - \mu_k(v^0, \sigma)\big)\Big).$$

If we assume that the k-th family is NGNL1, eq. (2.15) states that the shock wave of size $\mu_k(v^0, \sigma)$, and with left state $v^L = R_k\big(v^0, (\sigma - \mu_k(v^0, \sigma))\big)$, travels with the characteristic speed $\lambda_k(v^L)$ of its left state (cfr. figure 2.1). Similar meaning holds if the k-th family is NGNL2.

ii) The map $\sigma \mapsto T_k(v^0, \sigma)$ in (2.10) and its derivative satisfy

(2.16)
$$T_k(v^0, \sigma) = S_k\Big(R_k\big(v^0, \sigma - \mu_k(v^0, \sigma)\big), \mu_k(v^0, \sigma)\Big)$$
$$\frac{d}{d\sigma}T_k(v^0, \sigma) = \frac{d}{ds}S_k\Big(R_k\big(v^0, \sigma - \mu_k(v^0, \sigma)\big), s\Big)\Big|_{s=\mu_k(v^0, \sigma)} \qquad \forall \, \sigma.$$

iii) The map $\sigma \mapsto \zeta_k(v^0, \sigma) \doteq \sigma - \mu_k(v^0, \sigma)$ is strictly decreasing. Moreover, if we denote by $s \mapsto \zeta_k^{-1}(v^0, s)$ its inverse map and set

(2.17) $$\nu_k(v) \doteq \mu_k\big(\pi_i(v), \, \zeta_k^{-1}(\pi_k(v), \delta_k(v))\big),$$

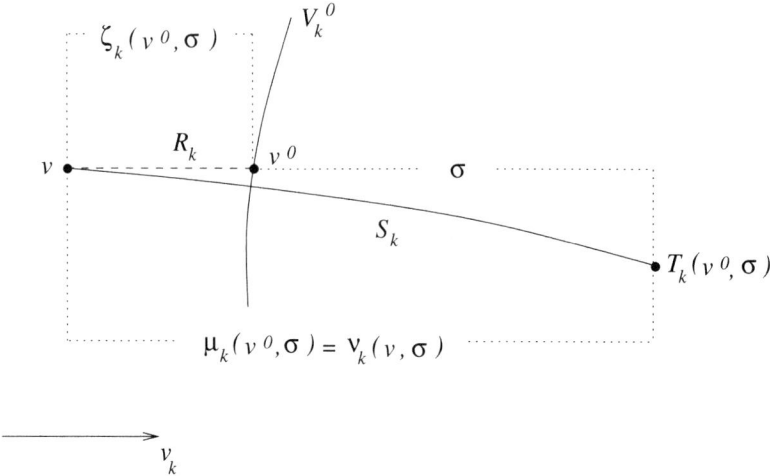

FIGURE 2.1. Illustration of (2.15), (2.17)

then, there holds

(2.18) $\quad T_k\big(\pi_k(v),\ \delta_k(v)+\nu_k(v)\big) = S_k\big(v,\ \nu_k(v)\big), \qquad \forall\, v \in \mathcal{V}.$

Because of (2.15), the quantity $\nu_k(v)$ defined at (2.17) satisfies

$$\lambda_k^s\big(v,\ \nu_k(v)\big) = \lambda_k(v),$$

i.e., if the k-th family is NGNL1, $\nu_k(v)$ represents the size of a left-sided contact discontinuity joining the left state v with the right state $S_k\big(v, \nu_k(v)\big)$, and traveling with the characteristic speed $\lambda_k(v)$. Similar meaning holds in the case where the k-th family is NGNL2. The map $\nu_k : \mathcal{V} \mapsto \mathbb{R}$ has the following expansion

(2.19) $\quad \nu_k(v) = -3\delta_k(v) + \mathcal{O}(1)\big|\delta_k(v)\big|^2 \qquad \forall\, v \in \mathcal{V}.$

We are now in the position to define the elementary curve associated to a NGNL characteristic family. Namely, consider the sets

(2.20) $\quad \mathcal{V}_k^- \doteq \{v \in \mathcal{V} :\ r_k \bullet \lambda_k(v) < 0\} \qquad \mathcal{V}_k^+ \doteq \{v \in \mathcal{V} :\ r_k \bullet \lambda_k(v) > 0\}$

and then, in the case the k-th family is NGNL1, define the *elementary curve of right states of the k-th family* $\Psi_k(v)$ as follows

(2.21)
$$\Psi_k(v,\sigma) = \begin{cases} S_k(v,\sigma) & \forall\,\sigma & \text{if}\quad v \in \mathcal{V}_k^0, \\[4pt] \begin{cases} S_k(v,\sigma) & \text{if}\quad \sigma < 0 \\ R_k(v,\sigma) & \text{if}\quad 0 \leq \sigma \leq -\delta_k(v) \\ T_k\big(\pi_k(v),\,\sigma + \delta_k(v)\big) & \text{if}\quad -\delta_k(v) < \sigma \leq \nu_k(v) \\ S_k(v,\sigma) & \text{if}\quad \sigma > \nu_k(v) \end{cases} & \text{if}\quad v \in \mathcal{V}_k^+, \\[4pt] \begin{cases} S_k(v,\sigma) & \text{if}\quad \sigma > 0 \\ R_k(v,\sigma) & \text{if}\quad -\delta_k(v) \leq \sigma \leq 0 \\ T_k\big(\pi_k(v),\,\sigma + \delta_k(v)\big) & \text{if}\quad \nu_k(v) \leq \sigma < -\delta_k(v) \\ S_k(v,\sigma) & \text{if}\quad \sigma < \nu_k(v) \end{cases} & \text{if}\quad v \in \mathcal{V}_k^-. \end{cases}$$

An entirely similar definition is given for the *elementary curve of left states of the k-th family* $\Psi_k(v)$ in the case the k-th family is NGNL2. On the other hand, in the case where the k-th family is LD, or GNL with

(2.22) $$r_k \bullet \lambda_k(v) > 0 \qquad \forall\, v \in \mathcal{V},$$

the corresponding elementary curve (of right states) is defined as usual by setting

(2.23) $$\Psi_k(v,\sigma) = \begin{cases} R_k(v,\sigma) & \text{if}\quad \sigma \geq 0 \\ S_k(v,\sigma) & \text{if}\quad \sigma < 0. \end{cases}$$

Observe that, by (2.5), (2.6), (2.10), (2.21)-(2.22), the elementary curves $\Psi_i(v,\sigma)$, $i = 1, 2$, have tangent e_i at the initial state v (no matter if the fields are LD, GNL or NGNL). Thus, applying the implicit function theorem it follows that, by possibly considering a further restriction of the domain $\mathcal{V}$, for any given pair of states $v^L, v^R \in \mathcal{V}$, there exists a unique intermediate state v^M, and wave sizes σ_1, σ_2, such that

(2.24)
$$\begin{aligned} v^M &= \Psi_1(v^L, \sigma_1) && \text{if} && \text{the first family is LD, or GNL, or NGNL1} \\ v^L &= \Psi_1(v^M, \sigma_1) && \text{if} && \text{the first family is NGNL2} \\ v^R &= \Psi_2(v^M, \sigma_2) && \text{if} && \text{the second family is LD, or GNL, or NGNL1} \\ v^M &= \Psi_2(v^R, \sigma_2) && \text{if} && \text{the second family is NGNL2} \end{aligned}$$

Then, if we set

(2.25) $$\bar{u}^1(x) \doteq \begin{cases} u(v^L) & \text{if}\ x < 0, \\ u(v^M) & \text{if}\ x > 0, \end{cases} \qquad \bar{u}^2(x) \doteq \begin{cases} u(v^M) & \text{if}\ x < 0, \\ u(v^R) & \text{if}\ x > 0, \end{cases}$$

the solution of the Riemann Problem with initial states $u(v^L)$, $u(v^R)$ is constructed by piecing together the solutions of the Riemann problem with initial data $\bar{u}^1$ on the quadrant of the (t, x)-plane where $x \leq 0$, with the solution of the Riemann problem with initial data $\bar{u}^2$ on the quadrant where $x \geq 0$.

By the above construction, in the particular case of a system satisfying the assumption **(A′)** (in which the first family is NGNL1 and the second family is NGNL2), the sets in (2.11), (2.20) take the form

(2.26)
$$\mathcal{V}_1^0 = \{v \in \mathcal{V} : v_1 = \widetilde{\gamma}_1(v_2)\} \qquad \mathcal{V}_2^0 = \{v \in \mathcal{V} : v_2 = \widetilde{\gamma}_2(v_1)\},$$
$$\mathcal{V}_1^- = \{v \in \mathcal{V} : v_1 > \widetilde{\gamma}_1(v_2)\} \qquad \mathcal{V}_2^- = \{v \in \mathcal{V} : v_2 < \widetilde{\gamma}_2(v_1)\},$$
$$\mathcal{V}_1^+ = \{v \in \mathcal{V} : v_1 < \widetilde{\gamma}_1(v_2)\} \qquad \mathcal{V}_2^+ = \{v \in \mathcal{V} : v_2 > \widetilde{\gamma}_2(v_1)\},$$

while the *elementary curve of right states of the first characteristic family* $\Psi_1(v)$, and the *elementary curve of left states of the second characteristic family* $\Psi_2(v)$, are defined by setting

(2.27)
$$\Psi_1(v,\sigma) = \begin{cases} S_1(v,\sigma) & \forall\, \sigma & \text{if } v \in \mathcal{V}_1^0, \\[4pt] \begin{cases} S_1(v,\sigma) & \text{if } \sigma < 0 \\ R_1(v,\sigma) & \text{if } 0 \leq \sigma \leq -\delta_1(v) \\ T_1(\pi_1(v),\, \sigma+\delta_1(v)) & \text{if } -\delta_1(v) < \sigma \leq \nu_1(v) \\ S_1(v,\sigma) & \text{if } \sigma > \nu_1(v) \end{cases} & \text{if } v \in \mathcal{V}_1^+, \\[4pt] \begin{cases} S_1(v,\sigma) & \text{if } \sigma > 0 \\ R_1(v,\sigma) & \text{if } -\delta_1(v) \leq \sigma \leq 0 \\ T_1(\pi_1(v),\, \sigma+\delta_1(v)) & \text{if } \nu_1(v) \leq \sigma < -\delta_1(v) \\ S_1(v,\sigma) & \text{if } \sigma < \nu_1(v) \end{cases} & \text{if } v \in \mathcal{V}_1^-, \end{cases}$$

(2.28)
$$\Psi_2(v,\sigma) = \begin{cases} S_2(v,\sigma) & \forall\, \sigma & \text{if } v \in \mathcal{V}_2^0, \\[4pt] \begin{cases} S_2(v,\sigma) & \text{if } \sigma > 0 \\ R_2(v,\sigma) & \text{if } -\delta_2(v) \leq \sigma \leq 0 \\ T_2(\pi_2(v),\, \sigma+\delta_2(v)) & \text{if } \nu_2(v) \leq \sigma < -\delta_2(v) \\ S_2(v,\sigma) & \text{if } \sigma < \nu_2(v) \end{cases} & \text{if } v \in \mathcal{V}_2^+, \\[4pt] \begin{cases} S_2(v,\sigma) & \text{if } \sigma < 0 \\ R_2(v,\sigma) & \text{if } 0 \leq \sigma \leq -\delta_2(v) \\ T_2(\pi_2(v),\, \sigma+\delta_2(v)) & \text{if } -\delta_2(v) < \sigma \leq \nu_2(v) \\ S_2(v,\sigma) & \text{if } \sigma > \nu_2(v) \end{cases} & \text{if } v \in \mathcal{V}_2^-. \end{cases}$$

Then, given v^L, $v^R \in \mathcal{V}$, according with (2.24) there will be a unique intermediate state $v^M \in \mathcal{V}$, and wave sizes σ_1, σ_2, such that

(2.23)
$$v^M = \Psi_1(v^L, \sigma_1), \qquad v^M = \Psi_2(v^R, \sigma_2).$$

The solution to the Riemann problem with initial data $\bar{u}^i$ as in (2.25) is defined as follows.

1 - Waves. Initial data $\bar{u}^1$. There are 3 possibilities:

1. $v^L \in \mathcal{V}_1^0$: the solution consists of a shock of the first characteristic family travelling with speed $\lambda_1^s(v^L, \sigma_1)$.

2. $v^L \in \mathcal{V}_1^+$: the solution consists of

   a) $0 \leq \sigma_1 \leq -\delta_1(v^L)$: a rarefaction wave of the first characteristic family, with characteristic speeds ranging over the interval
   $$[\lambda_1(v^L), \lambda_1(v^M)];$$

   b) $\sigma_1 < 0$ or $\sigma_1 \geq \nu_1(v^L)$: a shock of the first characteristic family travelling with speed $\lambda_1^s(v^L, \sigma_1)$;

   c) $-\delta_1(v^L) < \sigma_1 < \nu_1(v^L)$: a composed wave of the first characteristic family made of a rarefaction wave of size
   $$\widetilde{\sigma}_1 \doteq \sigma_1 - \mu_1(\pi_1(v^L), \sigma_1 + \delta_1(v^L)),$$

   with characteristic speeds ranging over the interval
   $$[\lambda_1(v^L), \lambda_1(\widetilde{v}^1)], \qquad \widetilde{v}^1 \doteq R_1(v^L, \widetilde{\sigma}_1),$$

   followed by a left contact discontinuity of size $\widehat{\sigma}_1 \doteq \sigma_1 - \widetilde{\sigma}_1$, travelling with speed
   $$\lambda_1^s(\widetilde{v}^1, \widehat{\sigma}_1) = \lambda_1(\widetilde{v}^1).$$

3. $v^L \in \mathcal{V}_1^-$: the solution consists of

   a) $-\delta_1(v^L) \leq \sigma_1 \leq 0$: a rarefaction wave as in 2-a);

   b) $\sigma_1 > 0$ or $\sigma_1 \leq \nu_1(v^L)$: a shock as in 2-b);

   c) $\nu_1(v^L) < \sigma_1 < -\delta_1(v^L)$: a composed wave as in 2-c), made of a rarefaction wave followed by a left contact discontinuity.

2 - Waves. Initial data $\bar{u}^2$. There are 3 possibilities:

1. $v^R \in \mathcal{V}_2^0$: the solution consists of a shock of the second characteristic family travelling with speed $\lambda_2^s(v^R, \sigma_2)$.

2. $v^R \in \mathcal{V}_2^+$: the solution consists of

   a) $-\delta_2(v^R) \leq \sigma_2 \leq 0$: a rarefaction wave of the second characteristic family, with characteristic speeds ranging over the interval
   $$[\lambda_2(v^M), \lambda_2(v^R)];$$

   b) $\sigma_2 > 0$ or $\sigma_2 \leq \nu_2(v^R)$: a shock of the second characteristic family travelling with speed $\lambda_2^s(v^R, \sigma_2)$;

   c) $\nu_2(v^R) < \sigma_2 < -\delta_2(v^R)$: a composed wave of the second characteristic family made of a right contact discontinuity of size
   $$\widehat{\sigma}_2 \doteq \mu_2(\pi_2(v^R), \sigma_2 + \delta_2(v^R)),$$

   travelling with speed
   $$\lambda_2^s(\widetilde{v}^2, \widehat{\sigma}_2) = \lambda_2(\widetilde{v}^2), \qquad \widetilde{v}^2 \doteq R_2(v^R, \sigma_2 - \widehat{\sigma}_2),$$

followed by a rarefaction wave of size $\widetilde\sigma_2 \doteq \sigma_2 - \widehat\sigma_2$, with characteristic speeds ranging over the interval

$$\left[\lambda_2(\widetilde v^2),\, \lambda_2(v^R)\right].$$

3. $v^R \in \mathcal{V}_2^-$: the solution consists of

   a) $0 \leq \sigma_2 \leq -\delta_2(v^R)$: a rarefaction wave as in 2-a);

   b) $\sigma_2 < 0$ or $\sigma_2 \geq \nu_2(v^R)$: a shock as in 2-b);

   c) $-\delta_2(v^R) < \sigma_2 < \nu_2(v^R)$: a composed wave as in 2-c), made of a right contact discontinuity followed by a rarefaction wave.

3. Outline of the Proof

We collect here the basic Steps in the proof of Theorem 1. Technical details will be worked out in the remaining sections.

Since we are dealing with a 2×2 system, it will be convenient to work always with a set of Riemann coordinates $v = (v_1, v_2)$. Throughout we shall often adopt the notation

$$\lfloor v \rfloor_i \doteq \left\lfloor \frac{v_i}{\varepsilon} \right\rfloor, \quad i = 1, 2, \qquad v = (v_1, v_2), \qquad \varepsilon > 0,$$

where $\lfloor \alpha \rfloor$ denotes the integer part of α, and let δ_k, π_k be the maps introduced in (2.12), (2.13) for a k-th NGNL family.

In the initial part of the proof, we fix $\varepsilon > 0$ and introduce a refined version of the wave-front tracking algorithm adopted in [**BC1**] which yields piecewise constant ε-approximate solutions, well defined for all $t > 0$, for a system satisfying the assumption (**A**). These are piecewise constant functions with jumps located along finitely many segments in the (t, x)-plane, and satisfying a set of approximate Rankine-Hugoniot conditions. The corresponding ε-Approximate Riemann Solver has the same features as the one constructed in [**BC1**] for systems with LD or GNL fields:

- Every Riemann Problem is solved with the same procedure and the solution depends continuously on the Riemann data v^L, v^R.

- Centered rarefaction waves are approximated by means of a piecewise constant rarefaction fan, inserting intermediate states on a fixed grid of step size ε.

- Shock waves of strength $|\sigma| < 2\sqrt\varepsilon$ connect a left and right state lying on the same rarefaction curve.

As observed in the Introduction, in order to produce a Riemann solver with such properties, that depends continuously on the Riemann data, we need to introduce a suitable approximation of the Hugoniot curves and of the elementary curves associated to the NGNL characteristic families. To this purpose, if the k-th family is NGNL and satisfies the assumption (**A**), we shall construct in Section 4 some maps (defined on a neighborhood of zero $\mathcal{I} \subset \mathbb{R}$)

(3.1)
$$\begin{aligned}(v,\sigma) &\mapsto \Psi_k^\varepsilon(v,\sigma),\\ (v,\sigma) &\mapsto \lambda_k^{\psi,\varepsilon}(v,\sigma),\\ (v,\sigma) &\mapsto \mu_k^\varepsilon(\pi_k(v),\sigma),\\ v &\mapsto \nu_k^\varepsilon(v),\end{aligned} \qquad v \in \mathcal{V},\ \sigma \in \mathcal{I},$$

that provide a careful approximation of the functions Ψ_k, λ_k^s, μ_k, ν_k involved in the definition of the exact solution of a Riemann problem. On the other hand, by the analysis in [**BC1**], in the case the k-th characteristic family is LD, or GNL and satisfies (2.22), one can use an approximation of the k-th elementary curve of the same type of the one adopted in [**BC1**], letting $\Psi_k^\varepsilon(v,\sigma)$ coincide with the k-th rarefaction curve $R_k(v,\sigma)$ for $\sigma \geq -\sqrt{\varepsilon}$, and with the k-th Hugoniot curve $S_k(v,\sigma)$ for $\sigma \leq -2\sqrt{\varepsilon}$, while for $\sigma \in [-2\sqrt{\varepsilon}, -\sqrt{\varepsilon}]$ it is defined as a smooth interpolation between the two curves. In a similar way, we let the traveling speed $\lambda_k^{\psi,\varepsilon}(v,\sigma)$ of a wave connecting the left and right states $\{v, \Psi_k^\varepsilon(v,\sigma)\}$ coincide with the averaged characteristic speed $\lambda_k^{r,\varepsilon}(v,\sigma)$ (computed along the rarefaction curve $R_k(v)$: see (3.3)-(3.4) below) for $\sigma \geq -2\sqrt{\varepsilon}$, and with the shock speed $\lambda_k^s(v,\sigma)$ for $\sigma \leq -5\sqrt{\varepsilon}$. An interpolation between $\lambda_k^{r,\varepsilon}$ and λ_k^s of the same type as for $\Psi_k^\varepsilon(v,\sigma)$ is used to define $\lambda_k^{\psi,\varepsilon}(v,\sigma)$ whenever $\sigma \in [-5\sqrt{\varepsilon}, -2\sqrt{\varepsilon}]$.

The detailed definitions of the *ε-approximate elementary curves* Ψ_k^ε, of the traveling speed $\lambda_k^{\psi,\varepsilon}$, and of the other maps in (3.1) associated to a k-th NGNL family is given in Section 4. Here, we collect only their basic properties. Throughout, we shall assume that a GNL family satisfies (2.22), while a NGNL family satisfies either one of (2.8)-(2.9) (i.e. is either NGNL1 or NGNL2). As we have done in Section 2, the map Ψ_k^ε represents the ε-approximate elementary curve of *right states* of the k-th family if the k-th family is LD, or GNL, or NGNL1, and the ε-approximate elementary curve of *left states* of the k-th family if the k-th family is NGNL2 (cfr. Definition 4.2 in Subsection 4.5).

P1. If the k-th family is LD, or GNL, then as $\varepsilon \to 0+$ there holds

$$\begin{aligned}\Psi_k^\varepsilon(v,\sigma) &\to \Psi_k(v,\sigma) & \sigma \in \mathcal{I},\\ \lambda_k^{\psi,\varepsilon}(v,\sigma) &\to \lambda_k^s(v,\sigma) & \sigma \in \mathcal{I} \cap \{\sigma \leq 0\},\end{aligned} \qquad v \in \mathcal{V},$$

the convergence being uniform in v, σ. The maps Ψ_k^ε, $\lambda_k^{\psi,\varepsilon}$ are continuously differentiable on their domain.

P2. If the k-th family is NGNL, then as $\varepsilon \to 0+$ there holds

$$\begin{aligned}\Psi_k^\varepsilon(v,\sigma) &\to \Psi_k(v,\sigma) & \sigma \in \mathcal{I},\ v \in \mathcal{V},\\ \mu_k^\varepsilon(v,\sigma) &\to \mu_k(v,\sigma) & \sigma \in \mathcal{I},\ v \in \mathcal{V},\\ \nu_k^\varepsilon(v) &\to \nu_k(v) & v \in \mathcal{V},\end{aligned}$$

and, if it is NGNL1, there holds

$$\lambda_k^{\psi,\varepsilon}(v,\sigma) \to \lambda_k^s(v,\sigma) \qquad \sigma \in \mathcal{I} \cap \left(\{\sigma \leq 0\} \cup \{\sigma \geq \nu_k^\varepsilon(v)\}\right), \quad v \in \mathcal{V}_k^+,$$

the convergence being uniform in v, σ. Of course, similar property holds for $\lambda_k^{\psi,\varepsilon}$ if $v \in \mathcal{V}_k^-$, or in the case where the k-th family is NGNL2. The maps Ψ_k^ε, $\lambda_k^{\psi,\varepsilon}$, μ_k^ε, ν_k^ε, admit left and right partial derivatives with respect to v_1, v_2, σ, and are piecewise continuously differentiable on their domain. Moreover, Ψ_k^ε, $\lambda_k^{\psi,\varepsilon}$ are continuous in any point.

P3.

$$\begin{aligned}\Psi_k^\varepsilon(v,\sigma) &= R_k(v,\sigma),\\ \lambda_k^{\psi,\varepsilon}(v,\sigma) &= \lambda_k^{r,\varepsilon}(v,\sigma),\end{aligned} \quad \text{if} \quad |\sigma| \leq 2\sqrt{\varepsilon}, \quad v \in \mathcal{V}, \tag{3.2}$$

where $\lambda_k^{r,\varepsilon}$ denotes the averaged speed defined as

$$\lambda_k^{r,\varepsilon}(v,\sigma) \doteq \sum_{\ell=\lfloor v \rfloor_k}^{\lfloor R_k(v,\sigma)\rfloor_k} \left[\frac{\text{meas}\big([\ell\varepsilon,(\ell+1)\varepsilon]\cap[v_k,v_k+\sigma]\big)}{\sigma} \cdot \widetilde{\lambda}_k^\ell(v)\right] \quad \sigma \geq 0, \tag{3.3}$$

with

$$\widetilde{\lambda}_k^\ell(v) \doteq \begin{cases} \displaystyle\sup_{0\leq s\leq \varepsilon} \lambda_k\big(R_k(v_i e_i,\,\ell\varepsilon+s)\big) \\ \quad (i\neq k) & \text{if } \begin{cases} \text{the } k\text{-th family is } LD, \\ \text{or } GNL, \text{ or } NGNL1\end{cases}\\[1em] \displaystyle\inf_{0\leq s\leq \varepsilon} \lambda_k\big(R_k(v_i e_i,\,\ell\varepsilon+s)\big)\\ \quad (i\neq k) & \text{if } \quad \text{the } k\text{-th family is } NGNL2\end{cases} \tag{3.4}$$

The definition of the averaged speed $\lambda_k^{r,\varepsilon}$ in the case $\sigma < 0$ is entirely similar.

P4. If the k-th family is NGNL1, then for any fixed $v \in \mathcal{V}_k^+$ and $\sigma \in [-\delta_k(v), \nu_k^\varepsilon(v)]$, setting

$$\widetilde{\sigma} \doteq \sigma - \mu_k^\varepsilon(\pi_k(v),\,\sigma+\delta_k(v)), \qquad \widehat{\sigma} \doteq \sigma - \widetilde{\sigma}, \tag{3.5}$$

the following hold. The equation

$$\Psi_k^\varepsilon(v,\sigma) - \Psi_k^\varepsilon\big[R_k(v,\sigma_1),\,\sigma_2\big] = 0 \tag{3.6}$$

in σ_1, σ_2 admits:
  i) a unique solution $(\sigma_1,\sigma_2) = (\widetilde{\sigma},\widehat{\sigma})$, if $\widehat{\sigma} > \nu_k^\varepsilon\big(R_k(v,\widetilde{\sigma})\big)$;
  ii) a set of solutions

$$\left\{ (\sigma_1,\sigma_2)\,:\,\widetilde{\sigma}\leq \sigma_1 \leq \widetilde{\sigma}+\varepsilon,\quad \sigma_2 = \sigma-\sigma_1\right\}, \tag{3.7}$$

if $\widehat{\sigma} = \nu_k^\varepsilon\big(R_k(v,\widetilde{\sigma})\big)$.
Moreover, one has $(R_k(v,\widetilde{\sigma}))_k \in \mathbb{Z}\varepsilon$ and, in the case $\widehat{\sigma} = \nu_k^\varepsilon\big(R_k(v,\widetilde{\sigma})\big)$, there holds

$$\lambda_k^{\lfloor R_k(v,\widetilde{\sigma})\rfloor_k}(v) = \lambda_k^{\psi,\varepsilon}\big(R_k(v,\widetilde{\sigma}),\,\widehat{\sigma}\big) = \lambda_k^{\psi,\varepsilon}\big(R_k(v,\sigma_1),\,\sigma_2\big), \tag{3.8}$$

for all (σ_1,σ_2) in (3.7).
Similar relations are verified if $v \in \mathcal{V}_k^-$, or in the case where the k-th family is NGNL2.

P5. For any fixed $v^L, v^R \in \mathcal{V}$, the following hold. If there is no NGNL family satisfying (2.9), then the system

$$v^R - \Psi_2^\varepsilon\big(\Psi_1^\varepsilon(v^L,\sigma_1),\,\sigma_2\big) = 0 \tag{3.9}_1$$

in σ_1, σ_2 admits a unique solution in $\mathcal{I} \times \mathcal{I}$. If both families are NGNL and satisfy (2.9), then the system

$$(3.9)_2 \qquad v^L - \Psi_1^\varepsilon\bigl(\Psi_2^\varepsilon(v^R, \sigma_2),\, \sigma_1\bigr) = 0$$

in σ_1, σ_2 admits a unique solution in $\mathcal{I} \times \mathcal{I}$. If the second family is NGNL2, but the first one is either LD, or GNL, or NGNL1, then the system

$$(3.9)_3 \qquad \Psi_1^\varepsilon(v^L, \sigma_1) - \Psi_2^\varepsilon(v^R, \sigma_2) = 0$$

in σ_1, σ_2 admits a unique solution in $\mathcal{I} \times \mathcal{I}$. If the first family is NGNL and satisfies (2.9), but the second one is either LD, or GNL, or NGNL1, then the system

$$(3.9)_4 \qquad \begin{cases} v^R - \Psi_2^\varepsilon(v^M, \sigma_2) = 0 \\ v^L - \Psi_1^\varepsilon(v^M, \sigma_1) = 0 \end{cases}$$

in σ_1, σ_2, v^M admits a unique solution in $\mathcal{I} \times \mathcal{I} \times \mathcal{V}$. Moreover, one can find some constant $C > 0$ independent of ε such that, for all $v^L, v^R \in \mathcal{V}$, if σ_1, σ_2 are the values determined by the system $(3.9)_\ell$ associated to (1.1), then one has

$$(3.10) \qquad \frac{1}{C}\bigl|v^R - v^L\bigr| \le |\sigma_1| + |\sigma_2| \le C\bigl|v^R - v^L\bigr|.$$

REMARK 3.1. Assume that the first family is NGNL1. Fix some state $v^L \in \mathcal{V}_1^-$, and let $\alpha \mapsto \widetilde{F}_1(\alpha)$ be a scalar, continuous, piecewise linear function such that

$$\widetilde{F}_1'(\alpha) = \widetilde{\lambda}_1^\ell(v^L) \qquad \text{if} \qquad \alpha \in]\ell\varepsilon,\, (\ell+1)\varepsilon[,$$

where $\widetilde{\lambda}_1^\ell(v^L)$ is the discrete characteristic speed defined at (3.4) (see figure 3.1). Consider the function

$$\omega(t,x) = \begin{cases} v^L & \text{if } x < \lambda_1^{\psi,\varepsilon}(v^L, \sigma) \cdot t, \\ v^R & \text{if } \lambda_1^{\psi,\varepsilon}(v^L, \sigma) \cdot t < x \le 0, \end{cases}$$

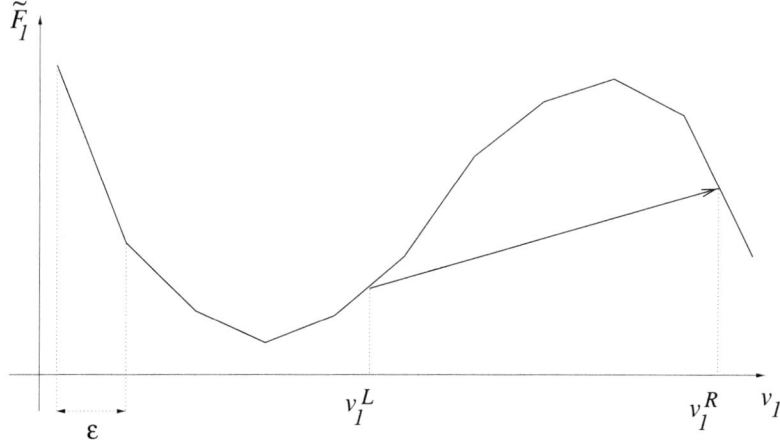

FIGURE 3.1. Illustration of Remark 3.1

where $v^R = \Psi_1^\varepsilon(v, \sigma)$. Then, because of the property **P3**, as long as $0 \leq \sigma \leq 2\sqrt{\varepsilon}$, the first component $\omega_1(t, x)$ of ω provides a weak, entropy-admissible solution to the scalar conservation law

$$(3.11) \qquad (v_1)_t + \left[\widetilde{F}_1(v_1)\right]_x = 0.$$

The second component is constant $\omega_2(t, x) = v_2^L$. This property motivates the definition (3.3) of averaged characteristic speed for a small shock. Of course, the same property holds for all shock waves σ with strength $|\sigma| \leq 2\sqrt{\varepsilon}$ that belong to any GNL, or LD, or NGNL family satisfying either one of (2.8)-(2.9).

REMARK 3.2. Assume that the first family is NGNL1. Fix $v^L \in \mathcal{V}_1^+$, $\sigma \in [-\delta_1(v^L), 2\sqrt{\varepsilon}]$, and assume that there exists some $\widetilde{\sigma} \in [0, \sigma]$ such that (see figure 3.2)

$$(3.12) \qquad v_1^L + \widetilde{\sigma} \in \mathbb{Z}\varepsilon, \qquad \widetilde{\lambda}_1^{\lfloor R_1(v^L, \widetilde{\sigma}) \rfloor_1}(v^L) = \lambda_1^{r,\varepsilon}\bigl[R_1(v^L, \widetilde{\sigma}), \sigma - \widetilde{\sigma}\bigr].$$

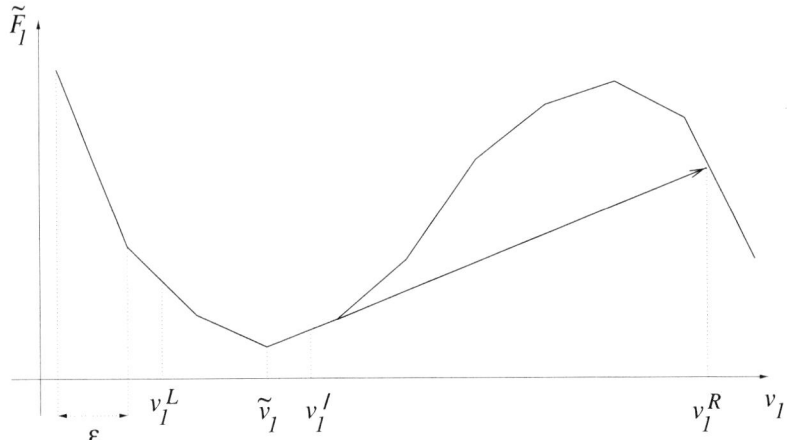

FIGURE 3.2. Illustration of (3.12), (3.13)

Let $\sigma' \in [\widetilde{\sigma}, \widetilde{\sigma} + \varepsilon]$, and set

$$\widehat{\sigma} \doteq \sigma - \widetilde{\sigma}, \qquad \widehat{\sigma}' \doteq \sigma - \sigma', \qquad \widetilde{v} \doteq R_1(v^L, \widetilde{\sigma}), \qquad v' \doteq R_1(v^L, \sigma'), \qquad v^R \doteq \Psi_1^\varepsilon(v, \sigma).$$

Notice that, from the properties of $\lambda_1^{r,\varepsilon}$ and by (3.12), it follows

$$\widetilde{\lambda}_1^{\lfloor \widetilde{v} \rfloor_1}(v^L) = \lambda_1^{r,\varepsilon}(\widetilde{v}, \widehat{\sigma}) = \lambda_1^{r,\varepsilon}(v', \widehat{\sigma}').$$

By **P3**, this relation can be rewritten as

$$(3.13) \qquad \widetilde{\lambda}_1^{\lfloor \widetilde{v} \rfloor_1}(v^L) = \lambda_1^{\psi,\varepsilon}(\widetilde{v}, \widehat{\sigma}) = \lambda_1^{\psi,\varepsilon}(v', \widehat{\sigma}'),$$

which is the analogous of (3.8) in **P4**. Then, setting $\ell_1 = \lfloor v^L \rfloor_1 + 1$ and $\ell_2 = \lfloor \widetilde{v} \rfloor_1$, for any given $\xi > 0$, consider the function (see figure 3.3)

$$\omega^{\sigma',\xi}(t,x) = \begin{cases} v^L & \text{if } x < \widetilde{\lambda}_1^{\lfloor v^L \rfloor_1}(v^L) \cdot t, \\ (\ell\varepsilon, v_2^L) & \text{if } \widetilde{\lambda}_1^{\ell-1}(v^L) \cdot t < x < \widetilde{\lambda}_1^{\ell}(v^L)) \cdot t, \quad \ell = \ell_1, \ldots, \ell_2, \\ v' & \text{if } \widetilde{\lambda}_1^{\lfloor \widetilde{v} \rfloor_1}(v^L) \cdot t < x < \xi + \widetilde{\lambda}_1^{\lfloor \widetilde{v} \rfloor_1}(v^L) \cdot t, \\ v^R & \text{if } \xi + \lambda_1^{\psi,\varepsilon}(v', \widehat{\sigma}') \cdot t < x \le 0, \end{cases}$$

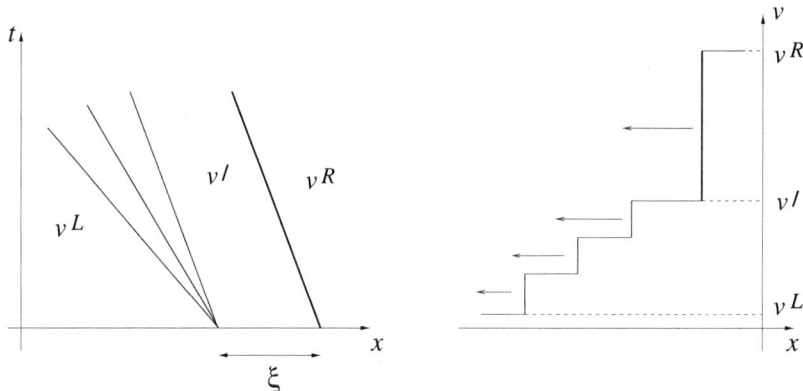

FIGURE 3.3. Illustration of Remark 3.2

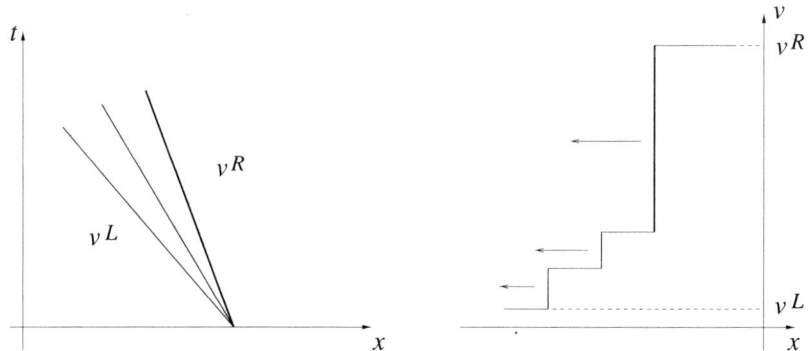

FIGURE 3.4. Illustration of Remark 3.2

that provides a piecewise constant approximation of a composed wave made of a centered rarefaction followed by a shock. As in the previous Remark, one can easily check that the first component of $\omega^{\sigma',\xi}(t,x)$ is a weak, entropy-admissible solution of the scalar conservation law (3.11), while the second component is constant $\omega_2^{\sigma',\xi}(t,x) = v_2^L$. Moreover, with the above notations, thanks to (3.13), we

find that the $\mathbb{L}^1$-limit
$$\omega(t,\cdot) \doteq \lim_{\xi\to 0+} \omega^{\sigma',\xi}(t,\cdot) \qquad t\geq 0,$$
is (see figure 3.4)
$$\omega(t,x) = \begin{cases} v^L & \text{if } x < \widetilde{\lambda}_1^{\lfloor v^L\rfloor_1}(v^L)\cdot t, \\ (\ell\varepsilon, v_2^L) & \text{if } \widetilde{\lambda}_1^{\ell-1}(v^L)\cdot t < x < \widetilde{\lambda}_1^{\ell}(v^L))\cdot t, \quad \ell = \ell_1,\ldots,\ell_2, \\ v^R & \text{if } \lambda_1^{\psi,\varepsilon}(\widetilde{v},\widehat{\sigma})\cdot t < x \leq 0. \end{cases}$$

and the first component of $\omega(t,x)$ is a weak, entropy-admissible solution of (3.11) as well. Clearly, the same property holds in the case we consider a piecewise constant function that approximates a composed wave σ, of strength $|\sigma|\leq 2\sqrt{\varepsilon}$, belonging to any NGNL family satisfying either one of (2.8)-(2.9). This observation shows the crucial importance of property **P4** to guarantee the stability and the continuous dependence on the initial data of approximate solutions constructed for a system satisfying the assumption (**A**).

DEFINITION 3.1. *Let (v^L, v^R) be a jump located at $x = x(t)$ and traveling with speed $\dot{x}$.*

1. *We say that (v^L, v^R) satisfies the Ψ^ε-Rankine-Hugoniot condition $(\Psi^\varepsilon\text{-RH})$ of a k-th GNL family verifying (2.22), if one of the following two cases occurs:*

   i) *There exists $\sigma \in \big[0, (\lfloor v^L\rfloor_k + 1)\varepsilon - v_k^L\big]$ such that*

   $$(3.14) \qquad v^R = R_k(v^L,\sigma), \qquad \dot{x} = \widetilde{\lambda}_k^{\lfloor v^L\rfloor_k}(v^L),$$

   where $\widetilde{\lambda}_k^{\lfloor v^L\rfloor_k}(v^L)$ is the discrete characteristic speed defined at (3.4).

   ii) *There exists $\sigma \leq 0$ such that*

   $$(3.15) \qquad v^R = \Psi_k^\varepsilon(v^L,\sigma), \qquad \dot{x} = \lambda_k^{\psi,\varepsilon}(v^L,\sigma).$$

2. *We say that (v^L, v^R) satisfies the Ψ^ε-Rankine-Hugoniot condition $(\Psi^\varepsilon\text{-RH})$ of a k-th LD family if there exists σ such that*

   $$(3.16) \qquad v^R = R_k(v^L,\sigma), \qquad \dot{x} = \lambda_k(v^L).$$

3. *We say that (v^L, v^R) satisfies the Ψ^ε-Rankine-Hugoniot condition $(\Psi^\varepsilon\text{-RH})$ of a k-th NGNL1 family if one of the following two cases occurs:*

   i) *If $v^L \in \mathcal{V}_k^+$ and there exists $\sigma \in \big[0, (\lfloor v^L\rfloor_k + 1)\varepsilon - v_k^L\big]$ such that (3.14) holds, or $v^L \in \mathcal{V}_k^-$ and there exists $\sigma \in \big[\lfloor v^L\rfloor_k\varepsilon - v_k^L, 0\big]$ so that (3.14) is verified.*

   ii) *If $v^L \in \mathcal{V}_k^+$ and there exists $\sigma \in \{s: s\leq 0\}\cup\{s: s\geq \nu_k^\varepsilon(v^L)\}$ such that (3.14) holds, or $v^L \in \mathcal{V}_k^-$ and there exists $\sigma \in \{s: s\geq 0\}\cup\{s: s\leq \nu_k^\varepsilon(v^L)\}$ so that (3.14) is verified.*

4. *We say that (v^L, v^R) satisfies the Ψ^ε-Rankine-Hugoniot condition $(\Psi^\varepsilon\text{-RH})$ of a k-th NGNL2 family if one of the following two cases occurs:*

i) If $v^R \in \mathcal{V}_k^+$ and there exists $\sigma \in \big[\lfloor v^R \rfloor_k \varepsilon - v_k^R,\, 0\big]$ such that

(3.17) $$v^L = R_k(v^R, \sigma), \qquad \dot{x} = \widetilde{\lambda}_k^{\lfloor v^R \rfloor_k}(v^R)$$

(here $\widetilde{\lambda}_2^{\lfloor v^R \rfloor_2}(v^R)$ is the discrete characteristic speed defined at (3.4)), or $v^R \in \mathcal{V}_k^-$ and there exists $\sigma \in \big[0,\, (\lfloor v^R \rfloor_k + 1)\varepsilon - v_k^R\big]$ so that (3.17) is verified.

ii) If $v^R \in \mathcal{V}_k^+$ and there exists $\sigma \in \{s : s \leq 0\} \cup \{s : s \geq \nu_k^\varepsilon(v^R)\}$ such that

(3.18) $$v^L = \Psi_k^\varepsilon(v^R, \sigma), \qquad \dot{x} = \lambda_k^{\psi,\varepsilon}(v^R, \sigma),$$

or $v^R \in \mathcal{V}_k^-$ and there exists $\sigma \in \{s : s \geq 0\} \cup \{s : s \leq \nu_k^\varepsilon(v^R)\}$ so that (3.18) is verified.

DEFINITION 3.2. We say that a piecewise constant function $v = v(t, x)$ is an ε-*approximate solution* if all of its jumps satisfy the Ψ^ε-RH conditions. Any line $x = x(t)$ across which v has a jump satisfying the Ψ^ε-RH conditions of the i-th family is an ε-*admissible wave-front* of the i-th family.

REMARK 3.3. By introducing the modified Rankine-Hugoniot conditions of Definition 3.1, we impose that each admissible jump of the ε-approximate solution corresponds to one of the following two types of waves:

- A shock wave of the k-th family that travels with the "modified approximate shock speed" $\lambda_k^{\psi,\varepsilon}$, whose left and right states are connected by the "modified approximate shock curve" Ψ_k^ε. In particular, the size σ of an admissible shock wave of a k-th NGNL family, joining two states v^L, v^R that lie on opposite sides w.r.t. the set $\mathcal{V}_k^0$ at (2.11), satisfies the bound $|\sigma| \geq |\nu_k^\varepsilon(v^L)|$. The errors in the size and in the speed are small and go to zero as $\varepsilon \to 0$.
- A rarefaction wave-front of strength $\leq \varepsilon$ (which approximates a piece of a rarefaction fan) that travels with the discrete characteristic speed defined at (3.4).

Notice that, because of the property **P3**, whenever a shock has strength less than $2\sqrt{\varepsilon}$, its left and right states are connected by a rarefaction curve and the shock travels with the averaged characteristic speed defined at (3.3)-(3.4).

The modified Rankine-Hugoniot conditions and the above definition of approximate solution determine a new way of approximately solving a Riemann problem. More precisely, given any pair of states v^L, $v^R \in \mathcal{V}$, consider the initial data

(3.19) $$\overline{u}(x) \doteq \begin{cases} u(v^L) & \text{if } x < 0, \\ u(v^R) & \text{if } x > 0. \end{cases}$$

We seek a self-similar, piecewise constant function $\omega = \omega(t, x)$, expressed in Riemann coordinates (v_1, v_2), that satisfies the Ψ^ε-RH conditions along each discontinuity line and attains the value (3.19) at time $t = 0$. The solution to this problem is provided by the following

ε-Approximate Riemann solver. Thanks to the property **P5**, given v^L, $v^R \in \mathcal{V}$, there exists a unique intermediate state $v^M \in \mathcal{V}$, and wave sizes $\sigma_1, \sigma_2 \in \mathcal{J}$, such

that
(3.20)
$$v^M = \Psi_1^\varepsilon(v^L, \sigma_1) \quad \text{if} \quad \text{the first family is LD, or GNL, or NGNL1}$$
$$v^L = \Psi_1^\varepsilon(v^M, \sigma_1) \quad \text{if} \quad \text{the first family is NGNL2}$$
$$v^R = \Psi_2^\varepsilon(v^M, \sigma_2) \quad \text{if} \quad \text{the second family is LD, or GNL, or NGNL1}$$
$$v^M = \Psi_2^\varepsilon(v^R, \sigma_2) \quad \text{if} \quad \text{the second family is NGNL2}$$

Then, as we have done in Section 2 for the exact solution of a Riemann problem, we construct the ε-approximate solution of the Riemann problem with initial data (3.19) by piecing together the ε-approximate solution of the Riemann problem with initial data

$$\overline{u}^1(x) \doteq \begin{cases} u(v^L) & \text{if } x < 0, \\ u(v^M) & \text{if } x > 0, \end{cases}$$

on the quadrant of the (t, x)-plane where $x \leq 0$, and the ε-approximate solution of the Riemann problem with initial data

$$\overline{u}^2(x) \doteq \begin{cases} u(v^M) & \text{if } x < 0, \\ u(v^R) & \text{if } x > 0, \end{cases}$$

on the quadrant where $x \geq 0$.

Observe that, if we consider in particular a system satisfying the assumption **(A')** (in which both family are NGNL and there hold (1.21), (1.22)), because of (3.20) the intermediate state $v^M \in \mathcal{V}$, and the wave sizes σ_1, σ_2, are uniquely determined by

(3.21) $$v^M = \Psi_1^\varepsilon(v^L)(\sigma_1), \qquad v^M = \Psi_2^\varepsilon(v^R)(\sigma_2).$$

In this case the ε-approximate solution $\omega(t, x)$ to each Riemann problem with initial data $\overline{u}^i$, $i = 1, 2$, is constructed as follows (expressed in Riemann coordinates).

1 - Waves. Initial data $\overline{u}^1$. There are 3 possibilities:

1) $v^L \in \mathcal{V}_1^0$: the solution consists of a shock of the first characteristic family travelling with speed $\lambda_1^{\psi,\varepsilon}(v^L, \sigma_1)$, i.e. (see figure 3.5)

(3.22) $$\omega(t, x) = \begin{cases} v^L & \text{if } x < \lambda_1^{\psi,\varepsilon}(v^L, \sigma_1) \cdot t, \\ v^M & \text{if } \lambda_1^{\psi,\varepsilon}(v^L, \sigma_1) \cdot t < x \leq 0. \end{cases}$$

2) $v^L \in \mathcal{V}_1^+$: the solution consists of

   a) $0 \leq \sigma_1 \leq -\delta_1(v^L)$: a piecewise constant rarefaction fan of the first characteristic family, i.e. (see figure 3.6)
   (3.23)
   $$\omega(t, x) = \begin{cases} v^L & \text{if } x < \widetilde{\lambda}_1^{\lfloor v^L \rfloor_1}(v^L) \cdot t, \\ (\ell\varepsilon, v_2^L) & \text{if } \widetilde{\lambda}_1^{\ell-1}(v^L) \cdot t < x < \widetilde{\lambda}_1^\ell(v^L) \cdot t, \quad \ell = \ell_1, \ldots, N_m, \\ v^M & \text{if } \widetilde{\lambda}_1^{N_m}(v^L) \cdot t < x \leq 0, \end{cases}$$

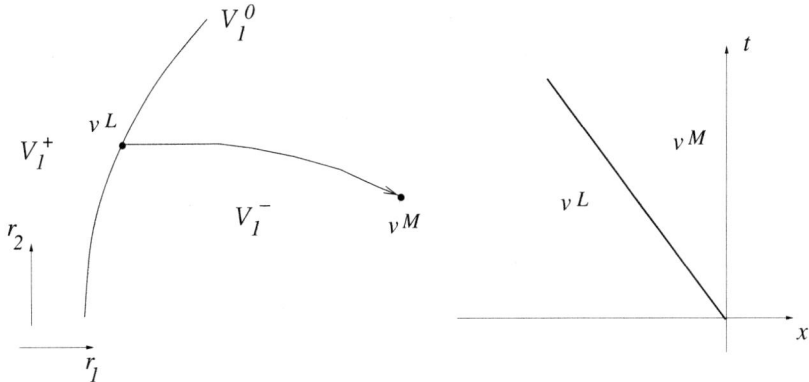

FIGURE 3.5. The formula (3.22)

where

$$\ell_1 = \lfloor v^L \rfloor_1 + 1, \qquad N_m = \begin{cases} \lfloor v^M \rfloor_1 & \text{if} \quad \lfloor v^M \rfloor_1 \varepsilon \neq v_1^M, \\ \lfloor v^M \rfloor_1 - 1 & \text{otherwise,} \end{cases}$$

while $\widetilde{\lambda}_1^\ell(v^L)$ are defined as in (3.4);

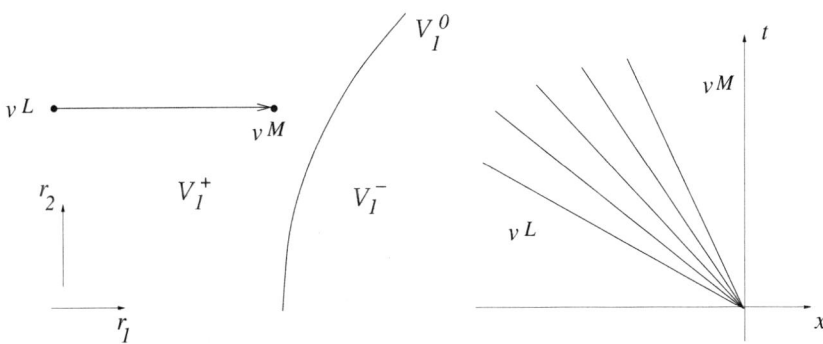

FIGURE 3.6. The formula (3.23)

b) $\sigma_1 < 0$ or $\sigma_1 \geq \nu_1^\varepsilon(v^L)$: a shock of the first characteristic family as in (3.22), travelling with speed $\lambda_1^{\psi,\varepsilon}(v^L, \sigma_1)$;

c) $-\delta_1(v^L) < \sigma_1 < \nu_1^\varepsilon(v^L)$: a composed wave of the first characteristic family, made of a piecewise constant rarefaction fan as in (3.23), of size

$$\widetilde{\sigma}_1 \doteq \sigma_1 - \mu_1^\varepsilon\big(\pi_1(v^L),\ \sigma_1 + \delta_1(v^L)\big),$$

followed by a shock $\widehat{\sigma}_1 \doteq \sigma_1 - \widetilde{\sigma}_1$, travelling with speed

$$\lambda_1^{\psi,\varepsilon}\big(\widetilde{v}^1, \widehat{\sigma}_1\big) \leq \widetilde{\lambda}_1^{\lfloor \widetilde{v}^1 \rfloor_1}(v^L), \qquad \widetilde{v}^1 \doteq R_1\big(v^L, \widetilde{\sigma}_1\big).$$

It follows that, setting $\ell_1 = \lfloor v^L \rfloor_1 + 1$ and $\ell_2 = \lfloor \widetilde{v}^1 \rfloor_1 - 1$, the solution is (see figure 3.7)

(3.24)
$$\omega(t,x) = \begin{cases} v^L & \text{if } x < \widetilde{\lambda}_1^{\lfloor v^L \rfloor_1}(v^L) \cdot t, \\ (\ell\varepsilon, v_2^L) & \text{if } \widetilde{\lambda}_1^{\ell-1}(v^L) \cdot t < x < \widetilde{\lambda}_1^{\ell}(v^L)) \cdot t, \quad \ell = \ell_1, \ldots, \ell_2, \\ \widetilde{v}^1 & \text{if } \widetilde{\lambda}_1^{\lfloor \widetilde{v}^1 \rfloor_1 - 1}(v^L) \cdot t < x < \lambda_1^{\psi,\varepsilon}(\widetilde{v}^1, \widehat{\sigma}_1) \cdot t, \\ v^M & \text{if } \lambda_1^{\psi,\varepsilon}(\widetilde{v}^1, \widehat{\sigma}_1) \cdot t < x \le 0, \end{cases}$$

with $\widetilde{\lambda}_1^{\ell}(v^L)$ defined as in (3.4);

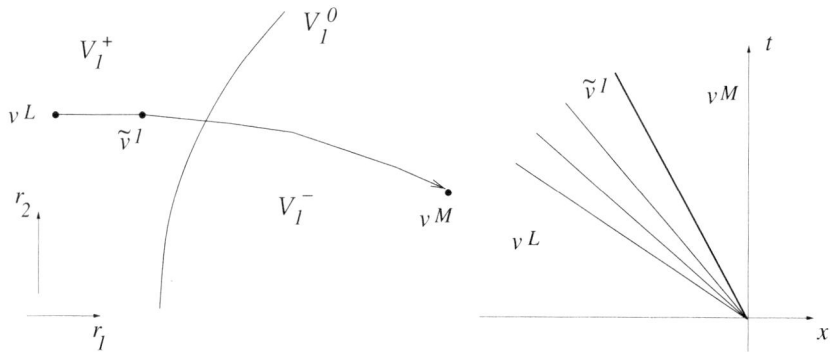

FIGURE 3.7. The formula (3.24)

3) $v^L \in \mathcal{V}_1^-$: the solution consists of

   a) $-\delta_1(v^L) \le \sigma_1 \le 0$: a piecewise constant rarefaction fan as in 2-a);

   b) $\sigma_1 > 0$ or $\sigma_1 \le \nu_1^\varepsilon(v^L)$: a shock as in 2-b);

   c) $\nu_1^\varepsilon(v^L) < \sigma_1 V - \delta_1(v^L)$: a composed wave as in 2-c), made of a piecewise constant rarefaction fan followed by a shock.

2 - Waves. Initial data $\overline{u}^2$. There are 3 possibilities:

1) $v^R \in \mathcal{V}_2^0$: the solution consists of a shock of the second characteristic family travelling with speed $\lambda_2^{\psi,\varepsilon}(v^R, \sigma_2)$, i.e. (see figure 3.8)

(3.25) $\qquad \omega(t,x) = \begin{cases} v^M & \text{if } 0 \le x < \lambda_2^{\psi,\varepsilon}(v^R, \sigma_2) \cdot t, \\ v^R & \text{if } x > \lambda_2^{\psi,\varepsilon}(v^R, \sigma_2) \cdot t. \end{cases}$

2) $v^R \in \mathcal{V}_2^+$: the solution consists of

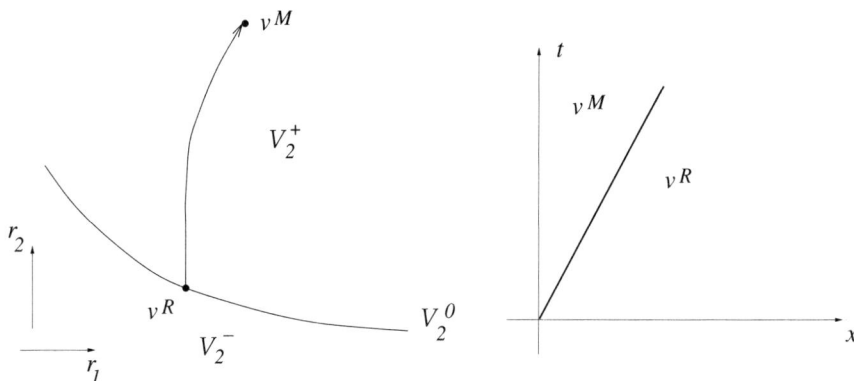

FIGURE 3.8. The formula (3.25)

a) $-\delta_2(v^R) \leq \sigma_2 \leq 0$: a piecewise constant rarefaction fan of the second characteristic family, i.e. (see figure 3.9)

(3.26)
$$\omega(t,\, x) = \begin{cases} v^M & \text{if } 0 \leq x < \widetilde{\lambda}_2^{\lfloor v^M \rfloor_2}(v^R) \cdot t, \\ (v_2^R,\, \ell\varepsilon) & \text{if } \widetilde{\lambda}_2^{\ell-1}(v^R) \cdot t < x < \widetilde{\lambda}_2^{\ell}(v^R) \cdot t, \quad \ell = \ell_1, \ldots, N_r, \\ v^R & \text{if } x > \widetilde{\lambda}_2^{N_r}(v^R) \cdot t, \end{cases}$$

where

$$\ell_1 = \lfloor v^M \rfloor_2 + 1, \qquad N_r = \begin{cases} \lfloor v^R \rfloor_1 & \text{if } \lfloor v^R \rfloor_1 \varepsilon \neq v_1^R, \\ \lfloor v^R \rfloor_1 - 1 & \text{otherwise,} \end{cases}$$

while $\widetilde{\lambda}_2^{\ell}(v^L)$ are defined as in (3.4);

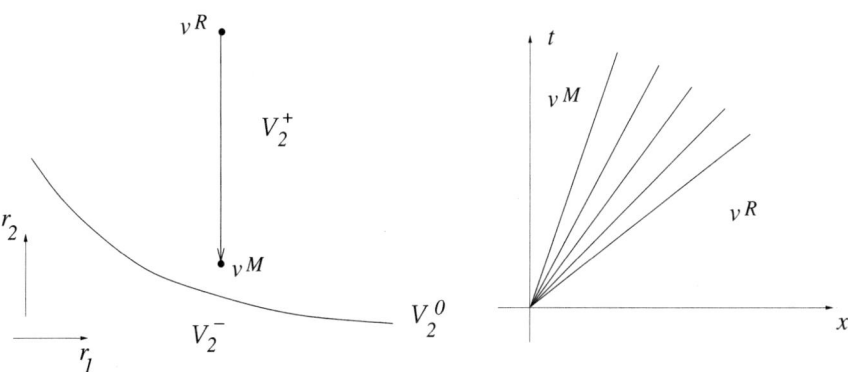

FIGURE 3.9. The formula (3.26)

b) $\sigma_2 > 0$ or $\sigma_2 \leq \nu_2^\varepsilon(v^R)$: a shock of the second characteristic family as in (3.25), travelling with speed $\lambda_2^{\psi,\varepsilon}(v^R, \sigma_2)$;

c) $\nu_2^\varepsilon(v^R) < \sigma_2 < -\delta_2(v^R)$: a composed wave of the second characteristic family, made of a shock of size

$$\widehat{\sigma}_2 \doteq \mu_2^\varepsilon\bigl(\pi_2(v^R),\ \sigma_2 + \delta_2(v^R)\bigr),$$

travelling with speed

$$\lambda_2^{\psi,\varepsilon}\bigl(\widetilde{v}^2, \widehat{\sigma}_2\bigr) \geq \widetilde{\lambda}_2^{\lfloor \widetilde{v}^2 \rfloor_2 - 1}(v^R), \qquad \widetilde{v}^2 \doteq R_2\bigl(v^R,\ \sigma_2 - \widehat{\sigma}_2\bigr),$$

followed by a piecewise constant rarefaction fan of size $\widetilde{\sigma}_2 \doteq \sigma_2 - \widehat{\sigma}_2$. Then, setting $\ell_1 = \lfloor \widetilde{v}^2 \rfloor_2 + 1$ and $\ell_2 = \lfloor v^R \rfloor_2$, the solution is (see figure 3.10)

(3.27)
$$\omega(t,x) = \begin{cases} v^M & \text{if } 0 \leq x < \lambda_2^{\psi,\varepsilon}\bigl(\widetilde{v}^2, \widehat{\sigma}_2\bigr) \cdot t, \\ \widetilde{v}^2 & \text{if } \lambda_2^{\psi,\varepsilon}\bigl(\widetilde{v}^2, \widehat{\sigma}_2\bigr) \cdot t < x < \widetilde{\lambda}_2^{\lfloor \widetilde{v}^2 \rfloor_2}(v^R) \cdot t, \\ (v_1^R, \ell\varepsilon) & \text{if } \widetilde{\lambda}_2^{\ell-1}(v^R) \cdot t < x < \widetilde{\lambda}_2^\ell(v^R) \cdot t, \qquad \ell = \ell_1, \ldots, \ell_2, \\ v^R & \text{if } x > \widetilde{\lambda}_2^{\lfloor v^R \rfloor_2}(v^R) \cdot t, \end{cases}$$

with $\widetilde{\lambda}_2^\ell(v^R)$ defined as in (3.4);

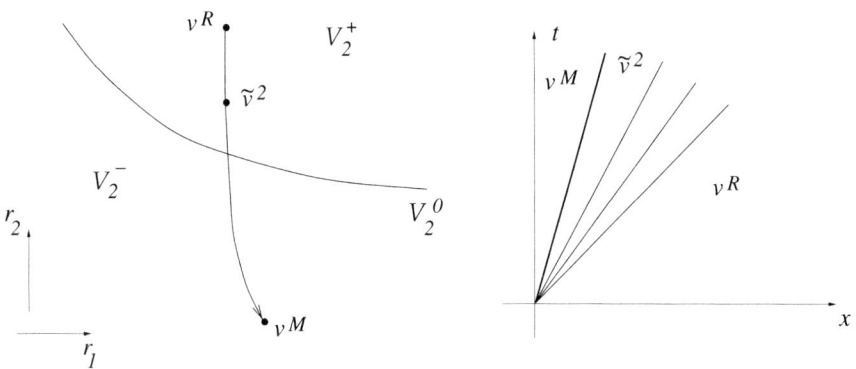

FIGURE 3.10. The formula (3.27)

3) $v^R \in \mathcal{V}_2^-$: the solution consists of

a) $0 \leq \sigma_2 \leq -\delta_2(v^R)$: a piecewise constant rarefaction fan as in 2-a);

b) $\sigma_2 < 0$ or $\sigma_2 \geq \nu_2(v^R)$: a shock as in 2-b);

c) $-\delta_2(v^R) \leq \sigma_2 < \nu_2^\varepsilon(v^R)$: a composed wave as in 2-c), made of a shock followed by a piecewise constant rarefaction fan.

REMARK 3.4. The ε-approximate solution $u = u(\omega(t,x))$ of the Riemann problem with initial data $\overline{u}^k$, $k \in \{1,2\}$, for a general 2×2 system satisfying the assumption (**A**), is constructed in an entirely similar way as above if the k-th family is NGNL satisfying either one of (2.8)-(2.9). On the other hand, in the case where the k-th family is LD or GNL, the ε-approximate solution $\omega(t,x)$ of the Riemann problem with initial data $\bar{u}^k$ is constructed as in [**BC1**]. In this case, there are only two possibilities.

1) The solution consists of a single jump of the k-th family of size σ_k (determined according with (3.20)) if either the k-th family is LD, or the k-th family is GNL and $\sigma_k < 0$. Hence $\omega(t,x)$ is defined in a similar way as in (3.22), (3.25).

2) The solution consists of a piecewise constant rarefaction fan of the k-th family of size σ_k (determined according with (3.20)) if the k-th family is GNL and $\sigma_k > 0$. Hence $\omega(t,x)$ is defined in a similar way as in (3.23), (3.26).

At this stage, for every fixed $\varepsilon > 0$, we have defined an algorithm for approximately solving the Riemann problem within the class of piecewise constant functions. Consider now a piecewise constant initial data $\bar{v}$ with bounded support and small total variation. An ε-approximate solution of the corresponding Cauchy problem can then be constructed (in the v-coordinates) by applying the above Riemann Solver at each point where two or more wave-fronts interact. More precisely, we first solve the Riemann problems at time $t = 0$ determined by the jumps in $\bar{v}$ using the ε-approximate Riemann Solver. The ε-approximate solution $v(t,x)$ is then prolonged until a time t_1 is reached, when the first set of interactions take place. Since $v(t_1, \cdot)$ is still a piecewise constant function, the corresponding Riemann problems can be solved applying again the above algorithm. This determines the solution v up to the time t_2 where the second set of interactions take place, etc...

Our goal is to prove that the ε-approximate solution generated by this algorithm are well defined for all $t \geq 0$. The core of the argument consists in showing that, as time increases,

1. the total variation of $v(t, \cdot)$ remains uniformly bounded;
2. the number of wave-fronts in v remains finite.

Regarding the first point, we shall proceed using the standard technique introduced by Glimm [**G**] based on the introduction of a nonlinear interaction potential $Q(v)$, and consisting in deriving a-priori estimates on the functional $\text{Tot.Var.}(v) + Q(v)$ that bounds the total variation of v.

Let $v : \mathbb{R} \to \mathbb{R}^2$ be a piecewise constant function with bounded support. Call $x_1 < \cdots < x_n$ the points where v suffers a jump. Assume that at each point x_α the corresponding Riemann problem with data $\omega_\alpha^L = v(x_\alpha-)$, $\omega_\alpha^R = v(x_\alpha+)$ is solved by our algorithm with a single ε-admissible wave-front. Call σ_i^α the size of this wave, where the index $i \in \{1,2\}$ refers to its family. This means that, if we are dealing in particular with a system satisfying the assumption (**A**′), by (3.21) we have

(3.28) $\quad \Psi_1^\varepsilon(\omega_\alpha^L, \sigma_1^\alpha) = \omega_\alpha^R, \quad$ if $\quad i = 1, \quad\quad \Psi_2^\varepsilon(\omega_\alpha^R, \sigma_2^\alpha) = \omega_\alpha^L, \quad$ if $\quad i = 2.$

Then, we define the *total strength* of waves in v as

(3.29) $$V^\varepsilon(v) \doteq \sum_{\alpha=1}^{N} |\sigma_i^\alpha|,$$

while the *interaction potential* is

$$(3.30) \quad Q^\varepsilon(v) \doteq K_1 \left[\sum_{x_\alpha < x_\beta} |\sigma_2^\alpha \cdot \sigma_1^\beta| + \sum_{i=1,2} \sum_{\sigma_i^\alpha \cdot \sigma_i^\beta < 0} |\sigma_i^\alpha \cdot \sigma_i^\beta| \right] + \sum_{i=1,2} \sum_{\sigma_i^\alpha \cdot \sigma_i^\beta > 0} |\sigma_i^\alpha \cdot \sigma_i^\beta|.$$

In the above expression of the interaction potential, K_1 denotes a suitable constant > 1 (whose precise value will be specified later: see (6.15) in Section 6.1) in the case where at least one of the family is NGNL, while $K_1 = 1$ in the classical case where both families are GNL or LD (when we can use the same interaction potential defined in [**BC1**]). Notice that, according with this definition, two fronts σ^α, σ^β, located at points $x_\alpha < x_\beta$, will be considered as *approaching* iff either σ^α belongs to the second characteristic family while σ^β belongs to the first one, or else they belong to the same family (no matter which angle they make). We will let $\mathcal{A}$ denote the set of approaching fronts contained in a solution v at time t.

As observed in the Introduction, a major technical difficulty that arises in establishing an a-priori bound on the Glimm-type functional $V^\varepsilon(v(t)) + Q^\varepsilon(v(t))$, stems from the fact that the front-tracking algorithm described in this section yields a class of "bad interactions" (involving shocks of an i-th NGNL family) which produce an increase of $Q^\varepsilon(v(t))$ of a magnitude which is not comparable with *any* order of the strength of the incoming fronts. To overcome this problem, we will perform in Section 6 an accurate analysis of the future history of the wave-fronts originated by such bad interactions and we will show that, although $Q^\varepsilon(v(t))$ increases after these interactions, it is still possible to derive a uniform bound on $Q^\varepsilon(v(t))$ and on $V^\varepsilon(v(t)) + Q^\varepsilon(v(t))$ for all $t \geq 0$.

To this purpose, we first introduce in Subsection 6.1 some rather technical notations concerning:
- the wave-fronts producing this particular type of bad interactions;
- the wave-fronts originated by these interactions;
- the interval of time $[t_0, \widehat{t}_0]$ within which $Q^\varepsilon(v(t))$ decreases by a quantity comparable with the growth caused by a bad interaction occurring at t_0 (see Definition 6.1);
- a Glimm-type functional E_i used to control the increase of $Q^\varepsilon(v(t))$ produced by a bad interaction of the i-th NGNL family (see (6.10)-(6.11));
- three Glimm-type functionals F_i, G_i, H_i used to measure the decrease of $Q^\varepsilon(v(t))$ due to (good) interactions involving wave-fronts of the i-th family (see (6.12)-(6.14)).

Then, relying on the basic interaction estimates on the difference between the strengths and the speeds of the incoming and outgoing waves at an interaction point collected in Section 5 (mostly of which are proved in [**AM2**] and in [**BC1**]), we establish in Lemma 6.1 of Subsection 6.2 the main estimates on the change in the values of V^ε and Q^ε across an interaction time. Next, we derive in Subsection 6.3 several lemmas (Lemmas 6.5-6.11) providing a priori bounds on the growth E_i of the interaction potential caused by the presence of bad interactions. Finally, we apply the estimates obtained in Subsections 6.2-6.3 to prove the main result of Section 6, Lemma 6.12, which provides a uniform a-priori bound on the increase of $Q^\varepsilon(v(t))$ and of $V^\varepsilon(v(t)) + Q^\varepsilon(v(t))$ on an arbitrary interval of time, and thus yields :

PROPOSITION 3.1. *There exist positive constants K_1, δ_1, c_0 independent of ε, for which, in the above setting, the following holds. Let $\bar v$ be a piecewise constant initial data with bounded support such that*

$$(3.31)_\delta \qquad V^\varepsilon(\bar v) + Q^\varepsilon(\bar v) < \delta,$$

with $\delta \leq \delta_1$, and let $v = v(t,x)$ be the corresponding ε-approximate solution constructed on some interval $[0,T[$, with $\varepsilon \leq c_0^{-1}$. Then, $v(t,\cdot)$ is a piecewise constant function that satisfies the bound

$$(3.32)\quad V^\varepsilon\big(v(t,\cdot)\big) + Q^\varepsilon\big(v(t,\cdot)\big) \leq (1 + c_0 \cdot \varepsilon)\Big[V^\varepsilon(\bar v) + Q^\varepsilon(\bar v)\Big], \qquad \forall\, 0 \leq t < T.$$

Now, fix $\delta_1 > 0$ according with Proposition 3.1 and, for any $\delta \leq \delta_1$, define the domain

$$(3.33)\quad \mathcal{D}^{\varepsilon,\delta} \doteq \big\{v \in \mathbb{L}^1(\mathbb{R};\mathbb{R}^2) \;:\; v \text{ is piecewise constant},\; V^\varepsilon(v) + Q^\varepsilon(v) < \delta\big\}.$$

Then, taking $\varepsilon < \varepsilon_0 \doteq c_0^{-1}$, if v is an ε-approximate solution with initial data $v(0,\cdot) \in \mathcal{D}^{\varepsilon,\delta}$, $\delta \leq \delta_1$, and if v can be constructed on some initial interval $[0,T[$, from (3.32) we derive the uniform bound

$$(3.34)\qquad V^\varepsilon\big(v(t,\cdot)\big) + Q^\varepsilon\big(v(t,\cdot)\big) \leq 2\delta, \qquad \forall\, 0 < t < T.$$

In particular, this means that the total variation of $v(t,\cdot)$ remains uniformly bounded for all $t \in [0,T]$. Then, thanks to the Property **P5**, every time an interaction occurs the corresponding Riemann Problem can be uniquely solved applying our algorithm. Thus, in order to prove that v can actually be prolonged beyond time $t = T$, we only need to show that the total number of wave-fronts and of interaction points remains finite. This result is accomplished in Section 7. Namely, we first use the estimates on the decrease of the interaction potential provided by Lemma 6.12 of Subsection 6.4, together with some additional interaction estimates established in Subsection 6.2 (Lemmas 6.2-6.4), to derive in Subsection 7.1 an a-priori bound on the total number of wave-fronts with strength $|\sigma| \geq 2\sqrt{\varepsilon}$.

PROPOSITION 3.2. *There exist $\varepsilon_0' \leq \varepsilon_0$, and $\delta_2 < \delta_1$, independent of $\varepsilon \leq \varepsilon_0'$, for which the following holds. Let $\bar v \in \mathcal{D}^{\varepsilon,\delta}$, $\delta \leq \delta_2$, and let $v = v(t,x)$ be the corresponding ε-approximate solution constructed on some interval $[0,T[$, with $\varepsilon \leq \varepsilon_0'$. Then, all of the shocks of v with strength $|\sigma| > 2\sqrt{\varepsilon}$ are located along a finite number of polygonal lines.*

From Proposition 3.2 and by property **P3** it follows that, if $v(0,\cdot) \in \mathcal{D}^{\varepsilon,\delta_2}$ and we construct the corresponding ε-solution v on some interval $[0,T[$, then outside a finite number of polygonal lines v behaves as the solution of a system with coinciding shock and rarefaction curves. This, in particular, implies that, whenever two small waves interact,

1. the outgoing waves have the same size of the incoming ones;
2. the possible generation of new wave-fronts can occur only when are involved in the interaction wave-fronts of different family or wave-fronts of the same family having sizes of opposite sign.

Thanks to these two key properties we can establish in Subsection 7.2:

PROPOSITION 3.3. *Let v be as in Proposition 3.2. Then the set of all points where two or more wave-fronts interact has no limit point in the (t, x)-plane.*

Then, using Propositions 3.1-3.3, we derive in Subsection 7.3 :

PROPOSITION 3.4. *Let $\delta_2 < 1/2$ be a constant chosen according with Proposition 3.2. Then, for any $0 < \varepsilon \leq \varepsilon_0'$, and for every initial data $v(0, \cdot) = \overline{v} \in \mathcal{D}^{\varepsilon, \delta}$, with $\delta \leq \delta_2$, the corresponding ε-approximate solution $v = v(t, x)$ is well defined for all $t > 0$ and contains a finite number of wave-fronts and interaction points in the (t, x)-plane. Moreover*

$$(3.35) \qquad V^\varepsilon\big(v(t, \cdot)\big) + Q^\varepsilon\big(v(t, \cdot)\big) < 2\,\delta, \qquad \forall\, t \geq 0.$$

To denote the unique, globally defined, ε-approximate solution $v(t, \cdot)$ with initial data $\overline{v} \in \mathcal{D}^{\varepsilon, \delta_2}$, we shall often use the semigroup notation

$$(3.36) \qquad S_t^\varepsilon \overline{v} \doteq v(t, \cdot).$$

The next part of the proof of Theorem 1 consists in showing that the map

$$(3.37) \qquad S^\varepsilon : [0, \infty[\, \times \mathcal{D}^{\varepsilon, \delta_2} \mapsto \mathcal{D}^{\varepsilon, 2\delta_2}, \qquad (t, \overline{v}) \to S_t^\varepsilon \overline{v},$$

is globally Lipschitz in the $\mathbb{L}^1$ norm, with a Lipschitz constant independent of ε. To estimate the distance between two approximate solutions v, w, the basic idea is, as in [**BC1**], to construct a one-parameter family of solutions v^θ connecting v, w, and then study how the length of the path $\gamma_t : \theta \to v^\theta(t, \cdot)$ varies in time. More precisely, given two piecewise constant initial data $\overline{v}, \overline{w} \in \mathbb{L}^1$, with small total variation, we consider continuous paths $\gamma : \theta \to \overline{v}^\theta$ that connect $\overline{v}$ with $\overline{w}$ by merely shifting the position of the jumps. Then, by studying the rates at which the jumps in the corresponding solution $\gamma_t(\theta) \doteq S_t^\varepsilon \overline{v}^\theta$ are shifted, for any fixed time $t > 0$, we will derive an a-priori estimate on the norm of a generalized tangent vector. In turn, this provides a bound on the length of γ_t and hence on the distance between $v(t, \cdot) \doteq S_t^\varepsilon \overline{v}$ and $w(t, \cdot) \doteq S_t^\varepsilon \overline{w}$.

In particular, our paths are *pseudopolygonal paths*, i.e. finite concatenation of *elementary paths* of the form

$$(3.38) \qquad \gamma(\theta) = \sum_{\alpha=1}^{N} \omega_\alpha \cdot \chi_{]x_{\alpha-1}^\theta,\, x_\alpha^\theta[}, \qquad x_\alpha^\theta = \overline{x}_\alpha + \xi_\alpha \theta, \qquad \theta \in]a, b[,$$

with $x_{\alpha-1}^\theta < x_\alpha^\theta$ for all $\theta \in]a, b[$ and $\alpha = 1, \ldots, N$. Here χ_I is the characteristic function of the interval I, $\omega_1, \ldots, \omega_N \in \mathbb{R}$ are constant states and ξ_α is the shift rates of the jump at x_α. Using entirely similar arguments to those developed in [**BC**], one can show that a basic property of our algorithm is that it preserves pseudopolygonal paths. In fact, if we consider a pseudopolygonal path $\gamma_0 : \theta \mapsto \overline{v}^\theta$ with $\overline{v}^\theta \in \mathcal{D}^{\varepsilon, \delta_2}$ for all $\theta \in [0, 1]$, then, for all $t > 0$, the path

$$(3.39) \qquad \gamma_t : \theta \mapsto v^\theta(t, \cdot) \doteq S_t^\varepsilon \overline{v}^\theta$$

is also a pseudopolygonal path. Moreover, there are finitely many parameters values $0 = \theta_0 < \theta_1 < \cdots < \theta_N = 1$ such that the wave-front configuration of the solution

v^θ on $[0,t]$ remains the same as θ ranges on each of the open interval $J_h =]\theta_{h-1}, \theta_h[$. Since, by definition, the $\mathbb{L}^1$-length of a continuous path $\gamma : [a,b] \to \mathbb{L}^1(\mathbb{R})$ is

$$
(3.40) \quad \|\gamma\|_{\mathbb{L}^1} \doteq \sup \left\{ \sum_{i=1}^N \|\gamma(\theta_i) - \gamma(\theta_{i-1})\|_{\mathbb{L}^1} \ :\ N \geq 1,\ a = \theta_0 < \theta_1 < \cdots < \theta_N = b \right\},
$$

the length of the pseudopolygonal path γ_t at (3.39) will be just the sum of the lengths of the elementary paths obtained by restricting γ_t to each subinterval J_h. Then, to estimate the length of γ_τ for any fixed $\tau > 0$, it is sufficient to consider the case where $\gamma_0 :\,]a,b[\to \mathbb{L}^1$ is an elementary path and the wave-front configurations of the corresponding ε-approximate solutions $v^\theta(t,x) \doteq S_t^\varepsilon \gamma_0(\theta)$, on the strip $[0,\tau] \times \mathbb{R}$, are the same for all $\theta \in]a,b[$.

If $v^\theta(t,\cdot)$ is the piecewise constant function

$$
(3.41) \quad v^\theta(t,x) = \sum_{\alpha=1}^{N(t)} \omega_\alpha \cdot \chi_{]x^\theta_{\alpha-1}(t),\, x^\theta_\alpha(t)[}(x), \qquad x^\theta_\alpha(t) = x_\alpha(t) + \xi_\alpha \theta,
$$

and we let $\Delta v^\theta(t, x_\alpha) \doteq \omega_{\alpha+1} - \omega_\alpha$ denote the jump of $v^\theta(t,\cdot)$ at x_α, the length of $\gamma_t : \theta \mapsto v^\theta(t,\cdot)$ is measured by

$$
(3.42) \quad \|\gamma_t\|_{\mathbb{L}^1} = \int_a^b \sum_\alpha \left|\Delta v^\theta(t, x_\alpha)\right| \left|\frac{dx^\theta_\alpha}{d\theta}\right| d\theta.
$$

Thus, in order to relate the length of γ_τ with the length of the path γ_0 of initial conditions, we will study how the integrand in (3.42)

$$
(3.43) \quad V_\xi^\varepsilon\left(v^\theta(t,\cdot)\right) \doteq \sum_\alpha \left|\Delta v^\theta(t, x_\alpha)\right| \left|\frac{dx^\theta_\alpha}{d\theta}\right| = \sum_\alpha |\omega_{\alpha+1} - \omega_\alpha||\xi_\alpha|
$$

varies in time, for any fixed $\theta \in]a,b[$. Clearly, this sum can change only at times $\bar{t}$ where some interaction takes place between wave-fronts of v^θ. In general, an interaction occurring at time $\bar{t}$ may well produce an increase of the sum at (3.43). However, by carefully studying the relation between $\|\gamma_{\bar{t}+}\|_{\mathbb{L}^1}$ and $\|\gamma_{\bar{t}-}\|_{\mathbb{L}^1}$, one can show that, as in the case of systems satisfying the standard hypothesis of genuine non-linearity or linearly degeneracy, also for systems satisfying our basic assumption (**A**) it is possible to control such increase of $\|\gamma_t\|_{\mathbb{L}^1}$ in terms of an interaction potential. Hence, as in [**BC1**], we shall define a weighted length $\|\cdot\|_\varepsilon$ equivalent to the standard $\mathbb{L}^1$-length and then will prove in Section 8 that, for every elementary path $\gamma_t : \theta \mapsto v^\theta(t,\cdot)$ of ε-approximate solutions, there holds $\|\gamma_t\|_\varepsilon \leq L\|\gamma_0\|_\varepsilon$, for some constant L independent of ε.

Let γ be the elementary path at (3.38) and assume that the Riemann problem with left data ω_α and right data $\omega_{\alpha+1}$ is solved by our algorithm with a single ε-admissible wave-front σ_i^α of the i-th family, $i \in \{1,2\}$ (in the particular case of a system satisfying the assumption (**A'**) this means that (3.28) holds with ω_α^L replaced by ω_α and ω_α^R replaced by $\omega_{\alpha+1}$). We define the *weighted length* of γ by setting

$$
(3.44) \quad \|\gamma\|_\varepsilon \doteq \sum_\alpha (b-a)|\sigma_i^\alpha \xi_\alpha| \cdot W_i^\alpha,
$$

(3.45)
$$W_i^\alpha \doteq \left[1 + K_3\left[\sum_{\substack{j \neq i \\ (\bar{x}_\alpha - \bar{x}_\beta)\cdot(i-j)<0}} |\sigma_j^\beta| + \sum_{\sigma_i^\beta \cdot \sigma_i^\alpha < 0} |\sigma_i^\beta|\right] + K_2 \sum_{\sigma_i^\beta \cdot \sigma_i^\alpha > 0} |\sigma_i^\beta|\right] e^{K_4 Q^\varepsilon(\gamma)},$$

where K_2, K_3, K_4, are positive constants to be determined later and $Q^\varepsilon(\gamma) \doteq Q^\varepsilon(\gamma(\theta))$ for any $\theta \in]a, b[$. As for the interaction potential, we will choose the constants K_2, K_3 in the weight (3.45) so that $K_3 > K_2$ in the case where at least one of the family is NGNL (cfr. Lemmas 8.9-8.10), while we can take $K_2 = K_3$ in the classical case where both families are GNL or LD (recovering the same weights used in [**BC1**]). Whenever γ is a pseudopolygonal path, its *weighted length* $\|\gamma\|_\varepsilon$ is defined as the sum of the weighted lengths of its elementary paths. Notice that, because of (3.10), there exists some constant $C_0 > 0$ independent of ε, such that

(3.46)
$$C_0^{-1}\|\gamma\|_{\mathbb{L}^1} \leq \|\gamma\|_\varepsilon \leq C_0\|\gamma\|_{\mathbb{L}^1}.$$

In order to establish an a-priori bound on the weighted length $\|\gamma_t\|_\varepsilon$ of an elementary path $\gamma_t : \theta \mapsto v^\theta(t, \cdot)$ of ε-approximate solutions, we will first derive in Lemma 8.1 of Subsection 8.1 the basic estimates on the change of the distance after an interaction time, for two nearby solutions obtained one from the other by shifting the position of the jumps in the initial data. Since the weighted length $\|\gamma_t\|_\varepsilon$ may well increase after the same type of bad interactions that produce a growth of the interaction potential, we then use the same technique developed in Section 6 to show that one can counterbalance such an increase of $\|\gamma_t\|_\varepsilon$ with the decrease of $\|\gamma_t\|_\varepsilon$ determined by the (good) interactions occurring at later times (and involving waves originated from the bad interactions). For this reason, we will use throughout Section 8 the same notations introduced in Subsection 6.1 concerning the wave-fronts of bad type and the fronts originated from bad interactions. Moreover, we will give two type of definitions of the time interval within which the weighted length $\|\gamma_t\|_\varepsilon$ decreases by a quantity comparable with the growth caused by a bad interaction (see Definitions 8.1-8.2), and we will define in (8.35)-(8.39) four functionals $\mathcal{E}_i, \mathcal{F}_i, \mathcal{G}_i, \mathcal{H}_i$ that are the analog of the Glimm-type interaction functional introduced in Section 6.1. A further functional $\mathcal{M}_i$ is also defined in (8.40) to measure the decrease of $\|\gamma_t\|_\varepsilon$ caused by (good) interactions in which the (front)·(shift) products have opposite signs. All such notations are collected in Section 8.2. Relying on the estimates provided by Lemma 8.1, we then establish in Lemma 8.2 of Subsection 8.3 the main estimate on the change in the values of the weighted length $\|\gamma_t\|_\varepsilon$ of an elementary path γ_t across an interaction time. Next, we derive in Subsection 8.4 several lemmas (Lemmas 8.3-8.8) providing a priori bounds on the growth $\mathcal{E}_i$ of the weighted length caused by the presence of bad interactions. Finally, we apply the estimates obtained in Subsections 8.3 to prove the main results of Section 8, Lemmas 8.9-8.11, which provide a uniform a-priori bound on the increase of $\|\gamma_t\|_\varepsilon$ on an arbitrary interval of time, and thus yield:

PROPOSITION 3.5. *In the above setting, there exist $\varepsilon_0'' \leq \varepsilon_0'$, $\delta_3 \leq \delta_2$, and positive constants K_2, K_3, K_4, L_1, independent of ε, such that the following holds. Let $\gamma_0 : \theta \mapsto \bar{v}^\theta \in \mathcal{D}^{\varepsilon,\delta_3}$, $\theta \in]a,b[$, and $\gamma_T : \theta \mapsto v^\theta(T, \cdot) \doteq S_T^\varepsilon \bar{v}^\theta$, $T > 0$, be elementary paths of ε-approximate solutions, with $\varepsilon \leq \varepsilon_0''$. Assume that on the strip $[0,T] \times \mathbb{R}$ the wave-front configuration of the ε-solution v^θ remains constant for* *all θ*. Then, there holds

(3.47)
$$\|\gamma_t\|_\varepsilon \leq L_1 \|\gamma_0\|_\varepsilon, \qquad \forall\ 0 \leq t \leq T.$$

PROPOSITION 3.6. *There exist constants $\delta^* \in]0, \delta_3]$, $L_2 > 0$, independent of ε, such that, setting*

$$\mathcal{D}^\varepsilon \doteq \left\{ v \in \mathbb{L}^1(\mathbb{R}; \mathbb{R}^2)\ :\ v \text{ is piecewise constant},\ V^\varepsilon(\overline{v}) + Q^\varepsilon(\overline{v}) < \delta^* \right\},$$

the following holds:

(i)

(3.48)
$$S^\varepsilon_t \overline{v} \in \mathcal{D}^{\varepsilon, 2\delta^*}, \qquad \forall\ \overline{v} \in \mathcal{D}^\varepsilon,$$

(ii)

(3.49)
$$\left\| S^\varepsilon_t \overline{v} - S^\varepsilon_t \overline{w} \right\|_{\mathbb{L}^1} \leq L_2 \|\overline{v} - \overline{w}\|_{\mathbb{L}^1}, \qquad \forall\ t \geq 0,\ \overline{v}, \overline{w} \in \mathcal{D}^\varepsilon.$$

Relying on Proposition 3.6, we can now construct a Standard Riemann Semigroup generated by (1.1) as a suitable limit of the flows S^ε. Consider the sequence of maps $S^{\varepsilon_n} : \mathcal{D}^{\varepsilon_n} \mapsto \mathcal{D}^{\varepsilon_n, 2\delta^*}$, with $\varepsilon_n = 2^{-n}$, and define the map S on the domain

(3.50)
$$\mathcal{D} \doteq \Big\{ \overline{v} \in \mathbb{L}^1(\mathbb{R}; \mathbb{R}^2)\ :\ \exists\ \{\overline{v}_n\}_{n \in \mathbb{N}},\ \overline{v}_n \in \mathcal{D}^{\varepsilon_n}\ \forall\ n,$$
$$\lim_n \overline{v}_n = \overline{v},\ \limsup_n \left[V^{\varepsilon_n}(\overline{v}_n) + Q^{\varepsilon_n}(\overline{v}_n) \right] \leq \delta^*/2 \Big\},$$

by setting

(3.51)
$$S_t \overline{v} \doteq \lim_{n \to +\infty} S^{\varepsilon_n}_t \overline{v}_n,$$

where $\{\overline{v}_n\}_{n \in \mathbb{N}}$ is any sequence with $\overline{v}_n \in \mathcal{D}^{\varepsilon_n}$ converging to $\overline{v}$ in $\mathbb{L}^1$, and such that

$$\limsup_n \left[V^{\varepsilon_n}(\overline{v}_n) + Q^{\varepsilon_n}(\overline{v}_n) \right] \leq \delta^*/2.$$

Notice that, thanks to (3.32), the domain $\mathcal{D}$ at (3.51) is positively invariant with respect to S. Then, we conclude the proof of Theorem 1 by establishing in Section 9:

PROPOSITION 3.7. *The semigroup S in (3.51) is well defined and satisfies the conditions (i)-(v) stated in Theorem 1.*

4. THE ALGORITHM

In this section, for any fixed $\varepsilon > 0$, if the k-th family is NGNL and satisfies the assumption (**A**), we first construct a suitable approximation $F^{k,\varepsilon}$ of the flux function F which, in turn, determines a corresponding definition of the *ε-approximate Hugoniot curve* of the k-th family S^ε_k. Next, we modify the curve S^ε_k by defining a curve $S^{\psi,\varepsilon}_k$ that coincides with the rarefaction curve R_k for small amplitudes, and with the ε-approximate Hugoniot curves S^ε_k for large ones. A suitable interpolation between the two curves is used for intermediate values. Relying on $S^{\psi,\varepsilon}_k$, we then define the *ε-approximate mixed curve* T^ε_k describing the states connected by an approximate k-th wave, which lie on opposite sides w.r.t. the curve Γ_k at (1.4). Finally, we give the definition of the *ε-approximate elementary curves* Ψ^ε_i, $i = 1, 2$

(used to construct the ε-approximate solution of a Riemann problem) that coincide with the ones introduced in [**BC1**] in the case where the i-th family is GNL or LD.

A more detailed discussion of the basic properties of such maps and the proofs of their convergence to the corresponding maps F, S_k, T_k, Ψ_k (involved in the definition of the exact solution of a Riemann problem for a 2×2 system satisfying the assumption (**A**)) can be found in [**AM2**]. Throughout the following, we shall often adopt the notation

$$\lfloor v \rfloor_i \doteq \left\lfloor \frac{v_i}{\varepsilon} \right\rfloor, \qquad i = 1, 2, \tag{4.1}$$
$$\lfloor\!\lfloor v \rfloor\!\rfloor_i \doteq R_i\big(v_j\, e_j,\, \lfloor v \rfloor_i \varepsilon\big) \quad (j \neq i),$$

where $\{e_1, e_2\}$ is the canonical basis of $\mathbb{R}^2$, and $\lfloor a \rfloor$ represents the integer part of $a \in \mathbb{R}$. As usual, the Landau symbol $\mathcal{O}(1)$ always denote a quantity whose absolute value satisfies a uniform bound, depending only on the system (1.1) and not on ε or on the functions of the variable v appearing in the sequel. We also recall the notation

$$\delta_k(v) \doteq v_k - \widetilde{\gamma}_k(v_j) \qquad (j \neq k), \tag{4.2}$$

$$\pi_{k,i}(v) \doteq \begin{cases} v_i & \text{if } i \neq k, \\ \widetilde{\gamma}_k(v_j) \quad (j \neq k) & \text{if } i = k, \end{cases} \tag{4.3}$$

introduced in Section 2 for a k-th NGNL family.

4.1. Approximate flux function.

Here, in connection with a k-th NGNL family satisfying the assumption (**A**), we introduce an ε-approximation of the flux function $u \mapsto F(u)$. Given $u = u(v)$, $v = (v_1, v_2) \in \mathcal{V}$, assume that

$$h_k \varepsilon \leq v_k < (h_k + 1)\varepsilon$$

for some integers h_k, and set

$$\overline{h}' \doteq \min\{0, h_k\}, \qquad \overline{h}'' \doteq \max\{0, h_k\}. \tag{4.4}$$

Then, the ε-approximate flux $u \mapsto F^{k,\varepsilon}(u)$ is defined as

$$F^{k,\varepsilon}\big(u(v)\big) \doteq F\big(u(\pi_k(v))\big) + \operatorname{sgn}\big(\delta_k(v)\big) \left(\sum_{\ell = \overline{h}'}^{\overline{h}''} \widetilde{\lambda}_k^\ell(v) \left(u(w_k^{\ell+1}) - u(w_k^\ell) \right) \right), \tag{4.5}$$

where $\widetilde{\lambda}_k^\ell(v)$ denotes the discrete speed defined in (3.4), and

$$w_k^{\overline{h}'} \doteq \begin{cases} v & \text{if } h_k \varepsilon \leq \widetilde{\gamma}_k(v_j) \quad (j \neq k), \\ \pi_k(v) & \text{otherwise,} \end{cases}$$

$$w_k^\ell \doteq R_k\big(v_j\, e_j,\, \ell\varepsilon\big) \qquad \text{if } \ell \neq \overline{h}',\ \overline{h}'' + 1, \tag{4.6}$$
$$(j \neq k)$$

$$w_k^{\overline{h}''+1} \doteq \begin{cases} \pi_k(v) & \text{if } h_k \varepsilon \leq \widetilde{\gamma}_k(v_j) \quad (j \neq k), \\ v & \text{otherwise.} \end{cases}$$

Notice that the map $u \mapsto F^{k,\varepsilon}(u)$ defined by (4.4)-(4.6), is continuous on Ω, continuously differentiable on

(4.7) $$\Omega_k^\varepsilon \doteq \{\, u = u(v_1, v_2) \in \Omega \ : \ v_k \notin \mathbb{Z}\varepsilon \,\},$$

and, at any point $u \in \Omega \setminus \Omega_k^\varepsilon$, it admits left and right directional derivatives along the characteristic fields $r_i(u)$, $i = 1, 2$. The derivatives of $u \mapsto F^{k,\varepsilon}(u)$ along the fields r_i are written

(4.8) $$DF^{k,\varepsilon}(u) \cdot r_i(u) \doteq \lim_{s \to 0} \frac{F^{k,\varepsilon}(u + s\, r_i(u)) - F^{k,\varepsilon}(u)}{s}.$$

The right and left directional derivatives of $F^{k,\varepsilon}$ along r_i are denoted $DF^{k,\varepsilon+}(u) \cdot r_i(u)$, and $DF^{k,\varepsilon-}(u) \cdot r_i(u)$, respectively. We call $\lambda_i^{k,\varepsilon}(u)$, $\lambda_i^{k,\varepsilon+}(u)$, and $\lambda_i^{k,\varepsilon-}(u)$, the eigenvalues of $A^{k,\varepsilon}(u) \doteq DF^{k,\varepsilon}(u)$, $A^{k,\varepsilon+}(u) \doteq DF^{k,\varepsilon+}(u)$, and $A^{k,\varepsilon-}(u) \doteq DF^{k,\varepsilon-}(u)$. Similar notations are used for the corresponding left and right eigenvectors. The right eigenvectors $r_i^{k,\varepsilon}$, $r_i^{k,\varepsilon\pm}$ are chosen so that

(4.9) $$\langle l_i(u),\, r_i^{k,\varepsilon}(u) \rangle = \langle l_i(u),\, r_i^{k,\varepsilon\pm}(u) \rangle = 1, \qquad i = 1, 2,$$

and the left eigenvector $l_i^{k,\varepsilon}$, $l_i^{k,\varepsilon\pm}$ are normalized as in (2.3) by
(4.10)
$$\langle l_i^{k,\varepsilon}(u),\, r_j^{k,\varepsilon}(u) \rangle = \langle l_i^{k,\varepsilon\pm}(u),\, r_j^{k,\varepsilon\pm}(u) \rangle = \begin{cases} 1 & \text{if } i = j \\ 0 & \text{if } i \neq j \end{cases} \qquad i,j = 1,2.$$

REMARK 4.1. A straight computation shows that, at any point $u = u(v) \in \Omega_k^\varepsilon$, the derivative of $F^{k,\varepsilon}$ along the field $r_k(u)$ take the form

(4.11) $$DF^{k,\varepsilon}(u) \cdot r_k(u) = \widetilde{\lambda}_k^{\lfloor v \rfloor_k}(v)\, r_k(u).$$

Similar formulas hold for the left and right directional derivatives of $F^{k,\varepsilon}$ along the fields $r_k(u)$ at any point $u \in \Omega \setminus \Omega_k^\varepsilon$. In particular, at any point $u = u(v_1, v_2)$ with $v_k \in \mathbb{Z}\varepsilon$, one has

(4.12) $$\begin{aligned} DF^{k,\varepsilon+}(u) \cdot r_k(u) &= \widetilde{\lambda}_k^{\lfloor v \rfloor_k}(v)\, r_i(u), \\ DF^{k,\varepsilon-}(u) \cdot r_k(u) &= \widetilde{\lambda}_k^{\lfloor v \rfloor_k - 1}(v)\, r_k(u). \end{aligned}$$

Condition (4.9) and the expression (4.11) of $DF^{k,\varepsilon}$ imply that, at any point $u = u(v) \in \Omega$, one has

(4.13) $$\begin{aligned} \lambda_k^{k,\varepsilon+}(u) &= \widetilde{\lambda}_k^{\lfloor v \rfloor_k}(v), \\ \lambda_k^{k,\varepsilon-}(u) &= \begin{cases} \widetilde{\lambda}_k^{\lfloor v \rfloor_k}(v) & \text{if } v_k \notin \mathbb{Z}\varepsilon, \\ \widetilde{\lambda}_k^{\lfloor v \rfloor_k - 1}(v) & \text{otherwise,} \end{cases} \\ r_k^{k,\varepsilon}(u) &= r_k^{k,\varepsilon\pm}(u) = r_k(u). \end{aligned}$$

The derivatives of $F^{k,\varepsilon}$ converge uniformly to the corresponding derivatives of F. Namely, one has

LEMMA 4.1. *For $i1, 2$, $\varepsilon > 0$, and any $u = u(v) \in \Omega_k^\varepsilon$, there holds*

$$\left| DF^{k,\varepsilon}(u) \cdot r_i(u) - DF(u) \cdot r_i(u) \right| = \mathcal{O}(1) |\delta_k(v)| \varepsilon. \tag{4.14}$$

A proof of Lemma 4.1 can be found in [**AM2**, Proposition 3.1]. Uniform estimates as (4.14) hold for the right and left directional derivatives of $u \mapsto F^{k,\varepsilon}(u)$ along the characteristic fields $r_i(u)$, at any point $u \in \Omega \setminus \Omega_k^\varepsilon$.

REMARK 4.2. From Lemma 4.1 and by Remark 4.1 it follows that, for $i = 1, 2$, $u = u(v) \in \Omega$, and $\varepsilon > 0$, there holds

$$\begin{aligned} \left| \lambda_i^{k,\varepsilon\pm}(u) - \lambda_i(u) \right| &= \mathcal{O}(1) |\delta_k(v)| \varepsilon, \\ \left| r_i^{k,\varepsilon\pm}(u) - r_i(u) \right| &= \mathcal{O}(1) |\delta_k(v)| \varepsilon, \\ \left| l_i^{k,\varepsilon\pm}(u) - l_i(u) \right| &= \mathcal{O}(1) |\delta_k(v)| \varepsilon, \end{aligned} \tag{4.15}$$

and

$$\left| \lambda_k^{k,\varepsilon\pm}\big(u(R_j(v,\sigma))\big) - \lambda_k^{k,\varepsilon\pm}(u) \right| = \mathcal{O}(1) |\sigma|, \qquad j \neq k. \tag{4.16}$$

4.2. Approximate Hugoniot curves.

Let $F^{k,\varepsilon}$ be the ε-approximate flux constructed for a k-th NGNL family in the previous subsection. Then, given two points $v, v' \in \mathcal{V}$, and setting $u \doteq u(v)$, $u' \doteq u(v')$, we call $\lambda_i^{k,\varepsilon}(v, v')$, $i = 1, 2$ the eigenvalues of the averaged matrix

$$A^{k,\varepsilon}(u, u') \doteq \int_0^1 DF^{k,\varepsilon}\big(u' + \xi(u' - u)\big) \, d\xi \tag{4.17}$$

ordered so that

$$\lambda_1^{k,\varepsilon}(v, v') < 0 < \lambda_2^{k,\varepsilon}(v, v') \tag{4.18}$$

and $l_i^{k,\varepsilon}(v, v')$, $r_i^{k,\varepsilon}(v, v')$, respectively, the corresponding left and right eigenvectors of $A^{k,\varepsilon}(v, v')$ normalized by
(4.19)
$$\langle l_i(v,v'), r_i^{k,\varepsilon}(v,v') \rangle = 1 \qquad \langle l_i^{k,\varepsilon}(v,v'), r_j^{k,\varepsilon}(v,v') \rangle = \begin{cases} 1 & \text{if } i = j \\ 0 & \text{if } i \neq j \end{cases}.$$

DEFINITION 4.1. Let u^L, u^R be two points in Ω. We say that (u^L, u^R) satisfies the *ε-Rankine-Hugoniot condition (ε-RH) of the k-th family* if, for some scalar λ, there holds

$$F^{k,\varepsilon}(u^R) - F^{k,\varepsilon}(u^L) = \lambda \, (u^R - u^L) \qquad (-1)^k \lambda > 0. \tag{4.20$_k$}$$

Notice that the solutions of (4.20)$_k$ are precisely the non-trivial solutions of the scalar equation

$$\langle l_j^{k,\varepsilon}(u^L, u^R), u^R - u^L \rangle = 0, \qquad j \neq k. \tag{4.21$_k$}$$

Moreover, the map $u^R \mapsto l_j^{k,\varepsilon}(u^L, u^R)$ has the same regularity of the map $u^R \mapsto F^{k,\varepsilon}(u^L, u^R)$, i.e. it's continuous on Ω, continuously differentiable on Ω_k^ε and, at any point in $\Omega \setminus \Omega_k^\varepsilon$, admits left and right directional derivatives along the vector fields $r_j(u^R)$. Therefore, by applying the implicit function theorem, one can show the existence of a continuous, piecewise smooth curve of solutions of (4.21)$_k$

$\sigma \mapsto S_k^\varepsilon(u^L, \sigma)$ that will be denoted as the *ε-approximate shock curve* through u^L of the k-th characteristic family. Such a curve can be parameterized by

$$(4.22) \qquad S_{k,j}^\varepsilon(v, \sigma) = \begin{cases} v_k + \sigma & \text{if } j = k, \\ v_j + \widetilde{S}_k^\varepsilon(v, \sigma) & \text{if } j \neq k, \end{cases}$$

where $S_{k,j}^\varepsilon$, $j = 1, 2$ are the components of S_k^ε, and $\widetilde{S}_k^\varepsilon$ denotes some piecewise smooth map. Call $\lambda_k^{s,\varepsilon}(v, s) \doteq \lambda_k^{k,\varepsilon}(v, S_k^\varepsilon(v, s))$ the speed of the ε-approximate shock wave connecting the states $\{u(v), u(S_k^\varepsilon(v, s))\}$.

REMARK 4.3. Let $v \in \mathcal{V}$ be given. Then, from Definition 4.1 and by Remark 4.1 it follows that
(4.23)
$$S_k^\varepsilon(v, \sigma) = R_k(v_j e_j, v_k + \sigma) \quad \text{if} \quad (j \neq k) \qquad \begin{cases} \sigma \in [\lfloor v \rfloor_k \varepsilon - v_k, \ (\lfloor v \rfloor_k + 1)\varepsilon - v_k], & v_k \notin \mathbb{Z}\varepsilon, \\ \text{or} \\ \sigma \in [-\varepsilon, \varepsilon], & v_k \in \mathbb{Z}\varepsilon. \end{cases}$$

Moreover, if the k-th family is NGNL1, $v \in \mathcal{V}_k^+$, and

$$(4.24) \qquad \exists \ \sigma' > 0 \ : \ \lambda_k^{s,\varepsilon}(v, \sigma') = \widetilde{\lambda}_k^{\lfloor v \rfloor_k}(v),$$

then there holds
(4.25)
$$S_k^\varepsilon\big(R_k(v, -\sigma''), \sigma' + \sigma''\big) = S_k^\varepsilon(v, \sigma')$$
$$\dot{S}_k^\varepsilon\big(R_k(v, -\sigma''), \sigma' + \sigma''\big) = \dot{S}_k^\varepsilon(v, \sigma') \qquad \forall \ \sigma'' \in [\lfloor v \rfloor_k \varepsilon - v_k, \ (\lfloor v \rfloor_k + 1)\varepsilon - v_k].$$
$$\lambda_k^{s,\varepsilon}\big(R_k(v, -\sigma''), \sigma' + \sigma''\big) = \lambda_k^{s,\varepsilon}(v, \sigma')$$

Similar statements hold if $v \in \mathcal{V}_k^-$, or if the k-th family is NGNL2. Notice that (4.25) is the analog of (3.8) in **P4**.

REMARK 4.4. From Lemma 4.1 it follows that, for $v = (v_1, v_2) \in \mathcal{V}$, and $\varepsilon > 0$, there holds

$$(4.26) \qquad \begin{aligned} \big|S_k^\varepsilon(v, \sigma) - S_k(v, \sigma)\big| &= \mathcal{O}(1) \max\{|\delta_k(v)|, |\sigma|\} \cdot \varepsilon, \\ \big|\lambda_k^{s,\varepsilon}(v, \sigma) - \lambda_k^s(v, \sigma)\big| &= \mathcal{O}(1) \max\{|\delta_k(v)|, |\sigma|\} \cdot \varepsilon. \end{aligned}$$

Indeed, using the ε-RH conditions, one can derive the sharper estimate

$$\big|S_k^\varepsilon(v, \sigma) - S_k(v, \sigma)\big| = \mathcal{O}(1) |\sigma| \max\{|\delta_k(v)|, |\sigma|\} \cdot \varepsilon$$

and further estimates on the convergence of the maps S_k^ε, $\lambda_k^{s,\varepsilon}$, and of their derivatives. A proof is given in [**AM2**, Proposition 3.2].

4.3. Interpolations.

For a k-th NGNL family, and for any given $v \in \mathcal{V}$, we construct here two maps $\sigma \mapsto S_k^{\psi,\varepsilon}(v, \sigma)$, $\sigma \mapsto \lambda_k^{\psi,\varepsilon}(v, \sigma)$, having the basic properties **P2-P4** required by our algorithm for a "correct" definition of the approximate solution of a Riemann problem. Such maps are obtained as suitable interpolations between, respectively, the functions S_k^ε, $\lambda_k^{s,\varepsilon}$, and R_k, $\lambda_k^{r,\varepsilon}$. This is done in four steps.

Step 1.

Consider a smooth function $\psi : \mathbb{R} \to [0,1]$ such that

$$\begin{cases} \psi(\sigma) = 0 & \text{if } \sigma \leq 0, \\ \psi(\sigma) = 1 & \text{if } \sigma \geq 1, \\ \psi'(\sigma) \in [0,1] & \text{if } \sigma \in [0,1], \end{cases}$$

and, given $v \in \mathcal{V}$, let $\varphi : \mathbb{R}^+ \to [0,1]$ be the map defined by

$$(4.27) \qquad \varphi(\sigma) \doteq \psi\left(\frac{\sigma - 3\sqrt{\varepsilon}}{\sqrt{\varepsilon}}\right) \qquad \sigma \geq 0.$$

Then, for $v \in \mathcal{V}$, and $\sigma \geq 0$, define the interpolated curves
(4.28)
$$S_k^{\varphi,\varepsilon}(v,\sigma) = \begin{cases} \varphi(\sigma) \cdot S_k^{\varepsilon}(v,\sigma) + [1-\varphi(\sigma)] \cdot R_k(v,\sigma) & \text{if } 0 \leq \delta_k(v), \\ \varphi(\sigma + \delta_k(v)) \cdot S_k^{\varepsilon}(v,\sigma) + \\ \quad + [1-\varphi(\sigma+\delta_k(v))] \cdot R_k(v,\sigma) & \text{if } -\sqrt{\varepsilon} \leq \delta_k(v) \leq 0, \\ \varphi(\sigma - \sqrt{\varepsilon}) \cdot S_k^{\varepsilon}(v,\sigma) + \\ \quad + [1-\varphi(\sigma-\sqrt{\varepsilon})] \cdot R_k(v,\sigma) & \text{if } \delta_k(v) \leq -\sqrt{\varepsilon}, \end{cases}$$

together with the interpolated speeds
(4.29)
$$\lambda_k^{\varphi,\varepsilon}(v,\sigma) = \begin{cases} \varphi(\sigma)\cdot \lambda_k^{s,\varepsilon}(v,\sigma)+[1-\varphi(\sigma)]\cdot \lambda_k^{r,\varepsilon}(v,\sigma) & \text{if } 0 \leq \delta_k(v), \\ \varphi(\sigma+\delta_k(v))\cdot \lambda_k^{s,\varepsilon}(v,\sigma) + \\ \quad + [1-\varphi(\sigma+\delta_k(v))]\cdot \lambda_k^{r,\varepsilon}(v,\sigma) & \text{if } -\sqrt{\varepsilon} \leq \delta_k(v) \leq 0, \\ \varphi(\sigma-\sqrt{\varepsilon})\cdot \lambda_k^{s,\varepsilon}(v,\sigma) + \\ \quad + [1-\varphi(\sigma-\sqrt{\varepsilon})]\cdot \lambda_k^{r,\varepsilon}(v,\sigma) & \text{if } \delta_k(v) \leq -\sqrt{\varepsilon}, \end{cases}$$

where $\lambda_k^{r,\varepsilon}$ denotes the averaged characteristic speed defined at (3.3)-(3.4).
In the case $\sigma \leq 0$, the definitions of $S_k^{\varphi,\varepsilon}(v,\sigma)$, $\lambda_k^{\varphi,\varepsilon}(v,\sigma)$, are entirely similar. Notice that the curve defined by (4.28) coincides with the rarefaction curve through v whenever $0 \leq \sigma \leq 2\sqrt{\varepsilon}$, and with the ε-approximate shock curve through v whenever $\sigma \geq 5\sqrt{\varepsilon}$.

Step 2.
Let $\mathcal{V}_k^0$ be the set at (2.11). For any fixed $v^0 \in \mathcal{V}_k^0$, we define now two sequences of numbers $(\sigma_k^{r,\ell}(v^0))_{\ell \in \mathbb{Z}}$, $(\sigma_k^{s,\ell}(v^0))_{\ell \in \mathbb{Z}}$, by setting $\sigma_k^{r,0}(v^0) = \sigma_k^{s,0}(v^0) \doteq 0$ and

$$(4.30) \quad \begin{aligned} \sigma_k^{r,\ell}(v^0) &\doteq \inf\left\{\sigma > \varepsilon \; : \; \lambda_k^{r,\varepsilon}\big(R_k(\lfloor\!\lfloor v^0 \rfloor\!\rfloor_k, -\ell\varepsilon),\, \ell\varepsilon+\sigma\big) = \widetilde{\lambda}_k^{\lfloor v^0 \rfloor_k - \ell}(v^0)\right\}, \\ \sigma_k^{s,\ell}(v^0) &\doteq \inf\left\{\sigma > \varepsilon \; : \; \lambda_k^{s,\varepsilon}\big(R_k(\lfloor\!\lfloor v^0 \rfloor\!\rfloor_k, -\ell\varepsilon),\, \ell\varepsilon+\sigma\big) = \widetilde{\lambda}_k^{\lfloor v^0 \rfloor_k - \ell}(v^0)\right\}, \end{aligned}$$

if $\ell > 0$ (see figure 4.1). Similar definitions are given for $\sigma_k^{r,\ell}(v^0)$, $\sigma_k^{s,\ell}(v^0)$, in the case $\ell < 0$.

By (4.30), if the k-th family is NGNL1, the quantities $\ell\varepsilon + \sigma_k^{r,\ell}(v^0)$, $\ell\varepsilon + \sigma_k^{s,\ell}(v^0)$ represent the size of a wave connecting the left state $R_k(\lfloor\!\lfloor v^0 \rfloor\!\rfloor_k, -\ell\varepsilon)$

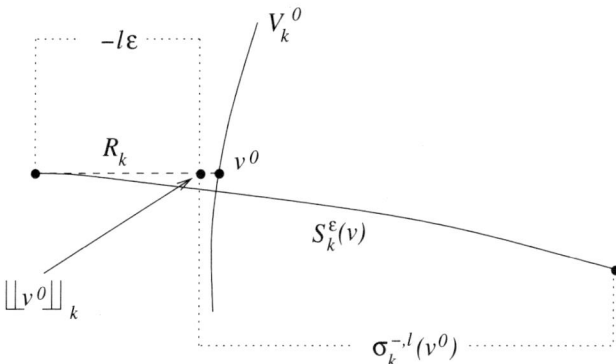

FIGURE 4.1. Illustration of (4.30)

with, respectively, the right state $R_k(\lfloor\lfloor v^0\rfloor\rfloor_k, \sigma_k^{r,\ell}(v^0))$, and the right state

$$S_k^\varepsilon\big(R_k(\lfloor\lfloor v^0\rfloor\rfloor_k, -\ell\varepsilon),\ \ell\varepsilon + \sigma_k^{s,\ell}(v^0)\big),$$

which travels with the characteristic speed of its left state (see figure 4.1). Of course $\sigma_k^{r,\ell}(v^0)$, $\sigma_k^{s,\ell}(v^0)$ have similar meaning if the k-th family is NGNL2. Such quantities will be used in the next subsection to construct the approximate mixed curve of the k-th family. Relying on the properties of the maps $\lambda_k^{s,\varepsilon}$, $\lambda_k^{r,\varepsilon}$, one can show (with the same type of arguments of the proof of Proposition 3.3 in [**AM2**]) that there exists some positive constant c, independent of ε and v^0, such that

$$(4.31) \qquad c\cdot\varepsilon \leq |\sigma_k^{r,\ell+1}(v^0) - \sigma_k^{r,\ell}(v^0))|, \qquad c\cdot\varepsilon \leq |\sigma_k^{s,\ell+1}(v^0) - \sigma_k^{s,\ell}(v^0))|,$$

for any $v^0 \in \mathcal{V}_k^0$, $\ell \in \mathbb{Z}$ and $\varepsilon > 0$. Next, for every $v \in \mathcal{V}$ such that $\delta_k(v) \leq 0$, we define the map

$$(4.32) \quad \lambda_k^{\widetilde{r},\varepsilon}(v,\sigma) \doteq \lambda_k^{r,\varepsilon}\big(v,\ \sigma + \sigma_k^{r,\ell(v)}(\pi_k(v)) - \sigma_k^{s,\ell(v)}(\pi_k(v))\big), \qquad \sigma \geq 0,$$

with $\ell(v) \doteq \lfloor(\pi_k(v))\rfloor_k - \lfloor v\rfloor_k$. Denote $\lambda_k^{\widetilde{\varphi},\varepsilon}(v,\sigma)$ the function defined as in (4.29) with $\lambda_k^{r,\varepsilon}$ replaced by $\lambda_k^{\widetilde{r},\varepsilon}$. Of course, similar definitions are given in the case $\delta_k(v) \geq 0$ and $\sigma < 0$.

Step 3.

For every $v \in \mathcal{V}$ with $\delta_k(v) \leq 0$, and $\sigma \geq 0$, define the map
(4.33)
$$\lambda_k^{\widetilde{\psi},\varepsilon}(v,\sigma) \doteq \psi\left(\frac{2\delta_k(v) + 3\sqrt{\varepsilon}}{\sqrt{\varepsilon}}\right)\lambda_k^{\varphi,\varepsilon}(v,\sigma) + \left[1 - \psi\left(\frac{2\delta_k(v) + 3\sqrt{\varepsilon}}{\sqrt{\varepsilon}}\right)\right]\lambda_k^{\widetilde{\varphi},\varepsilon}(v,\sigma).$$

Then, for any fixed $v^0 \in \mathcal{V}_k^0$, define the sequence $(\sigma_k^\ell(v^0))_{\ell\in\mathbb{Z}}$, by setting $\sigma_k^0(v^0) \doteq 0$ and

$$(4.34) \quad \sigma_k^\ell(v^0) \doteq \inf\left\{\sigma > \varepsilon\ :\ \lambda_k^{\widetilde{\psi},\varepsilon}\big(R_k(\lfloor\lfloor v^0\rfloor\rfloor_k, -\ell\varepsilon),\ \ell\varepsilon + \sigma\big) = \widetilde{\lambda}_k^{\lfloor v^0\rfloor_k - \ell}(v^0)\right\},$$

if $\ell > 0$. The same type of estimates in (4.31) hold for $(\sigma_k^\ell(v^0))_{\ell\in\mathbb{Z}}$. Notice that, by (4.33), the two maps $\lambda_k^{\widetilde{\varphi},\varepsilon}(v,\sigma)$, $\lambda_k^{\widetilde{\psi},\varepsilon}(v,\sigma)$, coincide for all $\sigma \geq 0$ whenever

$\delta_k(v) \leq -(3/2)\sqrt{\varepsilon}$. Thus, comparing the definitions (4.30) and (4.34), it follows that
$$\sigma_k^\ell(v^0) = \sigma_k^{s,\ell}(v^0) \qquad \forall\, \ell \geq \frac{3}{2\sqrt{\varepsilon}}.$$

Entirely similar definitions are given in the case $\delta_k(v) \geq 0$ and $\sigma < 0$.

Step 4.

For any $v \in \mathcal{V}$ with $\delta_k(v) \leq 0$, in connection with the sequence $\left(\sigma_k^\ell(\pi_k(v))\right)_\ell$ defined at (4.34), consider the map $\widehat{\varphi}_k(v, \cdot) : \mathbb{R} \to [0,1]$ defined by
(4.35)
$$\widehat{\varphi}_k(v, \sigma) \doteq \psi\left(\frac{\sigma - \sigma_k^{\ell(v)}(\pi_k(v)) + v_k - \lfloor \pi_k(v) \rfloor_k \, \varepsilon}{\sigma_k^{\ell(v)+1}(\pi_k(v)) - \sigma_k^{\ell(v)}(\pi_k(v))} \right), \qquad \ell(v) \doteq \lfloor \pi_k(v) \rfloor_k - \lfloor v \rfloor_k.$$

Then, define the interpolated curve
(4.36)
$$S_k^{\widehat{\varphi},\varepsilon}(v,\sigma) \doteq \widehat{\varphi}_k(v,\sigma) \cdot S_k^{\varphi,\varepsilon}\!\left(R_k(v,-\varepsilon),\, \sigma+\varepsilon\right) + \left(1-\widehat{\varphi}_k(v,\sigma)\right) \cdot S_k^{\varphi,\varepsilon}(v,\sigma), \qquad \sigma \geq 0,$$

and the interpolated speed
(4.37)
$$\lambda_k^{\widehat{\varphi},\varepsilon}(v,\sigma) \doteq \widehat{\varphi}_k(v,\sigma) \cdot \lambda_k^{\widetilde{\varphi},\varepsilon}\!\left(R_k(v,-\varepsilon),\, \sigma+\varepsilon\right) + \left(1-\widehat{\varphi}_k(v,\sigma)\right) \cdot \lambda_k^{\widetilde{\varphi},\varepsilon}(v,\sigma), \qquad \sigma \geq 0.$$

One can easily check that these maps satisfy a relation of the same type as (3.8) in **P4**. However, the map $S_k^{\widehat{\varphi},\varepsilon}(v,\sigma)$ never coincides with the approximate shock curve $S_k^\varepsilon(v,\sigma)$ even for large σ, while the map $\lambda_k^{\widehat{\varphi},\varepsilon}(v,\sigma)$ assumes different values from $\lambda_k^{r,\varepsilon}(v,\sigma)$ even for small σ. For this reason, we perform a last interpolation by setting, for all $v \in \mathcal{V}$, and $\sigma \geq 0$,
(4.38)
$$S_k^{\psi,\varepsilon}(v,\sigma) \doteq \psi\!\left(\frac{\sigma + [\delta_k(v)]^- - 4\sqrt{\varepsilon}}{\sqrt{\varepsilon}}\right) S_k^\varepsilon(v,\sigma) +$$
$$+ \left[1 - \psi\!\left(\frac{\sigma + [\delta_k(v)]^- - 4\sqrt{\varepsilon}}{\sqrt{\varepsilon}}\right)\right] \cdot \left[\psi\!\left(\frac{2\delta_k(v) + 3\sqrt{\varepsilon}}{\sqrt{\varepsilon}}\right) S_k^{\varphi,\varepsilon}(v,\sigma) + \right.$$
$$\left. + \left[1 - \psi\!\left(\frac{2\delta_k(v) + 3\sqrt{\varepsilon}}{\sqrt{\varepsilon}}\right)\right] S_k^{\widehat{\varphi},\varepsilon}(v,\sigma) \right],$$

(4.39)
$$\lambda_k^{\psi,\varepsilon}(v,\sigma) \doteq \psi\!\left(\frac{\sigma + [\delta_k(v)]^- - 4\sqrt{\varepsilon}}{\sqrt{\varepsilon}}\right) \lambda_k^{s,\varepsilon}(v,\sigma) +$$
$$+ \left[1 - \psi\!\left(\frac{\sigma + [\delta_k(v)]^- - 4\sqrt{\varepsilon}}{\sqrt{\varepsilon}}\right)\right] \cdot \left[\psi\!\left(\frac{2\delta_k(v) + 3\sqrt{\varepsilon}}{\sqrt{\varepsilon}}\right) \lambda_k^{\varphi,\varepsilon}(v,\sigma) + \right.$$
$$\left. + \left[1 - \psi\!\left(\frac{2\delta_k(v) + 3\sqrt{\varepsilon}}{\sqrt{\varepsilon}}\right)\right] \lambda_k^{\widehat{\varphi},\varepsilon}(v,\sigma) \right],$$

where
$$[\delta_k(v)]^- \doteq \min\{0, \delta_k(v)\}.$$

Similar definitions are given if $\sigma \leq 0$. On the other hand, in the case where the k-th family is LD, or GNL and satisfies (2.22), we define the modified shock curve $S_k^{\psi,\varepsilon}(v,\sigma)$, and the corresponding speed $\lambda_k^{\psi,\varepsilon}(v,\sigma)$ as in [**BC1**] by setting, for all $v \in \mathcal{V}$, and for any σ:

$$
(4.40) \quad \begin{aligned}
S_k^{\psi,\varepsilon}(v,\sigma) &\doteq \varphi(\sigma) \cdot S_k(v,\sigma) + [1 - \varphi(\sigma)] \cdot R_k(v,\sigma), \\
\lambda_k^{\psi,\varepsilon}(v,\sigma) &\doteq \varphi(\sigma) \cdot \lambda_k^s(v,\sigma) + [1 - \varphi(\sigma)] \cdot \lambda_k^{r,\varepsilon}(v,\sigma).
\end{aligned}
$$

The maps $S_k^{\psi,\varepsilon}(v,\sigma)$, $\lambda_k^{\psi,\varepsilon}(v,\sigma)$ defined in (4.38)-(4.40) are called, respectively, the *modified approximate shock curve* and the *modified approximate shock speed of the k-th family*.

From the above definitions it follows that, for any k-th characteristic family, and for $v \in \mathcal{V}$, one has

$$
(4.41) \quad \begin{aligned}
S_k^{\psi,\varepsilon}(v,\sigma) &= R_k(v,\sigma) \\
\lambda_k^{\psi,\varepsilon}(v,\sigma) &= \lambda_k^{r,\varepsilon}(v,\sigma)
\end{aligned} \quad \text{if} \quad |\sigma| \leq 2\sqrt{\varepsilon}.
$$

Moreover, if the k-th family is NGNL, for $v \in \mathcal{V}$, and $\sigma \geq 0$, there holds
(4.42)
$$
S_k^{\psi,\varepsilon}(v,\sigma) = \begin{cases}
R_k(v,\sigma) & \text{if} \quad \sigma \leq 2\sqrt{\varepsilon}, \\
S_k^{\varphi,\varepsilon}(v,\sigma) & \text{if} \quad \delta_k(v) \geq -\sqrt{\varepsilon} \quad \text{and} \quad \sigma + [\delta_k(v)]^- \leq 4\sqrt{\varepsilon}, \\
S_k^{\widehat{\varphi},\varepsilon}(v,\sigma) & \text{if} \quad \delta_k(v) \leq -(3/2)\sqrt{\varepsilon} \quad \text{and} \quad \sigma + [\delta_k(v)]^- \leq 4\sqrt{\varepsilon}, \\
S_k^{\varepsilon}(v,\sigma) & \text{if} \quad \sigma + [\delta_k(v)]^- \geq 5\sqrt{\varepsilon}.
\end{cases}
$$

Analogous relations are verified in the case $\sigma \leq 0$, and by the map $\lambda_k^{\psi,\varepsilon}(v,\sigma)$.

From (4.41) we deduce in particular that the maps $S_k^{\psi,\varepsilon}(v,\sigma)$, $\lambda_k^{\psi,\varepsilon}(v,\sigma)$ enjoy the property **P3** stated in Section 3, while the next result yields the property **P2**.

LEMMA 4.2. *If the k-th family is NGNL, for $\varepsilon > 0$, $v \in \mathcal{V}$, and any σ, the following estimates hold*

(4.43)
$$S_k^{\psi,\varepsilon}(v,\sigma) = S_k(v,\sigma) + \mathcal{O}(1) \, |\sigma| \max\{|\delta_k(v)|, |\sigma|\} \cdot \varepsilon,$$

$$\dot{S}_k^{\psi,\varepsilon}(v,\sigma) = \dot{S}_k + \mathcal{O}(1) \max\{|\delta_k(v)|, |\sigma|\} \cdot \varepsilon,$$

$$\partial_i S_k^{\psi,\varepsilon}(v,\sigma) = \partial_i S_k + \mathcal{O}(1) \max\{|\delta_k(v)|, |\sigma|\} \cdot \varepsilon, \qquad i = 1, 2,$$

$$\dot{S}_k^{\psi,\varepsilon}(v,\sigma) = \partial_k S_k^{\psi,\varepsilon}(v,\sigma) + \mathcal{O}(1) \, |\sigma| \cdot \Big[\max\{|\delta_k(v)|, |\sigma|\} \cdot \varepsilon + |\sigma| \cdot |\sigma - \nu_k(v)| \Big],$$

(4.44)
$$\lambda_k^{\psi,\varepsilon}(v,\sigma) = \lambda_k^s(v,\sigma) + \mathcal{O}(1)\max\{|\delta_k(v)|, |\sigma|\}\cdot\varepsilon,$$

$$\dot\lambda_k^{\psi,\varepsilon}(v,\sigma) = \dot\lambda_k^s(v,\sigma) + \mathcal{O}(1)\frac{\max\{|\delta_k(v)|, |\sigma|\}\cdot\varepsilon}{|\sigma|},$$

$$\partial_i\lambda_k^{\psi,\varepsilon}(v,\sigma) = \partial_i\lambda_k^s(v,\sigma) + \mathcal{O}(1)\frac{\max\{|\delta_k(v)|, |\sigma|\}\cdot\varepsilon}{|\sigma|}, \qquad i=1,2,$$

$$\dot\lambda_k^{\psi,\varepsilon}(v,\sigma) = \partial_k\lambda_k^{\psi,\varepsilon}(v,\sigma) + \mathcal{O}(1)\left[\frac{\max\{|\delta_k(v)|, |\sigma|\}\cdot\varepsilon}{|\sigma|} + |\sigma - \nu_k(v)|\right].$$

A proof of Lemma 4.2 and the derivation of further estimates on the second derivatives of $S_k^{\psi,\varepsilon}$, $\lambda_k^{\psi,\varepsilon}$, can be established relying on the properties of the approximate Hugoniot curve and of the approximate shock speed S_k^ε, λ_k^ε provided by Proposition 3.2 in [**AM2**], and using the Taylor expansions of the maps $S_k(v,\sigma)$, $\lambda_k^s(v,\sigma)$, $\lambda_k^{r,\varepsilon}(v,\sigma)$ (cfr. [**AM2**, Remark 2.2]).

REMARK 4.5. Concerning the property **P1**, observe that from the definition (4.40), and relying on the Taylor expansions of the maps $S_k(v,\sigma)$, $\lambda_k^s(v,\sigma)$, $\lambda_k^{r,\varepsilon}(v,\sigma)$ (cfr. [**BC1, B5**]), one can easily derive, for $\varepsilon > 0$, $v \in \mathcal{V}$, and any σ, the following estimates on the modified approximate shock curve and on the modified approximate shock speed of a k-th GNL or LD family:

$$S_k^{\psi,\varepsilon}(v,\sigma) = S_k(v,\sigma) + \mathcal{O}(1)|\sigma|\cdot\varepsilon,$$
$$\lambda_k^{\psi,\varepsilon}(v,\sigma) = \lambda_k^s(v,\sigma) + \mathcal{O}(1)\cdot\varepsilon.$$

4.4. Approximate mixed-curves.

We define now an approximation of the mixed-curve through a given point $v^0 \in \mathcal{V}_k^0$ for a k-th NGNL family satisfying the assumption (**A**). In connection with the sequence $(\sigma_k^\ell)_\ell \doteq (\sigma_k^\ell(v^0))_\ell$ introduced in (4.34) consider the map $\sigma \mapsto \mu_k^\varepsilon(v^0,\sigma)$ defined by

(4.45) $\qquad \mu_k^\varepsilon(v^0,\sigma) \doteq \ell\varepsilon + \sigma \qquad$ if $\qquad \begin{cases} \sigma_k^\ell \leq \sigma < \sigma_k^{\ell+1}, & 0 \leq \ell, \\ \sigma_k^{\ell-1} < \sigma \leq \sigma_i^\ell, & \ell < 0. \end{cases}$

Then, the curve $\sigma \mapsto T_k^\varepsilon(v^0,\sigma)$ defined by

(4.46) $\qquad T_k^\varepsilon(v^0,\sigma) \doteq S_k^{\psi,\varepsilon}\Big(R_k\big(\llbracket v^0\rrbracket_k,\, \sigma - \mu_k^\varepsilon(v^0,\sigma)\big),\, \mu_k^\varepsilon(v^0,\sigma)\Big) \qquad \forall\,\sigma$

is called the ε-*approximate mixed curve of the k-th family through* v^0. Such a curve converges with its derivative to the corresponding exact mixed curve T_k defined in (2.10). Indeed, the following holds

LEMMA 4.3. *For all* $v^0 \in \mathcal{V}_k^0$, $\varepsilon > 0$, *and any* σ, *there holds*

(4.47) $\qquad |\mu_k^\varepsilon(v^0,\sigma) - \mu_k(v^0,\sigma)| = \mathcal{O}(1)\,\varepsilon,$

(4.48) $\qquad |T_k^\varepsilon(v^0,\sigma) - T_k(v^0,\sigma)| = \mathcal{O}(1)\,\varepsilon,$

(4.49) $\qquad \left|\frac{d}{d\sigma}T_k^\varepsilon(v^0,\sigma) - \frac{d}{d\sigma}T_k(v^0,\sigma)\right| = \mathcal{O}(1)\,\varepsilon.$

Moreover, the second derivatives of T_k^ε w.r.t. σ are uniformly bounded.

A proof of Lemma 4.3 is given in [**AM2**, Proposition 3.3]. If the k-th family is NGNL1, consider now the map $\nu_k^\varepsilon : \mathcal{V} \to \mathbb{R}$, defined by setting
(4.50)
$$\nu_k^\varepsilon(v) \doteq \big(\pi_k(v)\big)_k - v_k \qquad \text{if} \quad v \in \mathcal{V}_k^+, \quad \lfloor \pi_k(v) \rfloor_k = \lfloor v \rfloor_k,$$

$$\nu_k^\varepsilon(v) \doteq \mu_1^\varepsilon\big(\pi_k(v), \sigma_k^{\ell(v)}(\pi_k(v))\big) + \big(\lfloor v \rfloor_k \varepsilon - v_k\big) \qquad \text{if} \quad v \in \mathcal{V}_k^+, \quad \lfloor \pi_k(v) \rfloor_k > \lfloor v \rfloor_k,$$

where $\ell(v) \doteq \lfloor \pi_k(v) \rfloor_k - \lfloor v \rfloor_k$. A similar definition is given in the case $v \in \mathcal{V}_k^-$, or if the k-th family is NGNL2. The map $\nu_k^\varepsilon(v)$ converges to the corresponding map $\nu_k(v)$ defined in (2.17) in connection with the exact mixed curve T_k. In fact, for $v \in \mathcal{V}$, $\varepsilon > 0$, there holds

(4.51) $$\nu_k^\varepsilon(v) = \nu_k(v) + \mathcal{O}(1)|\varepsilon|.$$

See [**AM2**, Remark 3.4] for a proof of (4.51). From (4.51), recalling (2.19), it follows

(4.52) $$\nu_k^\varepsilon(v) = -3\delta_k(v) + \mathcal{O}(1) \max\big\{|\varepsilon|, |\delta_k(v)|^2\big\}, \qquad v \in \mathcal{V}.$$

Notice that, using ν_k^ε, one can express the map $\widehat{\varphi}_k$ defined in (4.36) as

$$\widehat{\varphi}_k(v, \sigma) = \psi\left(\frac{\sigma - \nu_k^\varepsilon(v)}{\nu_k^\varepsilon(R_k(v, -\varepsilon)) - \nu_k^\varepsilon(v) - \varepsilon}\right), \qquad \delta_k(v) \leq 0.$$

Then, observing that by (4.29) one has

$$\lambda_k^{\varphi, \varepsilon}(v, \sigma) = \lambda_k^{\widetilde{\varphi}, \varepsilon}(v, \sigma) = \lambda_k^{\widetilde{\psi}, \varepsilon}(v, \sigma) = \lambda_k^{s, \varepsilon}(v, \sigma), \qquad \text{if} \quad \sigma \geq 5\sqrt{\varepsilon},$$

and comparing the definition (4.34) of $\big(\sigma_k^\ell(v^0)\big)_\ell$ with the definitions (4.37) and (4.39), we deduce that, if the k-th family is NGNL1, there holds

(4.53) $$\lambda_k^{\psi, \varepsilon}\big(v, \nu_k^\varepsilon(v)\big) = \lambda_k^{\widetilde{\psi}, \varepsilon}\big(v, \nu_k^\varepsilon(v)\big) = \lambda_k^{1, \varepsilon+}(v) \qquad \forall\, v \in \mathcal{V}_k^+.$$

Moreover, by (4.42) and (4.52) it follows
(4.54)
$$\begin{aligned} S_k^{\psi, \varepsilon}(v, \sigma) &= S_k^\varepsilon(v, \sigma) \\ \lambda_k^{\psi, \varepsilon}(v, \sigma) &= \lambda_k^{s, \varepsilon}(v, \sigma) \end{aligned} \quad \text{whenever} \quad \begin{aligned} &\delta_k(v) \geq 0, \quad \sigma \geq 5\sqrt{\varepsilon}, \\ &\text{or} \\ &\delta_k(v) < 0, \quad \sigma \geq \max\{\nu_k^\varepsilon(v), 9\sqrt{\varepsilon}\}. \end{aligned}$$

Similar definitions and relations hold in the case where $v \in \mathcal{V}_k^-$, or if the k-th family is NGNL2.

REMARK 4.6. *The maps $S_k^{\psi, \varepsilon}(v, \sigma)$, $\lambda_k^{\psi, \varepsilon}(v, \sigma)$, $\nu_k^\varepsilon(v)$, satisfy the property* **P4** *stated in Section 3. In particular, if the k-th family is NGNL1, $v = (v_1, v_2) \in \mathcal{V}_k^+$ and*

$$\exists\, \sigma' > 0 \; : \; \lambda_k^{\psi, \varepsilon}(v, \sigma') = \widetilde{\lambda}_k^{\lfloor v \rfloor_k}(v),$$

then it follows that $\sigma' = \nu_k^\varepsilon(v)$ and, for all $\sigma'' \in \left[v_k - \lfloor v \rfloor_k \varepsilon,\ v_k - (\lfloor v \rfloor_k + 1)\varepsilon\right]$, there holds

$$S_k^{\psi,\varepsilon}\big(R_k(v,\sigma''),\ \nu_k^\varepsilon(v) - \sigma''\big) = S_k^{\psi,\varepsilon}\big(v,\ \nu_k^\varepsilon(v)\big),$$

$$\dot{S}_k^{\psi,\varepsilon}\big(R_k(v,\sigma''),\ \nu_k^\varepsilon(v) - \sigma''\big) = \dot{S}_k^{\psi,\varepsilon}\big(v,\ \nu_k^\varepsilon(v)\big),$$

$$\lambda_k^{\psi,\varepsilon}\big(R_k(v,\sigma''),\ \nu_k^\varepsilon(v) - \sigma''\big) = \lambda_k^{\psi,\varepsilon}\big(v,\ \nu_k^\varepsilon(v)\big).$$

Similar relations hold for the map $\lambda_k^{\psi,\varepsilon}(v,\cdot)$ in the case $v \in \mathcal{V}_k^-$, or if the k-th family is NGNL2.

4.5. Approximate elementary curves.

We are now in the position to define the ε-*approximate elementary curves* that will be used to construct the ε-approximate solution of a Riemann problem for a 2×2 system satisfying the assumption **(A)**.

DEFINITION 4.2.

1. If the k-th characteristic family is LD, or GNL and satisfies (2.22), we define the ε-*approximate elementary curves of right states of the k-th family* $\Psi_k^\varepsilon(v)$ by setting

$$(4.55) \qquad \Psi_k^\varepsilon(v,\sigma) = S_k^{\psi,\varepsilon}(v,\sigma) \qquad \forall\, v,\, \sigma,$$

where $S_k^{\psi,\varepsilon}$ is the map defined in (4.40).

2. If the k-th characteristic family is NGNL1, we define the ε-approximate elementary curves of right states of the k-th family $\Psi_k^\varepsilon(v)$ by setting

(4.56)
$$\Psi_k^\varepsilon(v,\sigma) = \begin{cases} S_k^{\varphi,\varepsilon}(v,\sigma) & \forall\, \sigma & \text{if } v \in \mathcal{V}_k^0, \\[4pt] \begin{cases} S_k^{\varphi,\varepsilon}(v,\sigma) & \text{if } \sigma < 0 \\ R_k(v,\sigma) & \text{if } 0 \leq \sigma < -\delta_k(v) \\ T_k^\varepsilon\big(\pi_k(v),\ \sigma + \delta_k(v)\big) & \text{if } -\delta_k(v) \leq \sigma < \nu_k^\varepsilon(v) \\ S_k^{\psi,\varepsilon}(v,\sigma) & \text{if } \sigma \geq \nu_k^\varepsilon(v) \end{cases} & \text{if } v \in \mathcal{V}_k^+, \\[4pt] \begin{cases} S_k^{\varphi,\varepsilon}(v,\sigma) & \text{if } \sigma > 0 \\ R_k(v,\sigma) & \text{if } -\delta_k(v) < \sigma \leq 0 \\ T_k^\varepsilon\big(\pi_k(v),\ \sigma + \delta_k(v)\big) & \text{if } \nu_k^\varepsilon(v) < \sigma \leq -\delta_k(v) \\ S_k^{\psi,\varepsilon}(v,\sigma) & \text{if } \sigma \leq \nu_k^\varepsilon(v) \end{cases} & \text{if } v \in \mathcal{V}_k^-, \end{cases}$$

where $S_k^{\varphi,\varepsilon}$, $S_k^{\psi,\varepsilon}$, T_k^ε are the maps defined in (4.28), (4.38), (4.46), respectively.

3. If the k-th characteristic family is NGNL2, we define the ε-approximate elementary curves of left states of the k-th family $\Psi_k^\varepsilon(v)$ by setting

(4.57)
$$\Psi_k^\varepsilon(v,\sigma) = \begin{cases} S_k^{\varphi,\varepsilon}(v,\sigma) & \forall\,\sigma & \text{if}\quad v \in \mathcal{V}_k^0, \\ \begin{cases} S_k^{\varphi,\varepsilon}(v,\sigma) & \text{if}\quad \sigma > 0 \\ R_k(v,\sigma) & \text{if}\quad -\delta_k(v) \leq \sigma \leq 0 \\ T_k^\varepsilon\Big(\pi_k(v),\ \sigma + \delta_k(v)\Big) & \text{if}\quad \nu_k^\varepsilon(v) \leq \sigma < -\delta_k(v) \\ S_k^{\psi,\varepsilon}(v,\sigma) & \text{if}\quad \sigma < \nu_k^\varepsilon(v) \end{cases} & \text{if}\quad v \in \mathcal{V}_k^+, \\ \begin{cases} S_k^{\varphi,\varepsilon}(v,\sigma) & \text{if}\quad \sigma < 0 \\ R_k(v,\sigma) & \text{if}\quad 0 \leq \sigma < -\delta_k(v) \\ T_k^\varepsilon\Big(\pi_k(v),\ \sigma + \delta_k(v)\Big) & \text{if}\quad -\delta_k(v) \leq \sigma < \nu_k^\varepsilon(v) \\ S_k^{\psi,\varepsilon}(v,\sigma) & \text{if}\quad \sigma \geq \nu_k^\varepsilon(v) \end{cases} & \text{if}\quad v \in \mathcal{V}_k^-, \end{cases}$$

where $S_k^{\varphi,\varepsilon}$, $S_k^{\psi,\varepsilon}$, T_k^ε are the maps defined in (4.28), (4.38), (4.46), respectively.

REMARK 4.7. If the k-th family is LD, GNL, or NGNL1, for each $v \in \mathcal{V}$ the curve $\sigma \mapsto \Psi_k^\varepsilon(v,\sigma)$ describes all right states that are connected to v by a k-wave of an ε-approximate solution, while, in the case the k-th family is NGNL2, the curve $\sigma \mapsto \Psi_k^\varepsilon(v,\sigma)$ describes all left states that are connected to v by a k-wave of an ε-approximate solution. This means, in particular, that a k-wave front connecting two states v^L, v^R that lie on opposite sides w.r.t. the set $\mathcal{V}_k^0$ at (2.11), has strength $|\sigma| \geq |\nu_k^\varepsilon(v^L)|$, i.e. such a wave satisfies the Ψ^ε-RH conditions stated in Definition 3.1. Notice that, by construction, the maps $\sigma \mapsto \Psi_i^\varepsilon(v,\sigma)$, $i=1,2$, are everywhere continuous and piecewise twice continuously differentiable. Moreover, left and right first derivatives of $\Psi_i^\varepsilon(v,\sigma)$ are uniformly bounded with respect to σ and ε. These facts, together with Lemmas 4.2-4.3, and Remark 4.5, imply that the properties **P1-P2** stated in Section 3 are satisfied.

REMARK 4.8. In the particular case of a system satisfying the assumption $(\mathbf{A}')$ (in which both family are NGNL and there hold (1.21), (1.22)), according with Definition 4.2 we will define the ε-*approximate elementary curves of right states of the first family*

(4.58)
$$\Psi_1^\varepsilon(v,\sigma) = \begin{cases} S_1^{\varphi,\varepsilon}(v,\sigma) & \forall\,\sigma & \text{if}\quad v \in \mathcal{V}_1^0, \\ \begin{cases} S_1^{\varphi,\varepsilon}(v,\sigma) & \text{if}\quad \sigma < 0 \\ R_1(v,\sigma) & \text{if}\quad 0 \leq \sigma < -\delta_1(v) \\ T_1^\varepsilon\Big(\pi_1(v),\ \sigma + \delta_1(v)\Big) & \text{if}\quad -\delta_1(v) \leq \sigma < \nu_1^\varepsilon(v) \\ S_1^{\psi,\varepsilon}(v,\sigma) & \text{if}\quad \sigma \geq \nu_1^\varepsilon(v) \end{cases} & \text{if}\quad v \in \mathcal{V}_1^+, \\ \begin{cases} S_1^{\varphi,\varepsilon}(v,\sigma) & \text{if}\quad \sigma > 0 \\ R_1(v,\sigma) & \text{if}\quad -\delta_1(v) < \sigma \leq 0 \\ T_1^\varepsilon\Big(\pi_1(v),\ \sigma + \delta_1(v)\Big) & \text{if}\quad \nu_1^\varepsilon(v) < \sigma \leq -\delta_1(v) \\ S_1^{\psi,\varepsilon}(v,\sigma) & \text{if}\quad \sigma \leq \nu_1^\varepsilon(v) \end{cases} & \text{if}\quad v \in \mathcal{V}_1^-, \end{cases}$$

and the ε-approximate elementary curves of left states of the second family

(4.59)
$$\Psi_2^\varepsilon(v,\sigma) = \begin{cases} S_2^{\varphi,\varepsilon}(v,\sigma) & \forall \sigma & \text{if } v \in \mathcal{V}_2^0, \\ \begin{cases} S_2^{\varphi,\varepsilon}(v,\sigma) & \text{if } \sigma > 0 \\ R_2(v,\sigma) & \text{if } -\delta_2(v) \leq \sigma \leq 0 \\ T_2^\varepsilon\big(\pi_2(v), \sigma+\delta_2(v)\big) & \text{if } \nu_2^\varepsilon(v) \leq \sigma < -\delta_2(v) \\ S_2^{\psi,\varepsilon}(v,\sigma) & \text{if } \sigma < \nu_2^\varepsilon(v) \end{cases} & \text{if } v \in \mathcal{V}_2^+, \\ \begin{cases} S_2^{\varphi,\varepsilon}(v,\sigma) & \text{if } \sigma < 0 \\ R_2(v,\sigma) & \text{if } 0 \leq \sigma < -\delta_2(v) \\ T_2^\varepsilon\big(\pi_2(v), \sigma+\delta_2(v)\big) & \text{if } -\delta_2(v) \leq \sigma < \nu_2^\varepsilon(v) \\ S_2^{\psi,\varepsilon}(v,\sigma) & \text{if } \sigma \geq \nu_2^\varepsilon(v) \end{cases} & \text{if } v \in \mathcal{V}_2^-. \end{cases}$$

REMARK 4.9. For a system satisfying the assumption (**A'**), given $v^L, v^R \in \mathcal{V}$, and σ_1, σ_2 in a neighborhood of zero, consider the map

(4.60)
$$\Psi^\varepsilon\big(v^L, v^R; \sigma_1, \sigma_2\big) \doteq \Psi_1^\varepsilon\big(v^L, \sigma_1\big) - \Psi_2^\varepsilon\big(v^R, \sigma_2\big).$$

Then, because of the regularity of Ψ^ε, and since

$$\frac{\partial \Psi^\varepsilon}{\partial \sigma_1}\bigg|_{\sigma_1,\sigma_2=0}\big(v^L, v^R; \sigma_1, \sigma_2\big) = e_1,$$

$$\frac{\partial \Psi^\varepsilon}{\partial \sigma_2}\bigg|_{\sigma_1,\sigma_2=0}\big(v^L, v^R; \sigma_1, \sigma_2\big) = -e_2,$$

by using the implicit function theorem one can show that there exists a neighborhood of $0 \in \mathbb{R}^2$, still denoted $\mathcal{V}$, such that for every $v^L, v^R \in \mathcal{V}$, we can *uniquely determine* σ_1, σ_2 for which

(4.61)
$$\Psi^\varepsilon\big(v^L, v^R; \sigma_1, \sigma_2\big) = 0$$

holds, i.e. we can uniquely solve the system $(3.9)_3$. Clearly, one can repeat the same reasoning for a general 2×2 system (1.1) satisfying the assumption (**A**), showing that there exists a neighborhood of zero $\mathcal{V}$ where we can uniquely solve the system $(3.9)_\ell$, $\ell = 1, 2, 3, 4$, associated to (1.1). Notice that, thanks to the uniform bound on the derivatives of Ψ_i^ε, we can choose such a neighborhood with size independent of ε. Moreover, we can find some constant C independent of ε such that, for all $v^L, v^R \in \mathcal{V}$, if σ_1, σ_2 are the corresponding values determined by $(3.9)_\ell$, $\ell = 1, 2, 3, 4$, then one has

(4.62)
$$\frac{1}{C}\big|v^R - v^L\big| \leq |\sigma_1| + |\sigma_2| \leq C\big|v^R - v^L\big|.$$

This shows that the ε-approximate elementary curves above defined satisfy also the property **P5** stated in Section 3.

5. Basic Interaction Estimates

In this section we collect the basic estimates on the strengths and the speeds of the waves emerging after an interaction between wave-fronts of an ε-approximate solution of a system satisfying the assumption **(A)**, constructed as in Sections 3-4. Except for the Lemmas of the last subsection 5.4, all the other proofs can be found in [**AM2**, Section 4] and in [**BC1**, Section 4].

Throughout, with $\mathcal{O}(1)$ we always denote a quantity uniformly bounded by a constant which depends only on the system (1.1) and not on the parameter ε or on the size σ of the waves involved. Moreover, with a slight abuse of notation, we simply call σ a wave of size σ. We will also use the notation $\{v, \Psi_i^\varepsilon(v,\sigma)\}$ for the wave of the i-th family that connect the states v, $\Psi_i^\varepsilon(v,\sigma)$ where, according with Definition 4.2, v represents the left state in the case the i-th family is LD, or GNL, or NGNL1, while v represents the right state in the case where the i-th family is NGNL2. Two waves of the i-th family σ_α and σ_β are called *concordant* if $\text{sgn}(\sigma_\alpha) = \text{sgn}(\sigma_\beta)$, *discordant* otherwise. Following [**Li3**], we define the *angle* between two wave-fronts of the same family.

DEFINITION 5.1. Let $\{v', \Psi_i^\varepsilon(v',\sigma')\}$, $\{v'', \Psi_i^\varepsilon(v'',\sigma'')\}$, be two waves of the i-th family and assume that the wave of size σ' lies on the left of the wave of size σ''. Then we define the angle $\Theta_i(v',\sigma';\, v'',\sigma'')$ between $\{v', \Psi_i^\varepsilon(v',\sigma')\}$ and $\{v'', \Psi_i^\varepsilon(v'',\sigma'')\}$, as

$$(5.1) \qquad \Theta_i(v',\sigma';\, v'',\sigma'') \doteq \lambda_i^{\psi,\varepsilon}(v',\sigma') - \lambda_i^{\psi,\varepsilon}(v'',\sigma''), \qquad i=1,2.$$

Throughout the following we call v^L, v^R, respectively, the left and the right state near an interaction point, and denote σ_i^+, $i=1,2$, the size of the outgoing waves of the i-th family, which may consist of several wave-fronts (in the case the i-th outgoing wave is a piecewise constant rarefaction fan or, if the i-th family is NGNL, in the case the i-th outgoing wave is a composed wave). Moreover, when we consider the interaction between two incoming waves, we call v^M the middle state before the interaction takes place. Thus, in particular, if the interaction occurs between two waves σ', σ'' (σ' located on the left of σ'') of a k-th GNL, or NGNL1 family, then one has

$$v^M = \Psi_k^\varepsilon(v^L, \sigma'),$$

while, in the case where the interaction occurs between two waves σ', σ'' (σ'' located on the left of σ') of a k-th NGNL2 family, one has

$$v^M = \Psi_k^\varepsilon(v^R, \sigma').$$

Notice that, with a slight abuse of notation, sometimes we will say that the waves σ_1^+, σ_2^+ are generated by the interaction of two (possible composed) waves σ_1^-, σ_2^-, meaning that, if we let v^L, v^R denote, respectively, the state to the left of σ_1^- and the state to the right of σ_2^-, the Riemann problem (v^L, v^R) is solved by two waves of size σ_i^+, even though the two waves σ_i^- do not actually interact instantaneously together (whenever they don't consist of a single wave-front). Thus, in the particular case of a system satisfying the assumption **(A')**, because of (3.20) we have

$$\Psi_1^\varepsilon(v^L, \sigma_1^+) = \Psi_2^\varepsilon(v^R, \sigma_2^+).$$

On the other hand, if we say that two (single) wave-fronts σ', σ'' interact together, we precisely mean that their angle Θ defined as in (5.1) is positive.

5.1. Strength estimates for interaction between wave-fronts of the same family.

We first collect here the estimates on the change of wave-size whenever it occurs an interaction between two waves of the same family. We distinguish several cases, depending on the nature of the interacting waves, i.e. if they are shock or rarefaction waves, and on their relative position in the (t, x)-plane. For definiteness, we shall state the results only for interacting waves of the first family, assuming that the first family is either GNL or NGNL1, the other cases being entirely similar.

LEMMA 5.1. *Assume that the first family is either GNL or NGNL1, and consider two wave-fronts of the first family with sizes σ', σ'' that interact together (σ' located on the left of σ''). Then, one has*
(5.2)
$$\left|\sigma_1^+ - (\sigma' + \sigma'')\right| + \left|\sigma_2^+\right| \leq \begin{cases} 0 & \text{if } |\sigma'|, |\sigma''| \leq \sqrt{\varepsilon}, \\ C_1 |\sigma'\sigma''| \Theta_1(v^L, \sigma'; v^M, \sigma'') & \text{otherwise.} \end{cases}$$

for some constant $C_1 > 0$ independent of σ', σ'', ε,

A proof of Lemma 5.1 can be obtained with the same arguments of the proof of Lemma 3 in [**BC1**], and Proposition 4.1 in [**AM2**].

REMARK 5.1. Assume that the first family is NGNL1 and consider two concordant fronts of the first family interacting together: a rarefaction front σ' and a shock σ'' (σ' located on the left of σ''). Then $|\sigma'| \leq \varepsilon$ so that, thanks to (2.19) and to the estimate on ν_1^ε provided by (4.51), we have
$$\left|\sigma' + \sigma'' - \nu_1^\varepsilon(v^L)\right| = \mathcal{O}(1)\varepsilon.$$

It follows that, relying on the estimates on the modified approximate Hugoniot curve $S_1^{\psi,\varepsilon}$ provided by Lemma 4.2, and thanks to the properties of $S_1^{\psi,\varepsilon}$ stated in Remark 4.6, with the same arguments of the proof of Proposition 4.1 in [**AM2**] one can show that, besides (5.2), there holds

(5.3)
$$\sigma_1^+ = \sigma' + \sigma'',$$
$$\left|\sigma_2^+\right| \leq C_1 |\sigma''|\left|\sigma' + \sigma'' - \nu_1^\varepsilon(v^L)\right| \cdot \Theta_1(v^L, \sigma'; v^M, \sigma'').$$

LEMMA 5.2. *Assume that the first family is NGNL1 and consider a shock wave of size σ' interacting with a (discordant) rarefaction fan of size σ'' (located on the right of σ'), both of the first family. Suppose that the outgoing wave of the first family is a composed wave consisting of a piecewise constant rarefaction fan $\widetilde{\sigma}_1^+$ and of a shock wave $\widehat{\sigma}_1^+ \doteq \sigma_1^+ - \widetilde{\sigma}_1^+$. Then the following hold.*

i)
(5.4)
$$\left|\widetilde{\sigma}_1^+\right| \leq |\sigma''| + \varepsilon + \mathcal{O}(1)|\sigma'\sigma''|;$$

ii) *There exists some positive constant $C_2 < 1$ independent of σ', σ'', ε, such that, if $|\sigma''| \leq C_2 \varepsilon$, then $\widetilde{\sigma}_1^+$ is a single wave-front.*

A proof of Lemma 5.2 can be obtained with the same arguments of the proof of Proposition 4.1 in [**AM2**], relying on the estimates (4.31), (4.47), (4.51)-(4.52) on the maps μ_1^ε, ν_1^ε.

REMARK 5.2. Under the same assumptions and with the same notations of Lemma 5.2, if $|\sigma'| > \sqrt{\varepsilon}$, and $|\sigma''| \leq \varepsilon$, one can show that the outgoing rarefaction of the first family $\widetilde{\sigma}_1^+$ is always a single wave-front. This can be proved with same arguments of the proof of Proposition 4.1 in [**AM2**], and thanks to (4.51)-(4.52).

LEMMA 5.3. *Assume that the $i \in \{1, 2\}$-th family is either GNL, or LD, or NGNL. Then, for $v \in \mathcal{V}$, and wave sizes σ', σ'', there holds*

$$\left| \Psi_i^\varepsilon(v, \sigma' + \sigma'') - \Psi_i^\varepsilon\big(\Psi_i^\varepsilon(v, \sigma'),\ \sigma''\big) \right| \leq$$
(5.5)
$$\leq \begin{cases} 0 & \text{if } |\sigma'|, |\sigma''| \leq \sqrt{\varepsilon}, \\ \mathcal{O}(1) |\sigma'\sigma''| \cdot (|\sigma'| + |\sigma''|) & \text{otherwise.} \end{cases}$$

PROOF. Recalling the definitions of the maps $\Psi_i^\varepsilon, i = 1, 2$, and of the angle between two waves, one may easily verify that the estimate (5.5) can be obtained using Lemma 5.1 and the same type of arguments involved in its proof. □

5.2. Speed estimates for interaction between wave-fronts of the same family.

Next we collect the estimates on the change of wave-speed whenever it occurs an interaction between two waves of the same family. As in the former subsection, we shall consider only the case of interacting waves belonging to the first family, assuming that the first family is either GNL or NGNL1, the other case being entirely similar.

LEMMA 5.4. *Assume that the first family is either GNL or NGNL1 and consider two wave-fronts of the first family with size σ', σ'' that interact together. Suppose that the outgoing wave of the first family is a shock with size σ_1^+, and that one of the following cases occurs*

a) σ', σ'' *are both shock waves;*

b) σ' *is a rarefaction wave and σ'' is a shock wave located on the right of σ'.*

Let $\Lambda_1^+ = \lambda_1^{\psi,\varepsilon}[v^L, \sigma_1^+]$ denote the speed of the outgoing shock σ_1^+ of the first family. Then, one has

(5.6)
$$\left| \frac{\lambda_1^{\psi,\varepsilon}[v^M, \sigma''] - \Lambda_1^+}{\Theta_1(v^L, \sigma'; v^M, \sigma'')} \right| \begin{cases} = \left| \dfrac{\sigma'}{\sigma' + \sigma''} \right| & \text{if } |\sigma'|, |\sigma''| \leq \sqrt{\varepsilon}, \\ \leq \left| \dfrac{\sigma'}{\sigma' + \sigma''} \right| + \mathcal{O}(1) \min\{|\sigma'|, |\sigma''|\} & \text{otherwise,} \end{cases}$$

(5.7)
$$\left| \frac{\lambda_1^{\psi,\varepsilon}[v^L, \sigma'] - \Lambda_1^+}{\Theta_1(v^L, \sigma'; v^M, \sigma'')} \right| \begin{cases} = \left| \dfrac{\sigma''}{\sigma' + \sigma''} \right| & \text{if } |\sigma'|, |\sigma''| \leq \sqrt{\varepsilon}, \\ \leq \left| \dfrac{\sigma''}{\sigma' + \sigma''} \right| + \mathcal{O}(1) \min\{|\sigma'|, |\sigma''|\} & \text{otherwise.} \end{cases}$$

A proof of Lemma 5.4 can be obtained with the same arguments of the proof of Lemma 20 in [**BC1**], and Proposition 4.2 in [**AM2**].

LEMMA 5.5. *Assume that the first family is NGNL1 and consider a shock wave of size σ' interacting with a rarefaction front of size σ'' (located on the right of σ' and discordant with σ'), both of the first family. Let σ_i^+, $i = 1, 2$, be the sizes of the outgoing i-waves. If the outgoing wave of the first family consists of a rarefaction fan $\widetilde{\sigma}_1^+$ with speed $\widetilde{\Lambda}_1^+ \doteq \frac{1}{\widetilde{\sigma}_1^+} \sum_\ell \widetilde{\sigma}_{1,\ell}^+ \widetilde{\Lambda}_{1,\ell}^+$ (denoting with $\widetilde{\sigma}_{1,\ell}$ the fronts of $\widetilde{\sigma}_1^+$, and letting $\widetilde{\Lambda}_{1,\ell}^+$ be the corresponding speeds), followed by a shock wave $\widehat{\sigma}_1^+ \doteq \sigma_1^+ - \widetilde{\sigma}_1^+$ with speed $\widehat{\Lambda}_1^+$, then one has*

(5.8)
$$\left| \frac{\lambda_1^{\psi,\varepsilon}[v^M, \sigma''] - \widehat{\Lambda}_1^+}{\Theta_1(v^L, \sigma'; v^M, \sigma'')} \right| \begin{cases} = \left| 1 - \dfrac{\sigma''}{\widehat{\sigma}_1^+} - \dfrac{\widetilde{\sigma}_1^+}{\widehat{\sigma}_1^+} \cdot \dfrac{\left(\widetilde{\Lambda}_1^+ - \lambda_1^{\psi,\varepsilon}[v^L, \sigma']\right)}{\Theta_1(v^L, \sigma'; v^M, \sigma'')} \right| & \text{if } |\sigma'| \leq \sqrt{\varepsilon}, \\ \leq \left| 1 - \dfrac{\sigma''}{\widehat{\sigma}_1^+} \right| + \mathcal{O}(1) |\sigma''| & \text{otherwise}, \end{cases}$$

$$\left| \frac{\lambda_1^{\psi,\varepsilon}[v^L, \sigma'] - \widehat{\Lambda}_1^+}{\Theta_1(v^L, \sigma'; v^M, \sigma'')} \right| \begin{cases} = \left| -\dfrac{\sigma''}{\widehat{\sigma}_1^+} - \dfrac{\widetilde{\sigma}_1^+}{\widehat{\sigma}_1^+} \cdot \dfrac{\left(\widetilde{\Lambda}_1^+ - \lambda_1^{\psi,\varepsilon}[v^L, \sigma']\right)}{\Theta_1(v^L, \sigma'; v^M, \sigma'')} \right| & \text{if } |\sigma'| \leq \sqrt{\varepsilon}, \\ \leq \left| \dfrac{\sigma''}{\widehat{\sigma}_1^+} \right| + \mathcal{O}(1) |\sigma''| & \text{otherwise}, \end{cases}$$

(5.9)
$$\left| \frac{\lambda_1^{\psi,\varepsilon}[v^M, \sigma''] - \widetilde{\Lambda}_1^+}{\Theta_1(v^L, \sigma'; v^M, \sigma'')} \right| = 1 + \mathcal{O}(1) \left| \frac{\sigma''}{\sigma'} \right| \quad \text{if} \quad |\sigma'| > \sqrt{\varepsilon},$$

$$\left| \frac{\lambda_1^{\psi,\varepsilon}[v^L, \sigma'] - \widetilde{\Lambda}_1^+}{\Theta_1(v^L, \sigma'; v^M, \sigma'')} \right| = \mathcal{O}(1) \left| \frac{\sigma''}{\sigma'} \right| \quad \text{if} \quad |\sigma'| > \sqrt{\varepsilon}.$$

In the case where the outgoing wave of the first family consists of a single shock σ_1^+ with speed Λ_1^+, the estimates in (5.8) hold with $\widetilde{\sigma}_1^+ \doteq 0$, $\widehat{\sigma}_1^+ \doteq \sigma_1^+$, $\widehat{\Lambda}_1^+ \doteq \Lambda_1^+$.

A proof of Lemma 5.5 can be obtained with the same arguments of the proof of Proposition 4.2 in [**AM2**].

LEMMA 5.6. *Assume that the first family is either GNL or NGNL1 and consider two wave-fronts of the first family with sizes σ', σ'' (σ' located on the left of σ'') that interact together. Suppose that from the interaction emerges a shock of the first family of size σ_1^+, with speed $\Lambda_1^+ = \lambda_1^{\psi,\varepsilon}[v^L, \sigma_1^+]$. Then, there holds*
(5.10)
$$\left| (\sigma' + \sigma'') \cdot \Lambda_1^+ - \sigma' \cdot \lambda_1^{\psi,\varepsilon}[v^L, \sigma'] - \sigma'' \cdot \lambda_1^{\psi,\varepsilon}[v^M, \sigma''] \right| \leq \mathcal{O}(1) |\sigma' \sigma''| \, \Theta_1(v^L, \sigma'; v^M, \sigma'').$$

Moreover, in the case the first family is NGNL1, if σ' is a rarefaction front and σ'' is a concordant shock, then one has

(5.11)
$$\left| (\sigma' + \sigma'') \cdot \Lambda_1^+ - \sigma' \cdot \lambda_1^{\psi,\varepsilon}[v^L, \sigma'] - \sigma'' \cdot \lambda_1^{\psi,\varepsilon}[v^M, \sigma''] \right| \leq$$
$$\leq \mathcal{O}(1) |\sigma''| |\sigma' + \sigma'' - \nu_1^\varepsilon(v^L)| \cdot \Theta_1(v^L, \sigma'; v^M, \sigma'').$$

A proof of Lemma 5.6 can be obtained with the same arguments of the proof of Lemma 20 in [**BC1**], and Proposition 4.2 in [**AM2**], relying on the estimate (4.51) on the map ν_1^ε, and thanks to the properties of $\lambda_1^{\psi,\varepsilon}$ stated in Remark 4.6.

5.3. Interaction between waves of different families.

Here we collect the estimates on the change of wave-size and wave-speed whenever it occurs an interaction between two waves of different characteristic families.

LEMMA 5.7. *For a 2×2 system satisfying the assumption* (**A**), *consider two (possible composed) waves of different families with size σ_1^-, σ_2^- that interact together. Then one has*

(5.12)
$$|\sigma_1^+ - \sigma_1^-| + |\sigma_2^+ - \sigma_2^-| \leq \begin{cases} 0 & \text{if } |\sigma_1^-|, |\sigma_2^-| \leq \sqrt{\varepsilon}, \\ C_3 |\sigma_1^- \sigma_2^-| \cdot (|\sigma_1^-| + |\sigma_2^-|) & \text{otherwise,} \end{cases}$$

for some constant $C_3 > 0$ independent of σ_i^-, ε. Moreover, for $v \in \mathcal{V}$, and wave sizes σ_1^-, σ_2^-, there holds

(5.13)
$$\left| \Psi_1^\varepsilon(\Psi_2^\varepsilon(v, \sigma_2^-), \sigma_1^-) - \Psi_2^\varepsilon(\Psi_1^\varepsilon(v, \sigma_1^-), \sigma_2^-) \right| \leq$$

$$\leq \begin{cases} 0 & \text{if } |\sigma_1^-|, |\sigma_2^-| \leq \sqrt{\varepsilon}, \\ \mathcal{O}(1) |\sigma_1^- \sigma_2^-| \cdot (|\sigma_1^-| + |\sigma_2^-|) & \text{otherwise.} \end{cases}$$

A proof of Lemma 5.7 can be obtained with the same arguments of the proof of Lemma 2 in [**BC1**], relying on the estimates provided by Lemma 4.2.

LEMMA 5.8. *With the same notations and in the same setting of Lemma 5.7, suppose that the k-th family is NGNL, the k-th incoming wave σ_k^- is a shock wave, and the k-th outgoing wave σ_k^+ is a composed wave consisting of a piecewise constant rarefaction fan $\widetilde{\sigma}_k^+$ and of a shock wave $\widehat{\sigma}_k^+ \doteq \sigma_k^+ - \widetilde{\sigma}_k^+$. Then, the following hold.*

I)

(5.14) $\qquad |\widetilde{\sigma}_k^+| = \mathcal{O}(1) \left(|\sigma_j^-| + \varepsilon \right) \qquad j \neq k;$

II) *There exists some positive constant $C_4 < 1$ independent of σ_k^- and ε, such that, if $|\sigma_j^-| \leq C_4 \varepsilon$, $j \neq k$, then σ_k^+ contains at most a single rarefaction front $\widetilde{\sigma}_k^+$.*

A proof of Lemma 5.8 can be obtained with the same arguments of the proof of Lemma 2 in [**BC1**], relying on the estimates (4.31), (4.47), (4.51)-(4.52) on the maps μ_k^ε, ν_k^ε.

LEMMA 5.9. *With the same notations and in the same setting of Lemma 5.7, suppose that the k-th family is NGNL, the k-th incoming wave σ_k^- is a shock wave and the k-th outgoing wave σ_k^+ is a piecewise constant rarefaction fan, then the following holds.*

I)

(5.15) $\qquad |\sigma_k^+| = \mathcal{O}(1) \left[|\sigma_j^-| + \varepsilon \right] \qquad j \neq k.$

II) *The outgoing rarefaction σ_k^+ consists of at most two wave-fronts whenever*

(5.16)
$$|\sigma_j^-| < \sqrt{\varepsilon}, \quad j \neq k, \qquad \text{and} \qquad \left| \lfloor v^M \rfloor_k - \lfloor v^L \rfloor_k \right| + \left| \lfloor v^R \rfloor_k - \lfloor v^M \rfloor_k \right| \leq 1.$$

A proof of Lemma 5.9 can be obtained with the same arguments of the proof of Lemma 2 in [**BC1**], relying on the estimates (4.31), (4.47), (4.51)-(4.52) on the maps μ_k^ε, ν_k^ε.

LEMMA 5.10. *Under the assumptions of Lemma 5.7 and with the same notations, consider two wave-fronts of different families with size σ_1^-, σ_2^- that interact together and generate two (possibly composed) i-waves σ_i^+, $i = 1, 2$. Let Λ_i^+ denote the speed of* **any one** *of the outgoing i-fronts (contained in σ_i^+). Then, there holds*

(5.17) $$\left|\Lambda_i^+ - \Lambda_i^-\right| = \mathcal{O}(1)\left|\sigma_j^-\right| \qquad j \neq i, \qquad i = 1, 2.$$

A proof of Lemma 5.10 can be obtained with the same arguments of the proof of Lemma 18 in [**BC1**], relying on the estimates provided by Lemma 4.2.

5.4. Strength estimates for two coupled Riemann problems.

Having in mind to extend the previous estimates to the case of interactions involving an arbitrary number of incoming waves, we derive, as in [**BC1**], the following Lemmas which deal with the coupling of two Riemann problems. For notational convenience, we introduce the function

$$\widetilde{Q}(\sigma', \sigma'') \doteq \begin{cases} 0 & \text{if } |\sigma'|, |\sigma''| \leq \sqrt{\varepsilon}, \\ |\sigma'\sigma''| & \text{otherwise.} \end{cases}$$

LEMMA 5.11. *Let v^L, v^M, v^R be any three nearby states. Suppose that, for a 2×2 system satisfying the assumption* (**A**)*, the Riemann problems (v^L, v^M) and (v^M, v^R) are solved by waves of size σ'_1, σ'_2 and of size σ''_1, σ''_2, respectively. Then the Riemann problem (v^L, v^R) is solved by waves of size σ_i^+, $i = 1, 2$, satisfying the bound*

(5.18) $$\left|\sigma_1^+ - (\sigma'_1 + \sigma''_1)\right| + \left|\sigma_2^+ - (\sigma'_2 + \sigma''_2)\right|$$
$$\leq C_5 \Big[\widetilde{Q}(\sigma''_1, \sigma'_2) \cdot \max\{|\sigma''_1|, |\sigma'_2|\} + \widetilde{Q}(\sigma'_1, \sigma''_1) \cdot \max\{|\sigma'_1|, |\sigma''_1|\}$$
$$+ \widetilde{Q}(\sigma'_2, \sigma''_2) \cdot \max\{|\sigma'_2|, |\sigma''_2|\}\Big],$$

for some constant $C_5 > 1$ independent of ε.

PROOF. We shall prove the result only for a system satisfying the assumption (**A'**), the case of a general 2×2 system satisfying (**A**) being entirely similar. Consider the points

$$v' \doteq \Psi_1^\varepsilon(v^L, \sigma'_1) \qquad\qquad v'' \doteq \Psi_2^\varepsilon(v^R, \sigma''_2),$$
$$\omega' \doteq \Psi_1^\varepsilon(v', \sigma''_1) \qquad\qquad \omega''' \doteq \Psi_2^\varepsilon(v'', \sigma'_2),$$
$$\omega'' \doteq \Psi_1^\varepsilon(v^L, \sigma'_1 + \sigma''_1) \qquad\qquad \omega^{iv} \doteq \Psi_2^\varepsilon(v^R, \sigma'_2 + \sigma''_2).$$

Observe that by Remark 4.9 one has

$$v' = \Psi_2^\varepsilon(v^M, \sigma'_2) \qquad\qquad v'' = \Psi_1^\varepsilon(v^M, \sigma''_1).$$

Thus, by the implicit function theorem and using (5.5), (5.13), we derive

$$\left|\sigma_1^+ - (\sigma_1' + \sigma_1'')\right| + \left|\sigma_2^+ - (\sigma_2' + \sigma_2'')\right|$$
$$= \mathcal{O}(1) \left|\omega'' - \omega'^v\right|$$
$$= \mathcal{O}(1) \left[\left|\omega'' - \omega'\right| + \left|\omega' - \omega'''\right| + \left|\omega''' - \omega'^v\right|\right]$$
$$= \mathcal{O}(1) \Big[\widetilde{Q}\left(\sigma_1'', \sigma_2'\right) \cdot \max\{|\sigma_1''|, |\sigma_2'|\} +$$
$$+ \widetilde{Q}\left(\sigma_1', \sigma_1''\right) \cdot \max\{|\sigma_1'|, |\sigma_1''|\} + \widetilde{Q}\left(\sigma_2', \sigma_2''\right) \cdot \max\{|\sigma_2'|, |\sigma_2''|\}\Big].$$

$\square$

The next result provides some estimates on the rarefaction and shock components of the new waves of a NGNL family generated by the coupling of two Riemann problems. They can be easily derived by using the same arguments of the proofs of Lemmas 5.2, 5.8, 5.9 and 5.10. In the following, as previously done, we denote the rarefaction fan and the shock component of a composed wave σ of a NGNL family, respectively, by $\widetilde{\sigma}$ and by $\widehat{\sigma} \doteq \sigma - \widetilde{\sigma}$. We also use the notation $\widetilde{\sigma} \doteq \sigma$, $\widehat{\sigma} \doteq 0$, if σ is a rarefaction fan, and $\widetilde{\sigma} \doteq 0$, $\widehat{\sigma} \doteq \sigma$, if σ is a shock wave, both in the case of a GNL and of a NGNL family.

LEMMA 5.12. *For a 2×2 system satisfying the assumption* (A), *let v^L, v^M, v^R, and σ_i', σ_i'', σ_i^+, $i = 1, 2$, be as in Lemma 5.11. Call $\widetilde{\sigma}_{i,\ell}'$, $\widetilde{\sigma}_{i,\ell}''$ and $\widetilde{\sigma}_{i,\ell}^+$, respectively, the wave-fronts of $\widetilde{\sigma}_i'$, $\widetilde{\sigma}_i''$ and $\widetilde{\sigma}_i^+$, where we understand $\widetilde{\sigma}_{i,\ell}' = 0$ ($\widetilde{\sigma}_{i,\ell}'' = 0$, $\widetilde{\sigma}_{i,\ell}^+ = 0$) in the case $\widetilde{\sigma}_i' = 0$ ($\widetilde{\sigma}_i'' = 0$, $\widetilde{\sigma}_i^+ = 0$). Then there exists some constant C_6 independent of ε, σ_i', σ_i'', such that the following hold.*

I) *If $sgn(\sigma_1'') \neq sgn(\sigma_1^+)$, $sgn(\sigma_1') = sgn(\sigma_1^+)$, then one has*

(5.19)
$$|\widetilde{\sigma}_1^+| \leq |\widetilde{\sigma}_1'| + |\widetilde{\sigma}_1''| + \varepsilon +$$
$$+ C_6\Big[|\sigma_2'\sigma_1''|[|\sigma_2'| + |\sigma_1''|] + |\sigma_1'\sigma_1''|[|\sigma_1'| + |\sigma_1''|]\Big],$$

$$|\widehat{\sigma}_1^+| \leq |\widehat{\sigma}_1'| + C_6\Big[|\sigma_2'\sigma_1''|[|\sigma_2'| + |\sigma_1''|] + |\sigma_1'\sigma_1''|[|\sigma_1'| + |\sigma_1''|]\Big],$$

$$\sum_\ell |\widetilde{\sigma}_{1,\ell}^+|[|\sigma_1^+| - |\widetilde{\sigma}_{1,\ell}^+|] \leq \sum_\ell |\widetilde{\sigma}_{1,\ell}'|[|\sigma_1'| - |\widetilde{\sigma}_{1,\ell}'|] + |\sigma_1'|[|\sigma_1''| + \varepsilon] +$$
$$+ C_6\Big[|\sigma_2'\sigma_1''|[|\sigma_2'| + |\sigma_1''|] + |\sigma_1'\sigma_1''|[|\sigma_1'| + |\sigma_1''|]\Big].$$

II) *If $sgn(\sigma_1') = sgn(\sigma_1'') = sgn(\sigma_1^+)$, then one has*

(5.20)
$$|\widetilde{\sigma}_1^+| \le |\widetilde{\sigma}_1'| + |\widetilde{\sigma}_1''| + \\ + C_6\Big[|\sigma_2'\sigma_1''|[|\sigma_2'| + |\sigma_1''|] + |\sigma_1'\sigma_1''|[|\sigma_1'| + |\sigma_1''|]\Big],$$

$$|\widehat{\sigma}_1^+| \le |\widehat{\sigma}_1'| + |\widehat{\sigma}_1''| + \\ + C_6\Big[|\sigma_2'\sigma_1''|[|\sigma_2'| + |\sigma_1''|] + |\sigma_1'\sigma_1''|[|\sigma_1'| + |\sigma_1''|]\Big],$$

$$\sum_\ell |\widetilde{\sigma}_{1,\ell}^+|[|\sigma_1^+| - |\widetilde{\sigma}_{1,\ell}^+|] \le \sum_\ell |\widetilde{\sigma}_{1,\ell}'|[|\sigma_1'| - |\widetilde{\sigma}_{1,\ell}'|] + \sum_\ell |\widetilde{\sigma}_{1,\ell}''|[|\sigma_1''| - |\widetilde{\sigma}_{1,\ell}''|] + \\ + C_6\Big[|\sigma_2'\sigma_1''|[|\sigma_2'| + |\sigma_1''|] + |\sigma_1'\sigma_1''|[|\sigma_1'| + |\sigma_1''|]\Big].$$

III) If $sgn(\sigma_1') \ne sgn(\sigma_1'')$, $sgn(\sigma_1'') = sgn(\sigma_1^+)$, then one has

$$|\widetilde{\sigma}_1^+| \le |\widetilde{\sigma}_1''| + C_6\,|\sigma_2'||\sigma_1''|^2,$$

(5.21)
$$|\widehat{\sigma}_1^+| \le |\widehat{\sigma}_1''| + C_6\,|\sigma_2'||\sigma_1''|^2,$$

$$\sum_\ell |\widetilde{\sigma}_{1,\ell}^+|[|\sigma_1^+| - |\widetilde{\sigma}_{1,\ell}^+|] \le \sum_\ell |\widetilde{\sigma}_{1,\ell}''|[|\sigma_1''| - |\widetilde{\sigma}_{1,\ell}''|] + C_6\,|\sigma_2'||\sigma_1''|^2.$$

Moreover, there exists some constant $0 < C_7 < 1$ independent of ε, σ_i', σ_i'', so that:

IV) *If*

$$|\sigma_2'| + |\sigma_1''| \le C_7\,\varepsilon \qquad sgn(\sigma_1'') \ne sgn(\sigma_1^+),$$

then σ_1^+ contains at most one more wave-front than σ_1'.

V) *If*

$$|\sigma_2''| \le C_7\,\varepsilon \qquad sgn(\sigma_1'') = sgn(\sigma_1^+),$$

then σ_1^+ contains at most one more wave-front than σ_1' and σ_1''.

Entirely similar properties hold for the waves of the second family.

We conclude this section deriving the same type of estimates provided by Lemma 5.12, in the case of an interaction involving an arbitrary number of wave-fronts. As previously done, we let $\widetilde{\sigma}$, $\widehat{\sigma}$ denote, respectively, the rarefaction fan and the shock component of a composed wave σ of a NGNL family, and set $\widetilde{\sigma} \doteq \sigma$, $\widehat{\sigma} \doteq 0$, if σ is a rarefaction front, and $\widetilde{\sigma} \doteq 0$, $\widehat{\sigma} \doteq \sigma$, whenever σ is a shock of a GNL or of a NGNL family.

LEMMA 5.13. *For a 2×2 system satisfying the assumption* (**A**), *consider n_2 fronts $\sigma_{2,1}, \ldots, \sigma_{2,n_2}$ of the second characteristic family and n_1 fronts $\sigma_{1,1}, \ldots, \sigma_{1,n_1}$ of the first characteristic family interacting all together at time τ_0, and generating two i-waves σ_i^+, $i = 1, 2$. Denote $\widetilde{\sigma}_{i,\ell}^+$ the wave-fronts of $\widetilde{\sigma}_i^+$. Then, for $i = 1, 2$, the following hold.*

a) *There exists some constant $C_8 > 0$, independent of ε, $\sigma_{i,m}$, such that*

(5.22)
$$|\tilde{\sigma}_i^+| \leq \sum_{\sigma_{i,m}^- \cdot \sigma_i^+ < 0} |\sigma_{i,m}^-| + \varepsilon + C_8 \Bigg[\sum_{m,p} |\sigma_{1,m}^- \cdot \sigma_{2,p}^-| + \sum_{j=1}^{2} \sum_{\sigma_{j,m}^- \cdot \sigma_{j,p}^- < 0} |\sigma_{j,m}^- \cdot \sigma_{j,p}^-| +$$
$$+ \sum_{j=1}^{2} \sum_{\sigma_{j,m}^- \cdot \sigma_{j,p}^- > 0} |\sigma_{j,m}^- \cdot \sigma_{j,p}^-| \big[|\sigma_{j,m}^-| + |\sigma_{j,p}^-| \big] \Bigg],$$

(5.23)
$$|\sigma_i^+| \leq \sum_{\sigma_{i,m}^- \cdot \sigma_i^+ > 0} |\sigma_{i,m}^-| + C_8 \Bigg[\sum_{m,p} |\sigma_{1,m}^- \cdot \sigma_{2,p}^-| + \sum_{j=1}^{2} \sum_{\sigma_{j,m}^- \cdot \sigma_{j,p}^- < 0} |\sigma_{j,m}^- \cdot \sigma_{j,p}^-| +$$
$$+ \sum_{j=1}^{2} \sum_{\sigma_{j,m}^- \cdot \sigma_{j,p}^- > 0} |\sigma_{j,m}^- \cdot \sigma_{j,p}^-| \big[|\sigma_{j,m}^-| + |\sigma_{j,p}^-| \big] \Bigg],$$

(5.24)
$$\sum_{\ell} |\tilde{\sigma}_{i,\ell}^+| \big[|\sigma_i^+| - |\tilde{\sigma}_{i,\ell}^+| \big] \leq \varepsilon \cdot \sum_{\sigma_{i,m}^- \cdot \sigma_i^+ > 0} |\sigma_{i,m}^-| +$$
$$+ C_8 \Bigg[\sum_{m,p} |\sigma_{1,m}^- \cdot \sigma_{2,p}^-| + \sum_{j=1}^{2} \sum_{\sigma_{j,m}^- \cdot \sigma_{j,p}^- < 0} |\sigma_{j,m}^- \cdot \sigma_{j,p}^-| +$$
$$+ \sum_{j=1}^{2} \sum_{\sigma_{j,m}^- \cdot \sigma_{j,p}^- > 0} |\sigma_{j,m}^- \cdot \sigma_{j,p}^-| \big[|\sigma_{j,m}^-| + |\sigma_{j,p}^-| \big] \Bigg].$$

b) *There exists some constant $0 < C_9 < 1$ independent of ε, $\sigma_{i,m}$, such that, if*

(5.25)
$$\sum_{\sigma_{i,m}^- \cdot \sigma_i^+ < 0} |\sigma_{i,m}^-| + \sum_{\substack{p \\ j \neq i}} |\sigma_{j,p}^-| \leq C_9 \, \varepsilon,$$

then σ_i^+ consists of at most two wave-fronts.

PROOF. We shall prove the result for the outgoing fronts of the first family, the other case being entirely similar. By considering a slightly perturbed configuration, we first let $\sigma_{2,1}, \ldots, \sigma_{2,n_2}$ interact all together at some time $\tau' < \tau_0$, producing two waves of size σ_1', σ_2'. Afterwards we let $\sigma_{1,1}, \ldots, \sigma_{1,n_1}$ interact at some time $\tau'' > \tau'$, $\tau'' < \tau_0$, producing two waves of size σ_1'', σ_2''. Finally, we let σ_1', σ_2', σ_1'', σ_2'' interact at time τ_0. Then, working by induction on the number of interacting waves and using Lemma 5.3 one derives

$$|\sigma_1'| + \bigg| \sigma_2' - \sum_{m=1}^{n_2} \sigma_{2,m} \bigg| = \mathcal{O}(1) \sum_{m,p} |\sigma_{2,m}^- \sigma_{2,p}^-| \big[|\sigma_{2,m}^-| + |\sigma_{2,p}^-| \big]$$

and hence, in particular,

$$|\sigma_1'| = \mathcal{O}(1) \sum_{m,p} |\sigma_{2,m}^- \sigma_{2,p}^-| [|\sigma_{2,m}^-| + |\sigma_{2,p}^-|],$$

(5.26)

$$|\sigma_2'| = \mathcal{O}(1) \sum_m |\sigma_{2,m}^-|.$$

On the other hand, since $\sigma_{1,m}$ are all fronts of the same family, using again an inductive argument, and relying on Lemma 5.2 (in the case the first family is NGNL), and on Lemma 5.3, we deduce:

I)

(5.27)
$$|\tilde{\sigma}_1''| \leq \sum_{\sigma_{1,m}^- \cdot \sigma_1^+ < 0} |\sigma_{1,m}^-| + \varepsilon + \mathcal{O}(1) \sum_{m,p} |\sigma_{1,m}^- \sigma_{1,p}^-| [|\sigma_{1,m}^-| + |\sigma_{1,p}^-|],$$

(5.28)
$$|\sigma_1''| \leq \sum_{\sigma_{1,m}^- \cdot \sigma_1^+ > 0} |\sigma_{1,m}^-| + \mathcal{O}(1) \sum_{m,p} |\sigma_{1,m}^- \sigma_{1,p}^-| [|\sigma_{1,m}^-| + |\sigma_{1,p}^-|],$$

(5.29)
$$\sum_\ell |\tilde{\sigma}_{1,\ell}''|[|\sigma_1''| - |\tilde{\sigma}_{1,\ell}''|] \leq \varepsilon \cdot \sum_{\sigma_{1,m}^- \cdot \sigma_1^+ > 0} |\sigma_{1,m}^-| + \sum_{\sigma_{1,m}^- \cdot \sigma_{1,p}^- < 0} |\sigma_{1,m}^- \sigma_{1,p}^-| +$$
$$+ \mathcal{O}(1) \sum_{\sigma_{1,m}^- \cdot \sigma_{1,p}^- < 0} |\sigma_{1,m}^- \sigma_{1,p}^-|[|\sigma_{1,m}^-| + |\sigma_{1,p}^-|],$$

where $\tilde{\sigma}_{1,\ell}''$ denote the wave-fronts of $\tilde{\sigma}_1''$ and we understand $\tilde{\sigma}_{1,\ell}'' = 0$ if $\tilde{\sigma}_1'' = 0$.

II) There exists some constant $c > 0$ independent of ε, $\sigma_{1,m}$, such that, if

$$\sum_{\sigma_{1,m}^- \cdot \sigma_1^+ < 0} |\sigma_{1,m}^-| \leq c \cdot \varepsilon,$$

then σ_1'' consists of at most two wave-fronts.

Now call v^L the state to the left of σ_1', v^M the state between σ_2' and σ_1'', and v^R the state to the right of σ_2''. Then, applying Lemma 5.12, by (5.26)-(5.29) we derive (5.22)-(5.24). On the other hand, notice that if $\sum_{m=1}^{n_2} |\sigma_{2,m}| < \sqrt{\varepsilon}$, then by construction one has $\sigma_1' = 0$. Thus b) immediately follows if the first family is GNL, while in the case where the first family is NGNL, we recover b) from II) and Lemma 5.12, with similar arguments as in the proofs of Lemmas 5.2, 5.8. □

6. Bounds on the Total Variation and on the Interaction Potential

Throughout this section we fix $0 < \varepsilon < 1$, and consider a piecewise constant ε-approximate solution $v = v(t,x)$ of a system satisfying the assumption (A), constructed by the algorithm in Sections 3-4 on some initial time interval $[0, T[$, always working with Riemann coordinates $v = (v_1, v_2)$. A wave σ connecting the left state v^L with the right state v^R will be often equivalently denoted by (v^L, v^R)

or by $(v^{L_\sigma}, v^{R_\sigma})$. Two waves of the i-th family σ_α and σ_β are called *concordant* if $\text{sgn}(\sigma_\alpha) = \text{sgn}(\sigma_\beta)$, *discordant* otherwise. We denote $V(t) \doteq V^\varepsilon(v(t, \cdot))$ and $Q(t) \doteq Q^\varepsilon(v(t, \cdot))$, respectively, the total strength of waves and the interaction potential for $v(t, \cdot)$ defined as in (3.29), (3.30). For the one-sided limits and the jumps of V, Q we use notations such as

$$V(\tau+) \doteq \lim_{t \to \tau+} V(t), \qquad V(\tau-) \doteq \lim_{t \to \tau-} V(t), \qquad \Delta V(t) \doteq V(\tau+) - V(\tau-).$$

We shall use the interaction potential $Q(t)$ and the estimates collected in Section 5, to derive an a-priori bound on the total variation of $v(t, \cdot)$ for all $0 \leq t < T$. The standard argument followed to accomplish this goal consists in showing that, for a suitable constant K_0, the function

(6.1) $$t \mapsto V(t) + K_0 Q(t),$$

bounding the total variation, is non-increasing (see [**G, DP1, B2**]). However, for approximate solutions of a NGNL system defined with our algorithm, the quantity $Q(t)$ defined as in (3.30), after some particular type of interactions, may well increase of an amount not comparable with the possible decrease of $V(t)$. To see this, observe that the summation defining Q^ε contains all the products of wave-fronts belonging to the same family. Thus, $Q(t)$ shall increase after any interaction that produces a composed wave consisting, say, of a shock σ^+ and of a rarefaction $\tilde{\sigma}^+$, in the case where the quantity $|\sigma^+ \tilde{\sigma}^+|$ is not comparable with the amount of interaction $\sum_{i,j} |\sigma_i^- \sigma_j^-|$ of the incoming waves $\sigma_1^-, \ldots, \sigma_k^-$. On the other hand, although $V(t)$ may decrease as well after such interactions due to some cancellation, this, in general, is not sufficient to counterbalance the increase of $Q(t)$.

EXAMPLE 6.1. Suppose that the first characteristic family of (1.1) is NGNL1, and consider an ε-approximate solution $v = v(t, x)$ consisting of a shock σ' and of a rarefaction σ'' of opposite signs, both of the first family, which interact at some time t_0. Call v^L, v^R, respectively, the left and the right state near the interaction point, and σ_i^+, $i = 1, 2$, the size of the outgoing wave of the i-th family generated by the interaction. Suppose that v^L, v^R lie on opposite sides with respect to the curve $\mathcal{V}_1^0$ at (2.11), and that v^L belongs to a vertical line of the grid of step ε associated with the ε-Approximate Riemann solver (see fig. 6.1), i.e. $v_1^L = \lfloor v^L \rfloor_1 \varepsilon$ (using the notation $\lfloor \cdot \rfloor_i$ introduced at (4.1)). Moreover, assume that $0 < \sigma' \approx \sqrt{\varepsilon}$, $-\varepsilon^2 < \sigma'' < 0$, and that

$$\varepsilon + \nu_1^\varepsilon(\tilde{v}) < \sigma' + \sigma'' < \nu_1^\varepsilon(v^L), \qquad \tilde{v} \doteq R_1(v^L, \varepsilon).$$

Then, by (4.45), (4.50), (4.58), and by Definition 3.1, σ_1^+ must be a composed wave consisting of a rarefaction $\tilde{\sigma}_1^+ = \varepsilon$ and of a shock $\hat{\sigma}_1^+ \doteq \sigma_1^+ - \varepsilon$ (see fig. 6.2). Thus, the interaction estimates (5.5) yield

$$V(t_0+) - V(t_0-) \geq -|\sigma''| + \mathcal{O}(1) |\sigma'|^2 |\sigma''|$$
$$= \mathcal{O}(1) |\tilde{\sigma}_1^+| |\hat{\sigma}_1^+|^2,$$

$$Q(t_0+) - Q(t_0-) \geq |\tilde{\sigma}_1^+ \hat{\sigma}_1^+| - K_0 K_1 |\sigma' \sigma''| + \mathcal{O}(1) |\sigma' \sigma''| \cdot V(t_0-)$$
$$= |\tilde{\sigma}_1^+ \hat{\sigma}_1^+| + \mathcal{O}(1) |\tilde{\sigma}_1^+| |\hat{\sigma}_1^+|^2,$$

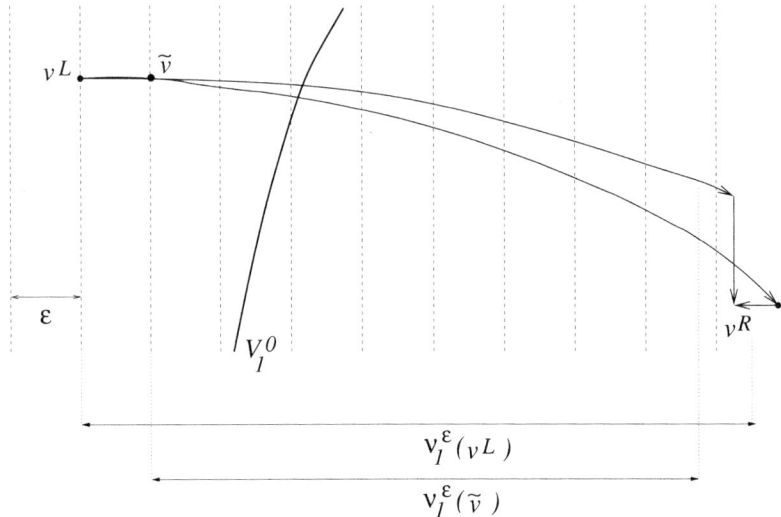

FIGURE 6.1. Example 6.1

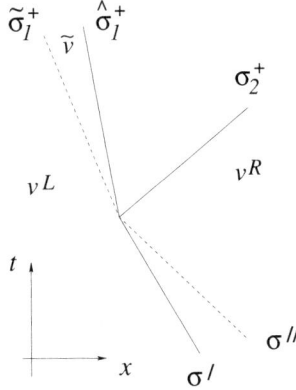

FIGURE 6.2. Example 6.1

which clearly shows how the function (6.1) increases at t_0 no matter which constants K_0, K_1 may be chosen.

The occurrence of interactions of this type is due both to the structure of the solution of the Riemann Problem for systems satisfying the basic assumption (**A**), and to the peculiar properties of the front tracking algorithm adopted here in which, as in [**BC1**],

- the ε-approximate Riemann solver provides solutions which depend continuously on the Riemann data v^L, v^R;
- every Riemann problem is solved by the same procedure;
- centered rarefactions waves are cut along a fixed grid of step ε.

To cope with this technical difficulty we shall keep track of the future history of all wave-fronts emerging from this particular type of *"bad"* interactions. In this way we will show that, although $Q(t)$ increases after such interactions, it is still possible to derive a uniform bound on $Q(t)$ for all $t \geq 0$. To this purpose some further rather technical notations is needed.

6.1. Wave-front basic notations.

1. Shock fronts producing a bad interaction. Let $\sigma_1, \ldots, \sigma_n$ be n wave-fronts that interact together at time t_0, and denote by $\sigma_i^+, i = 1, 2$ the sizes of the outgoing waves. We will say that $\mathcal{J} \doteq \{\sigma_1, \ldots, \sigma_n\}$ is a **family of fronts of bad type (FFB)** for the i-th NGNL characteristic family at time t_0 if the following holds:

I) Among the fronts $\sigma_1, \ldots, \sigma_n$ there is a shock σ of the i-th family, with strength $|\sigma| > \varepsilon$.

II) The total strength of the fronts in $\mathcal{J}$ that either belong to the i-th family and are discordant with σ_i^+, or else belong to the other family, is smaller than ε. More precisely, assume that $\sigma_{\ell_1}, \ldots, \sigma_{\ell_H}$ are such fronts, let C_9 be the constant in (5.25) and, recalling (4.50), (4.51), let $C_{10} > 1$ be a constant independent of ε so that there holds

$$(6.2) \qquad \left(1 + \frac{1}{C_{10}}\right)|v - v'| \leq \left|\nu_i^\varepsilon(v) - \nu_i^\varepsilon(v')\right| \leq C_{10}|v - v'| \qquad \forall\, v, v'.$$

Then, one has

$$(6.3) \qquad \sum_{h=1}^{H} |\sigma_{\ell_h}| < C_{11}\,\varepsilon, \qquad C_{11} \doteq \min\left\{C_9, \frac{1}{8C_{10}}\right\}.$$

III) The outgoing wave σ_i^+ of the i-th family is a composed wave with a rarefaction component of strength

$$|\widetilde{\sigma}_i^+| > \sum_{h=1}^{H} |\sigma_{\ell_h}|.$$

An interaction among the fronts of a FFB for the i-th NGNL family will be called an **interaction of bad type** for the i-th characteristic family. The collection of all FFB for the i-th family at time t will be denoted by $\mathcal{B}^i(t)$. In the case the i-th family is GNL or LD we will set $\mathcal{B}^i(t) \doteq \emptyset$ for all t. We also set

$$\mathcal{B}^i \doteq \bigcup_{t > 0} \mathcal{B}^i(t), \quad i = 1, 2, \qquad \mathcal{B}(t) \doteq \mathcal{B}^1(t) \cup \mathcal{B}^2(t), \qquad \mathcal{B} \doteq \mathcal{B}^1 \cup \mathcal{B}^2.$$

REMARK 6.1. According with the approximate Riemann solver described in Section 3, and by the definition of the approximate mixed-curve given in Section 4.4, it follows that the outgoing i-wave σ_i^+ emerging from an interaction of bad type for the i-th family is a composed wave with a single rarefaction front $\sigma_i^{+,r}$ and a shock $\sigma_i^{+,s}$ of strength $\geq \varepsilon$. Moreover, if we let σ_0 be the incoming i-front that has maximum strength, and let $\sigma'_\ell, \ell = 1, \ldots, L$, be all the other incoming i-fronts concordant with σ_i^+, then, there holds

$$(6.4) \qquad \sum_{\ell=1}^{L} |\sigma'_\ell| \leq C_{12}\,\varepsilon.$$

for some constant $C_{12} > 0$ independent of ε.

REMARK 6.2. Consider n_2 fronts of the second family $\sigma_{2,1}^-, \ldots, \sigma_{2,n_2}^-$, and n_1 fronts of the first family $\sigma_{1,1}^-, \ldots, \sigma_{1,n_1}^-$ that interact together at t_0. Let σ_i^+, $i = 1, 2$, denote the size of the outgoing waves and, if σ_i^+ contains a rarefaction fan $\widetilde{\sigma}_i^+$, call $\widetilde{\sigma}_{i,\ell}^+$ the wave-fronts of $\widetilde{\sigma}_i^+$, otherwise set $\widetilde{\sigma}_{i,\ell}^+ = 0$. By the above definition it follows, for $i = 1, 2$, that if $\mathcal{J}_i \doteq \{\sigma_{i,1}^-, \ldots, \sigma_{i,n_i}^-\}$ is not FFB at time t_0, i.e. if $\mathcal{J}_i \notin \mathcal{B}^i(t_0)$, then one has

$$\min\left\{\varepsilon, |\widetilde{\sigma}_i^+|\right\} = \mathcal{O}(1)\left[\sum_{\sigma_{i,m}^- \cdot \sigma_i^+ < 0} |\sigma_{i,m}^-| + \sum_{j \neq i} |\sigma_{j,p}^-|\right].$$

Hence, applying Lemma 5.13, we obtain
(6.5)
$$\sum_\ell |\widetilde{\sigma}_{i,\ell}^+|\left[|\sigma_i^+| - |\widetilde{\sigma}_{i,\ell}^+|\right] \leq C_{13}\left[\sum_{j=1}^2 \sum_{\sigma_{j,m} \cdot \sigma_{j,p} < 0} |\sigma_{j,m}\,\sigma_{j,p}| + \sum_{m,p} |\sigma_{1,m}\,\sigma_{2,p}| + \right.$$
$$\left. + \sum_{\substack{j \neq i \\ 1 \leq m \leq n_j}} |\sigma_{j,m}| \cdot \sum_{\substack{\sigma_{j,m} \cdot \sigma_{j,p} > 0 \\ m \neq p}} |\sigma_{j,m}\,\sigma_{j,p}|\right],$$

for some constant $C_{13} > 1$, independent of ε, $\sigma_{i,m}$.

2. Wave-fronts originated in the time interval $[t_0, \tau]$ from a given set of fronts $\mathcal{J}$ present at time t_0. For $i = 1, 2$, given any set $\mathcal{J}$ of wave-fronts that interact together at t_0, denote by $\mathcal{N}^i(\mathcal{J}, \tau)$, $\tau \geq t_0$, the collection of all wave-fronts of the i-th family present in $v(\tau, x)$ that are "originated" from $\mathcal{J}$. More precisely, we inductively define the sequence of interaction points $t_0 < t_1 < t_2 < \cdots < t_k$, and of sets $\mathcal{J}_{S,k}$, $\mathcal{J}_{R,k}$, $\mathcal{N}_S^i(\mathcal{J}, \tau)$, $\mathcal{N}_R^i(\mathcal{J}, \tau)$, $t_{k-1} < \tau \leq t_k$, as follows.

I) Let $\mathcal{J}^i$ be the collection of all wave-fronts of the i-th family in $\mathcal{J}$. Then, set
$$\mathcal{N}_S^i(\mathcal{J}, t_0) = \mathcal{N}_R^i(\mathcal{J}, t_0) \doteq \mathcal{J}^i.$$

II) Let $\mathcal{J}_{S,1}$ ($\mathcal{J}_{R,1}$) be the collection of all shocks (rarefaction wave-fronts) of the i-th family emerging from the interaction at t_0

III) Let t_1 be the first time when it occurs an interaction involving a wave front of $\mathcal{J}_{S,1}$ ($\mathcal{J}_{R,1}$). Then, set
$$\mathcal{N}_S^i(\mathcal{J}, \tau) \doteq \mathcal{J}_{S,1}, \qquad \mathcal{N}_R^i(\mathcal{J}, \tau) \doteq \mathcal{J}_{R,1} \qquad t_0 < \tau \leq t_1.$$

IV) For $k \geq 1$, define $\mathcal{J}_{S,k+1}$ ($\mathcal{J}_{R,k+1}$) as the collection of all waves in $\mathcal{J}_{S,k}$ ($\mathcal{J}_{R,k}$) not interacting at t_k, together with all waves of the i-th family emerging from the interactions occurring at t_k that involve waves of $\mathcal{J}_{S,k}$ ($\mathcal{J}_{R,k}$).

V) For $k \geq 1$, let t_{k+1} be the first time when it occurs an interaction involving a wave front of $\mathcal{J}_{S,k+1}$ ($\mathcal{J}_{R,k+1}$). Then, set
$$\mathcal{N}_S^i(\mathcal{J}, \tau) \doteq \mathcal{J}_{S,k+1}, \qquad \mathcal{N}_R^i(\mathcal{J}, \tau) \doteq \mathcal{J}_{R,k+1} \qquad t_k < \tau \leq t_{k+1}.$$

Moreover, set

$$\mathcal{N}^i(\mathcal{J}, \tau) \doteq \mathcal{N}^i_S(\mathcal{J}, \tau) \cup \mathcal{N}^i_R(\mathcal{J}, \tau), \qquad \mathcal{N}^i(\mathcal{J}) \doteq \bigcup_{\tau \geq t_0} \mathcal{N}^i(\mathcal{J}, \tau),$$

$$\mathcal{N}(\mathcal{J}, \tau) \doteq \mathcal{N}^1(\mathcal{J}, \tau) \cup \mathcal{N}^2(\mathcal{J}, \tau).$$

3. Wave-fronts interacting at time t with a given set of fronts $\mathcal{J}$. For $i = 1, 2$, given any set $\mathcal{J}$ of wave-fronts in v, denote by $\mathcal{I}^i(\mathcal{J}, t)$ the collection of all wave-fronts in v of the i-th family that interacts at time t with some wave in $\mathcal{J}$. Moreover, set $\mathcal{I}(\mathcal{J}, t) \doteq \mathcal{I}^1(\mathcal{J}, t) \cup \mathcal{I}^2(\mathcal{J}, t)$.

4. Wave-fronts canceled by an interaction occurring at time t. For $i = 1, 2$, denote by $\mathcal{C}^i(t)$ the collection of all wave-fronts in v of the i-th family that have an interaction at time t, and are discordant with the outgoing waves (of the same family) originated by such interactions.

5. Time interval $[t_0, \widehat{t}_0]$ within which the interaction potential decreases by a quantity comparable with the increase $Q(t_0+) - Q(t_0-) > 0$ caused by a bad interaction occurring at t_0.

DEFINITION 6.1. Assume that the first characteristic family is NGNL1, let $\mathcal{J}_0$ be a family of fronts of bad type for the first family that interact together at time t_0, and let σ_0 be the 1-front with maximum strength in $\mathcal{J}_0$. We will call **interaction time of good type (ITG)** for $(\mathcal{J}_0, t_0)$ the first time $\widehat{t}_0 > t_0$ at which it occurs an interaction that satisfies one of the following five conditions.

I) The interaction point $(\widehat{t}_0, x_{\widehat{t}_0})$ is connected with (t_0, x_0) by a polygonal line $x(\cdot)$ in the (t, x)-plane with a finite number of nodes

$$(t_0, x_0), (t_1, x_1), \ldots, (t_H, x_H) = (\widehat{t}_0, x_{\widehat{t}_0}),$$

having the following properties (see figure 6.3):

I.I) The nodes (t_h, x_h), $0 \leq h \leq H$, are all interaction points.

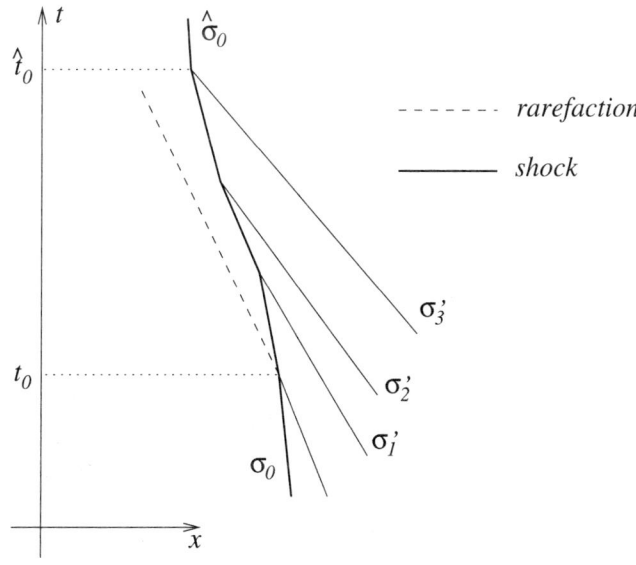

FIGURE 6.3. Definition 6.1, Case I

I.II) On each segment joining (t_h, x_h) and (t_{h+1}, x_{h+1}), $0 \leq h < H$, the solution v has a shock σ_{h+1} belonging to $\mathcal{N}_S^1(\mathcal{J}_0, t_{h+1})$, concordant with σ_0.

I.III) The wave $\sigma_{\widehat{t}_0}$ of the first family emerging from $(\widehat{t}_0, x_{\widehat{t}_0})$ is concordant with σ_0 and the total strength of waves of the first family that interact with waves of $\mathcal{N}_S^1(\mathcal{J}_0, t)$, $t_0 < t \leq \widehat{t}_0$, is such that either

$$(6.6) \qquad \sum_{\substack{\sigma_h' \in \mathcal{I}^1(\sigma_h, t_h),\ 0 < h \leq H \\ \sigma_h' \cdot \sigma_0 < 0}} |\sigma_h'| > \frac{\varepsilon}{8 C_{10}},$$

or

$$(6.7) \qquad \sum_{\substack{\sigma_h' \in \mathcal{I}^1(\sigma_h, t_h),\ 0 < h \leq H \\ \sigma_h' \cdot \sigma_0 > 0}} |\sigma_h'| > (10\, C_{10})^2 \cdot \varepsilon,$$

where C_{10} is the constant in (6.2). This property guarantees that the decrease of the interaction potential caused by interactions occurring in the time interval $]t_0, \widehat{t}_0]$ between waves of the first family is either of order $K_1(|\sigma_0| - \varepsilon/(8C_{10})) \cdot \varepsilon/(8C_{10}) \approx Q(t_0+) - Q(t_0-)$ (in the case the interactions take place between discordant fronts), or of order $|\sigma_0| \cdot 100\, C_{10}^2 \cdot \varepsilon \approx Q(t_0+) - Q(t_0-)$ (in the case the interactions take place between concordant fronts).

II) The interaction point $(\widehat{t}_0, x_{\widehat{t}_0})$ is connected with (t_0, x_0) by a polygonal line as in I) having properties I.I), I.II), and (see figure 6.4):

II.III) The total strength of waves of the second family that interact

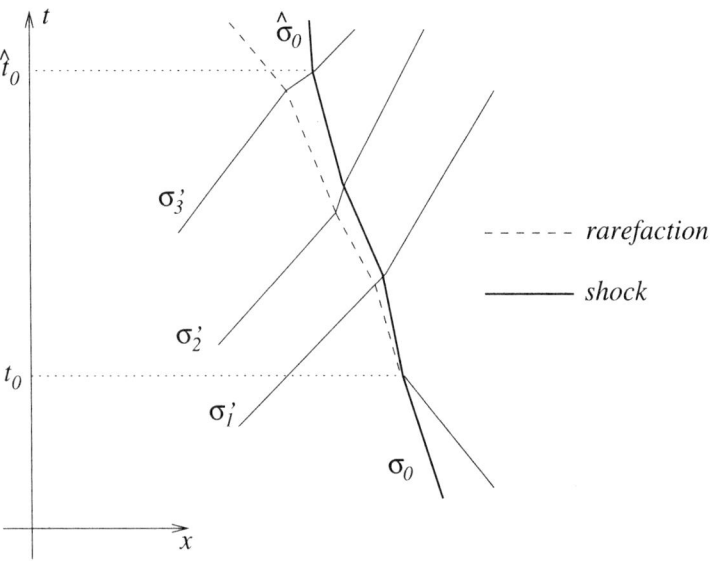

FIGURE 6.4. Definition 6.1, Case II

with waves of $\mathcal{N}_S^1(\mathcal{J}_0, t)$, $t_0 < t \leq \widehat{t}_0$, is such that

(6.8) $$\sum_{\substack{\sigma_h' \in \mathcal{I}^2(\sigma_h, t_h) \\ 0 < h \leq H}} |\sigma_h'| > \frac{\varepsilon}{8C_{10}}.$$

This property guarantees that the decrease of the interaction potential caused by interactions occurring in the time interval $]t_0, \widehat{t}_0\,]$ between waves of different families is at least of order $K_1|\sigma_0| \cdot \varepsilon/(8C_{10}) \approx Q(t_0+) - Q(t_0-)$.

III) The interaction point $(\widehat{t}_0, x_{\widehat{t}_0})$ is connected with (t_0, x_0) by a polygonal line $x(\cdot)$ as in I) having property I.I) and (see figure 6.5):

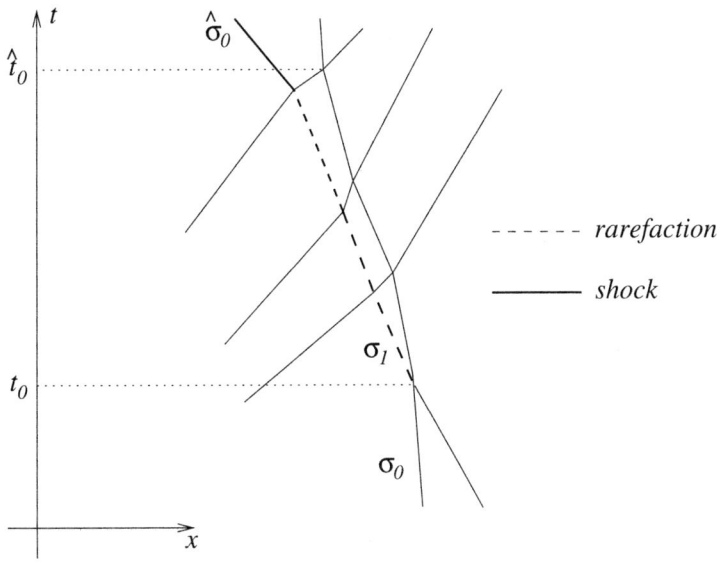

FIGURE 6.5. Definition 6.1, Case III

III.II) On each segment joining (t_h, x_h) and (t_{h+1}, x_{h+1}), $0 \leq h < H$, the solution v has a rarefaction front σ_{h+1} belonging to $\mathcal{N}_R^1(\mathcal{J}_0, t_{h+1})$, concordant with σ_0.

III.III) The wave $\sigma_{\widehat{t}_0}$ of the first family emerging from $(\widehat{t}_0, x_{\widehat{t}_0})$ is discordant with σ_0, and the total strength of waves of the second family that interact with fronts of $x(t)$, $t \in\,]t_0, \widehat{t}_0\,]$ is $\leq \sqrt{\varepsilon}$. This property guarantees that, on the time interval $]t_0, \widehat{t}_0\,]$, the interaction potential decreases by a quantity that is at least of order $K_1(|\sigma_0| - |\sigma_1|) \cdot |\sigma_1| \approx Q(t_0+) - Q(t_0-)$, due to the cancellation of the rarefaction front $\sigma_1 \in \mathcal{N}_R^1(\mathcal{J}_0, t_1)$ generated by the bad interaction at t_0.

IV) There is an interaction point $(\bar{t}, \bar{x})$, $\bar{t} \leq \widehat{t}_0$, connected with (t_0, x_0) by a polygonal line $x(\cdot)$ as in I) having properties I.I), III.II), and:

IV.III) The total strength of waves of the second family that interact with fronts of $x(\cdot)$, $t \in\,]t_0, \bar{t}\,]$ is $> \sqrt{\varepsilon}$.

Moreover, the total strength of waves of the second family originated by interactions of the fronts in $x(t)$, $t \in \,]t_0, \bar{t}\,]$, that are canceled in the time interval $]t_0, \widehat{t_0}]$ (and were approaching the fronts in $\mathcal{N}_S^1(\mathcal{J}_0, t)$, $t \in \,]t_0, \widehat{t_0}]$), is $> \sqrt{\varepsilon}/2$. This property guarantees that, on the time interval $]t_0, \widehat{t_0}\,]$, the interaction potential decreases by a quantity that is at least of order $K_1|\sigma_0| \cdot \sqrt{\varepsilon} \gg Q(t_0+) - Q(t_0-)$, due to the cancellation of the 2-fronts generated in the time interval $]t_0, \bar{t}\,]$.

v) The interaction point $(\widehat{t_0}, x_{\widehat{t_0}})$ is connected with (t_0, x_0) by a polygonal line $x(\cdot)$ as in I) having properties I.I), I.II), and (see figure 6.6):

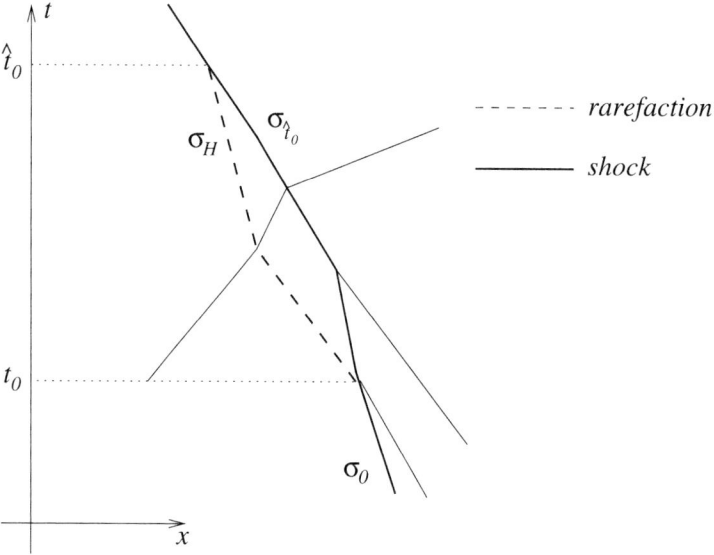

FIGURE 6.6. Definition 6.1, Case V

V.III) The wave $\sigma_H \in \mathcal{N}_S^1(\mathcal{J}_0, \widehat{t_0})$ entering the node $(\widehat{t_0}, x_{\widehat{t_0}})$ interacts at $\widehat{t_0}$ with some front $\sigma_{\widehat{t_0}} \in \mathcal{N}_R^1(\mathcal{J}_0, \widehat{t_0})$ on its left. This properties guarantees that the fronts $\sigma_1 \in \mathcal{N}_R^1(\mathcal{J}_0, t_1)$, $\sigma_1' \in \mathcal{N}_S^1(\mathcal{J}_0, t_1)$ that were generated by the bad interaction at time t_0, after subsequent interactions with other fronts eventually interact among themselves, and the interaction potential decreases by a quantity that is at least of order $|\sigma_1'||\sigma_1| \approx Q(t_0+) - Q(t_0-)$.

We will say that the interaction occurring at $\widehat{t_0}$ (satisfying one of the above five conditions) is an **interaction of good type** for $(\mathcal{J}_0, t_0)$. An entirely similar definition of interaction time (and interaction) of good type is given in the case we are considering a family of fronts of bad type for any k-th NGNL characteristic family that satisfies either one of (2.8)-(2.9) (i.e. that is either NGNL1 or NGNL2).

In connection with the above definition of interaction time of good type for a FFB $\mathcal{J}_0 \in \mathcal{B}^i(t_0)$, $i \in \{1,2\} \cap NGNL$, it is useful to introduce the map $\Xi_i(\mathcal{J}_0; \cdot):$

$[t_0, \infty) \mapsto \{0, 1\}$ defined by setting

(6.9) $$\Xi_i(\mathcal{J}_0; t) \doteq \begin{cases} 0 & \text{if} \quad t \geq \widehat{t}_0, \\ 1 & \text{if} \quad t \in [t_0, \widehat{t}_0 [. \end{cases}$$

We also set $\Xi_i(\mathcal{J}_0; t) \doteq 0$ for all $t \geq t_0$, if $\mathcal{J}_0$ is a set of fronts that have an interaction at t_0 which is not of bad type for the i-th family, i.e. if $\mathcal{J}_0 \notin \mathcal{B}^i(t_0)$. This means, in particular, that if the i-th family is GNL or LD, then one has $\Xi_i(\mathcal{J}_0; t) = 0$ for any set of fronts $\mathcal{J}_0$ interacting at t_0, and any time $t \geq t_0$.

REMARK 6.3. Recalling the properties of the approximate Riemann solver described in Section 3, and from Definition 6.1 it follows that, if the first family is NGNL1, given $\mathcal{J}_0 \in \mathcal{B}^1(t_0)$, and letting σ_1 be the shock generated by the bad interaction at t_0, either the set $\mathcal{N}_R^1(\mathcal{J}_0, t)$ consists of concordant fronts with σ_1 for all $t < \widehat{t}_0$, or at some time $\bar{t} < \widehat{t}_0$ there is a cancellation of the front in $\mathcal{N}_R^1(\mathcal{J}_0, t)$ that is adjacent to the ones in $\mathcal{N}_S^1(\mathcal{J}_0, t)$. In the latter case one of the following two situations occurs. Case 1: the total strength of 2-fronts which are originated by interactions with the fronts in $\mathcal{N}_R^1(\mathcal{J}_0, t)$, $t \in [t_0, \bar{t}]$, and that are canceled in the time interval $[t_0, \widehat{t}_0]$, is $> \sqrt{\varepsilon}/2$. Hence, condition IV) is verified. Case 2: the total strength of 2-fronts which are originated by interactions with the fronts in $\mathcal{N}_R^1(\mathcal{J}_0, t)$, $t \in [t_0, \bar{t}]$, and that interact with the front in $\mathcal{N}_S^1(\mathcal{J}_0, t)$ in the time interval $[t_0, \widehat{t}_0]$, is $> \sqrt{\varepsilon}/2$ and thus condition II) is verified.

Concerning the set $\mathcal{N}_S^1(\mathcal{J}_0, t)$, we will see that, if $x(\cdot)$ is a polygonal line as in Definition 6.1-I.I)-I.II) joining (t_0, x_0) with some interaction point (t_h, x_h), $t_0 < t_h < \widehat{t}_0$, then the wave $\sigma_{h+1} > 0$ of the first family emerging from (t_h, x_h), satisfies the bound (cfr. proof of Lemma 6.6):

$$\sigma_{h+1} > \nu_1^\varepsilon(v^{L_{\sigma_1}}) + \frac{\varepsilon}{5C_{10}}.$$

On the other hand, condition V) of Definition 6.1 guarantees that the fronts of the first family cannot interact from the left with the ones in $\mathcal{N}_S^1(\mathcal{J}_0, t)$ for all $t < \widehat{t}_0$ (as well as no 1-front can interact from the right with the fronts in $\mathcal{N}_R^1(\mathcal{J}_0, t)$ before $\widehat{t}_0$), while condition II) ensures that the total strength of 2-fronts that hit $x(\cdot)$ is $< \sqrt{\varepsilon}$. Therefore, by the properties of the approximate Riemann solver and of the mixed curves discussed in Section 3 (see Remark 3.3), we deduce that the left states of σ_1 and σ_h lie on the same vertical line of the grid of step ε, and stay on the same side w.r.t. the curve $\mathcal{V}_1^0$ in (2.11). Then, the above estimate ensures in particular that the wave of the first family emerging from (t_h, x_h) is concordant with σ_1 and consists precisely of the single shock-front σ_{h+1}. Hence, the set $\mathcal{N}_S^1(\mathcal{J}_0, t)$ consists of a single shock for all $t < \widehat{t}_0$.

By the above observations it follows also that, in the case $\widehat{t}_0$ satisfies condition V) (and none of the other conditions I)-IV)), if $x'(t)$, $t \in [t_0, \widehat{t}_0]$ is a polygonal line as in Definition 6.1-I.I)-III.II), starting with the rarefaction front σ_1' generated by the bad interaction at t_0, any front σ_h' of $x'(\cdot)$ is a rarefaction front (because on its right there is a shock $\sigma_h \in \mathcal{N}_S^1(\mathcal{J}_0)$ connecting two states that lie on opposite side w.r.t. $\mathcal{V}_1^0$). Moreover, the total strength of 2-fronts that hit $x'(\cdot)$ is $\leq \sqrt{\varepsilon}$, since otherwise $\widehat{t}_0$ would satisfy either condition II) or condition IV) of Definition 6.1. On the other hand, no 1-front can hit $x'(\cdot)$ before $\widehat{t}_0$. In fact, an interaction of a front in $x'(\cdot)$ with a 1-front coming from the left would produce the cancellation of the front in $x'(\cdot)$ (and hence condition iii) would be verified), while interactions of a front in $x'(\cdot)$ with 1-fronts coming from the right cannot occur (because of condition v)). Therefore, by the properties of the approximate Riemann solver we deduce that, in the case $\widehat{t}_0$ satisfies condition v) of Definition 6.1 (and none of the other conditions I)-IV)), the set $\mathcal{N}_R^1(\mathcal{J}_0, t)$, $t < \widehat{t}_0$ consists of a single rarefaction front of size $\sigma_t^r = \sigma_1'$, concordant with the shock in $\mathcal{N}_S^1(\mathcal{J}_0, t)$.

Similar properties hold in the case $\mathcal{J}_0 \in \mathcal{B}^k(t_0)$ for any k-th NGNL family that satisfies either one of (2.8)-(2.9).

6. Glimm-type interaction functionals.

Consider a family of fronts $\mathcal{J}$ interacting together at t_0, and let σ_i^+, $i = 1, 2$, be the sizes of the outgoing waves. In connection with $\mathcal{J}$, for any i-th characteristic family, we shall introduce here four functionals E_i, F_i, G_i, H_i, as follows. We first define the quantity

$$E_i(\mathcal{J}; t_0) \doteq \Big[\text{pot. int. of raref. } i\text{-fronts } \textit{with} \text{ all the other } i\text{-fronts generated at } t_0 \Big],$$

i.e. set

(6.10) $$E_i(\mathcal{J}; t_0) \doteq \sum_{\ell} |\widetilde{\sigma}_{i,\ell}^+| \big[|\sigma_i^+| - |\widetilde{\sigma}_{i,\ell}^+| \big].$$

Observe that, if we assume in particular that the fronts in $\mathcal{J}$ have an interaction of bad type at t_0 for the i-th family, i.e. if $\mathcal{J} \in \mathcal{B}^i(t_0)$, $i \in NGNL$, then, recalling that the i-th outgoing wave is a composed wave consisting of a single rarefaction front $\sigma_0^r \in \mathcal{N}_R^i(\mathcal{J}, t_0+)$, and of a shock $\sigma_0^s \in \mathcal{N}_S^i(\mathcal{J}, t_0+)$ (see Remark 6.1), we can rewrite (6.10) as $|\sigma_0^r \sigma_0^s|$. The quantity in (6.10) provides an upper bound for the increase of potential interaction occurring across t_0, due to the self-interactions among the new fronts of the i-th family generated at t_0. For any $\tau \in [t_0, \widehat{t}_0]$, we define also the quantity

(6.11) $$E_i(\mathcal{J}; \tau) \doteq \begin{cases} |\sigma_\tau^r \sigma_\tau^s| & \text{if } \widehat{t}_0 \text{ satisfies cond. v) of Def. 6.1,} \\ |\sigma_0^r \sigma_0^s| & \text{otherwise,} \end{cases}$$

where σ_τ^r, σ_τ^s denote, respectively, the single rarefaction front of $\mathcal{N}_R^i(\mathcal{J}, \tau)$, and the shock of $\mathcal{N}_S^i(\mathcal{J}, \tau)$ (cfr. Remark 6.3). In the case $\mathcal{J} \notin \mathcal{B}^i(t_0)$, we shall write $E_i(\mathcal{J}; \tau) \doteq E_i(\mathcal{J}; t_0)$, for all $\tau \geq t_0$.

Next, for any time $\tau \geq t_0$, we introduce the functionals

(6.12) $$F_i(\mathcal{J}; t_0, \tau) \doteq \frac{1}{4} \sum_{\substack{\sigma' \in \mathcal{N}_S^i(\mathcal{J}, t_k) \\ \sigma'' \in \mathcal{I}^i(\sigma', t_k) \\ \sigma' \cdot \sigma'' > 0 \\ t_0 \leq t_k \leq \tau}} |\sigma' \sigma''|,$$

(6.13)
$$G_i(\mathcal{J};t_0,\tau) \doteq \frac{K_1}{4} \sum_{\substack{\sigma' \in \mathcal{N}_S^i(\mathcal{J},t_k) \\ \sigma'' \in \mathcal{I}^i(\sigma',t_k) \\ \sigma' \cdot \sigma'' < 0 \\ t_0 \leq t_k \leq \tau}} |\sigma'\sigma''| + \frac{K_1}{8} \sum_{\substack{\sigma' \in \mathcal{N}_S^i(\mathcal{J},t_k) \\ \sigma'' \in \mathcal{I}^j(\sigma',t_k),\ j \neq i \\ t_0 \leq t_k \leq \tau}} |\sigma'\sigma''|,$$

(6.14)
$$H_i(\mathcal{J};t_0,\tau) \doteq \frac{K_1}{4} \sum_{\substack{\sigma' \in \mathcal{C}^i(t_k) \cap N_R^i(\mathcal{J},t_k) \\ \sigma'' \in \mathcal{N}_S^i(\mathcal{J}',t_k) \setminus \mathcal{I}(\sigma',t_k), \\ \mathcal{J}' \in \mathcal{B}^i(t),\ t<t_k \\ t_0 \leq t_k \leq \tau}} |\sigma'\sigma''| + \frac{K_1}{4} \sum_{\substack{\sigma' \in \mathcal{C}^j(t_k) \cap \mathcal{N}^i(\mathcal{N}_R^i(\mathcal{J},t_h),\ t_k) \\ j \neq i,\ t_0 \leq t_h \leq t_k \leq \tau \\ \sigma'' \in \mathcal{N}_S^i(\mathcal{J}',t_k) \setminus \mathcal{I}(\sigma',t_k), \\ \mathcal{J}' \in \mathcal{B}^i(t),\ t<t_k \\ (\sigma',\sigma'') \in \mathcal{A}}} |\sigma'\sigma''|.$$

Such functionals measure the decrease of potential interaction caused by the (good) interactions occurring in the time interval $[t_0,\tau]$ and involving fronts of the i-th family originated from $\mathcal{J}$. In the above summations t_k are the interaction times for the fronts of v, $\mathcal{A}$ denotes the set of approaching waves (cfr. definition in Section 3) present in $v(t_k)$, while $K_1 > 0$ denotes the constant appearing in the potential interaction Q^ε defined at (3.30), chosen so that

(6.15)
$$K_1 \geq \max\left\{4C_5,\ (40\,C_{10})^3 \cdot C_{12},\ 24\,C_{13}\right\},$$

where C_5 is the constant in Lemma 5.11, while C_{10}, , C_{12}, C_{13} are the constants at (6.2), (6.4) and (6.5). Notice that, in the case where the i-th family is LD or GNL, by definition one has $\mathcal{B}^i(t) = \emptyset$, for all t. Hence, in this case, the functional $H_i(\mathcal{J};t_0,\tau)$ in (6.14) is always zero for any $\tau \geq t_0$.

REMARK 6.4. Let $\mathcal{J}_0 = \{\sigma_{2,1},\ldots,\sigma_{2,n_2},\sigma_{1,1},\ldots,\sigma_{1,n_1}\}$ be a family of fronts interacting together at t_0, with $\sigma_{i,m}$ belonging to the i-th characteristic family. Observe that, by the definitions of sets of fronts originated from $\mathcal{J}_0$ given above at **2**, one has
$$\mathcal{N}_S^i(\mathcal{J}_0,\ t_0) = \mathcal{N}_R^i(\mathcal{J}_0,\ t_0) = \mathcal{J}_0^i = \{\sigma_{i,1},\ldots,\sigma_{i,n_i}\}.$$

Thus, if we compute the above functionals precisely at t_0, we obtain
(6.16)
$$F_i(\mathcal{J}_0;t_0,t_0) = \frac{1}{4} \sum_{\sigma_{i,m} \cdot \sigma_{i,p} > 0} |\sigma_{i,m}\,\sigma_{i,p}|,$$

$$G_i(\mathcal{J}_0;t_0,t_0) = \frac{K_1}{4} \sum_{\sigma_{i,m} \cdot \sigma_{i,p} < 0} |\sigma_{i,m}\,\sigma_{i,p}| + \frac{K_1}{8} \sum_{m,p} |\sigma_{1,m}\,\sigma_{2,p}|,$$

$$H_i(\mathcal{J}_0;t_0,t_0) = \frac{K_1}{4}\left[\sum_{\substack{\sigma_{i,m} \in \mathcal{C}^i(t_0) \\ \sigma' \in \mathcal{N}_S^i(\mathcal{J}',t_0) \setminus \mathcal{I}(\mathcal{J},t_0) \\ \mathcal{J}' \in \mathcal{B}^i(t),\ t<t_0}} |\sigma_{j,m}\,\sigma'| + \sum_{\substack{\sigma_{j,m} \in \mathcal{C}^j(t_0),\ j \neq i \\ \sigma' \in \mathcal{N}_S^i(\mathcal{J}',t_0) \setminus \mathcal{I}(\mathcal{J},t_0) \\ (\sigma_{j,m},\sigma') \in \mathcal{A} \\ \mathcal{J}' \in \mathcal{B}^i(t),\ t<t_0}} |\sigma_{j,m}\,\sigma'|\right].$$

Notice that, as observed above, in the case where the i-th family is GNL or LD, we have $H_i(\mathcal{J}_0; t_0, t_0) = 0$.

6.2. Main estimates.

Our first goal is to derive the basic estimates on the change in the values of $V(t)$ and $Q(t)$ across times where interactions involving arbitrary number of wavefronts take place. Throughout the following, in connection with any pair of waves σ_α, σ_β, we shall often use the notation

$$\hat{K}(\sigma_\alpha, \sigma_\beta) \doteq \begin{cases} |\sigma_\alpha \cdot \sigma_\beta| & \text{if } \sigma_\alpha, \sigma_\beta \text{ belong to the same family} \\ & \text{and are concordant,} \\ K_1 |\sigma_\alpha \cdot \sigma_\beta| & \text{otherwise.} \end{cases}$$

LEMMA 6.1. *For a 2×2 system satisfying the assumption* (**A**), *consider n_2 fronts $\mathcal{J}_2 \doteq \{\sigma_{2,1}, \ldots, \sigma_{2,n_2}\}$ of the second family and n_1 fronts $\mathcal{J}_1 \doteq \{\sigma_{1,1}, \ldots, \sigma_{1,n_1}\}$ of the first family interacting all together at t_0, that generate two i-waves σ_i^+, $i = 1, 2$. Moreover, assume that no other interaction occurs at time t_0 between waves in v and that*

$$(6.17) \qquad V(t_0-) + Q(t_0-) < C_{14} \doteq \min\left\{ \frac{1}{40 C_5}, \frac{1}{5 K_1} \right\},$$

where $K_1 > 1$ denotes the constant in (6.15), while C_5 is the constants of Lemma 5.11. Then, setting $\mathcal{J} \doteq \mathcal{J}_1 \cup \mathcal{J}_2$, the following hold.

$$(6.18) \quad \left|\sigma_1^+ - \sum_{m=1}^{n_1} \sigma_{1,m}\right| + \left|\sigma_2^+ - \sum_{m=1}^{n_2} \sigma_{2,m}\right| < \left[-\Delta Q(t_0) + \sum_{i \in NGNL} \Xi_i(\mathcal{J}; t_0) \, E_i(\mathcal{J}; t_0) \right]$$

$$\leq 8 \sum_{i=1}^{2} \Big[F_i(\mathcal{J}; t_0, t_0) + G_i(\mathcal{J}; t_0, t_0) \Big],$$

$$(6.19) \quad \Delta Q(t_0) \leq -3 \sum_{i=1}^{2} \Big[F_i(\mathcal{J}; t_0, t_0) + G_i(\mathcal{J}; t_0, t_0) + H_i(\mathcal{J}; t_0, t_0) \Big] + \sum_{i \in NGNL} \Xi_i(\mathcal{J}; t_0) \, E_i(\mathcal{J}; t_0),$$

$$(6.20) \quad \Delta(V + Q)(t_0) \leq -\frac{5}{3} \sum_{i=1}^{2} \Big[F_i(\mathcal{J}; t_0, t_0) + G_i(\mathcal{J}; t_0, t_0) + H_i(\mathcal{J}; t_0, t_0) \Big] + \sum_{i \in NGNL} \Xi_i(\mathcal{J}; t_0) \, E_i(\mathcal{J}; t_0).$$

REMARK 6.5. *According with Lemma 6.1 and definition (3.30), the change in the values of the interaction potential $Q(t)$ across an interaction time t_0 can be* estimated by

$$\Delta Q(t_0) \leq \Big[\text{pot. int. between } \textit{new fronts} \text{ generated by the interaction at } t_0,$$
$$\text{whenever such an interaction is of } \textit{bad} \text{ type} \Big] +$$
$$- \Big[\text{pot. int. between } \textit{fronts involved} \text{ in the interaction at } t_0 \Big] +$$
$$- \Big[\text{pot. int. between } \textit{fronts canceled} \text{ by the interaction at } t_0$$
$$\textit{and } \text{ all the other fronts present at } t_0, \text{ generated by}$$
$$\text{some } \textit{bad} \text{ interaction occuring at a } \textit{previous time } t < t_0 \Big].$$

The same type of estimate holds for the change in the values of $V + Q$ across t_0.

PROOF OF LEMMA 6.1. The proof is given in three steps.

Step 1. We shall work by induction on the number of adjacent fronts in $\mathcal{J}$. More precisely, for each $k = 1, \ldots, n_1 + n_2$, let v^L be the state to the left of $\sigma_{2,1}$, and call v_k^R the state to the right of the k-th front in $\mathcal{J}$. Then, consider the Riemann problem with initial states v^L, v_k^R, let σ_1^k, σ_2^k, be the sizes of the outgoing waves generated by this problem, and call $\widetilde{\sigma}_{i,\ell}^k$ the wave-fronts of the rarefaction component $\widetilde{\sigma}_i^k$ of σ_i^k (with the understanding that $\widetilde{\sigma}_{i,\ell}^k = 0$ if $\widetilde{\sigma}_i^k = 0$). The configuration of the fronts and states after the k-th inductive step is illustrated in figure 6.7. Let $\mathcal{J}^k$ be the set of the first k adjacent fronts in $\mathcal{J}$, and denote with V^k and Q^k, respectively, the total strength of waves and the potential interaction (defined as in (3.29)-(3.30)) evaluated for $v(t_0)$ replacing the first k fronts $\mathcal{J}^k$ with the waves σ_1^k, σ_2^k. Denote with E^k, F^k, G^k, H^k the interaction functionals for the set of fronts $\mathcal{J}^k$ at t_0 defined as in (6.10), (6.12)-(6.14), i.e. set

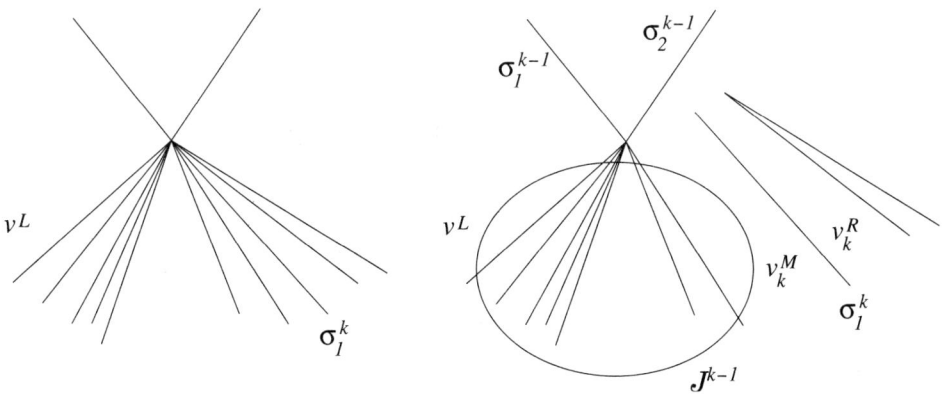

FIGURE 6.7

$$E^k \doteq \sum_{i=1}^{2} \sum_{\ell} |\widetilde{\sigma}_{i,\ell}^k| \big[|\sigma_i^k| - |\widetilde{\sigma}_{i,\ell}^k| \big],$$

$$F^k \doteq \begin{cases} \dfrac{1}{4} \displaystyle\sum_{\substack{1 \leq m < p \leq k \\ \sigma_{2,m} \cdot \sigma_{2,p} > 0}} |\sigma_{2,m}\, \sigma_{2,p}| & \text{if } k \leq n_2, \\[2ex] F^{n_2} + \dfrac{1}{4} \displaystyle\sum_{\substack{1 \leq m < p \leq k - n_2 \\ \sigma_{1,m} \cdot \sigma_{1,p} > 0}} |\sigma_{1,m}\, \sigma_{1,p}| & \text{otherwise,} \end{cases}$$

$$G^k \doteq \begin{cases} \dfrac{K_1}{4} \displaystyle\sum_{\substack{1 \leq m < p \leq k \\ \sigma_{2,m} \cdot \sigma_{2,p} < 0}} |\sigma_{2,m}\, \sigma_{2,p}| & \text{if } k \leq n_2, \\[2ex] G^{n_2} + \dfrac{K_1}{4} \Bigg[\displaystyle\sum_{\substack{1 \leq m < p \leq k - n_2 \\ \sigma_{1,m} \cdot \sigma_{1,p} < 0}} |\sigma_{1,m}\, \sigma_{1,p}| + \displaystyle\sum_{\substack{1 \leq m \leq n_1 \\ 1 \leq p \leq k - n_2}} |\sigma_{1,m}\, \sigma_{2,p}|\Bigg] & \text{otherwise,} \end{cases}$$

$$H^k \doteq \begin{cases} \dfrac{K_1}{4}\Bigg[\displaystyle\sum_{\substack{1 \leq m \leq k \\ \sigma_{2,m} \cdot \sigma_2^k < 0}} |\sigma_{2,m}|\Bigg]\Bigg[\displaystyle\sum_{\substack{\sigma' \in \mathcal{N}_S^2(\mathcal{J}', t_0) \setminus \mathcal{J} \\ \mathcal{J}' \in \mathcal{B}^2(t),\, t < t_0}} |\sigma'| + \displaystyle\sum_{\substack{\sigma' \in \mathcal{N}_S^1(\mathcal{J}', t_0) \setminus \mathcal{J} \\ \mathcal{J}' \in \mathcal{B}^1(t),\, t < t_0 \\ (\sigma_{2,m}, \sigma') \in \mathcal{A}}} |\sigma'|\Bigg] & \\[1ex] & \hspace{-2em} \text{if } k \leq n_2, \\[2ex] H^{n_2} + \dfrac{K_1}{4}\Bigg[\displaystyle\sum_{\substack{1 \leq m \leq k - n_2 \\ \sigma_{1,m} \cdot \sigma_1^k < 0}} |\sigma_{1,m}|\Bigg]\Bigg[\displaystyle\sum_{\substack{\sigma' \in \mathcal{N}_S^1(\mathcal{J}', t_0) \setminus \mathcal{J} \\ \mathcal{J}' \in \mathcal{B}^1(t),\, t < t_0}} |\sigma'| + \displaystyle\sum_{\substack{\sigma' \in \mathcal{N}_S^2(\mathcal{J}', t_0) \setminus \mathcal{J} \\ \mathcal{J}' \in \mathcal{B}^2(t),\, t < t_0 \\ (\sigma_{1,m}, \sigma') \in \mathcal{A}}} |\sigma'|\Bigg] & \\[1ex] & \hspace{-2em} \text{otherwise.} \end{cases}$$

Notice that, recalling the definition (6.10), one has

$$(6.21) \qquad E^{n_1+n_2} = \sum_{i=1}^{2} E_i(\mathcal{J}; t_0).$$

By induction on $k = 1, \ldots, n_1 + n_2$, we will prove that the following estimates hold

$$(6.22)_k \quad \sum_{i=1}^{2} |\sigma_i^k| \leq \Bigg[\sum_{m=1}^{k-n_2} |\sigma_{1,m}| + \sum_{m=1}^{\min\{k,n_2\}} |\sigma_{2,m}|\Bigg]\Bigg[1 + C_5 \sum_{m=1}^{k-1-n_2} |\sigma_{1,m}| + \sum_{m=1}^{\min\{k-1,n_2\}} |\sigma_{2,m}|\Bigg],$$

$$(6.23)_k \qquad V^k + Q^k \leq V(t_0-) + Q(t_0-) + E^k - 2F^k - 2G^k - 4H^k,$$

$$(6.24)_k \quad \Bigg|\sigma_1^k - \sum_{m=1}^{k-n_2} \sigma_{1,m}\Bigg| + \Bigg|\sigma_2^k - \sum_{m=1}^{\min\{k,n_2\}} \sigma_{2,m}\Bigg| < Q^1 - Q^k + E^k - 2F^k - 2G^k,$$

where C_5 denotes the constant in (5.18), and we understand the first summation in $(6.22)_k$, $(6.24)_k$ as equal to zero if $k \leq n_2$. By definition one has $V^1 = V(t_0-)$, $Q^1 = Q(t_0-)$, $E^1 = F^1 = G^1 = H^1 = 0$, and thus $(6.22)_1 - (6.24)_1$ hold. Let now $k > 1$ and suppose that $k > n_2$, i.e. that we are considering the case in which the k-th interacting wave belongs to the first family, the other case being entirely similar. Observe that

$$\left|\sigma_1^k - \sum_{m=1}^{k-n_2} \sigma_{1,m}\right| + \left|\sigma_2^k - \sum_{m=1}^{n_2} \sigma_{2,m}\right| \leq \left|\sigma_1^{k-1} - \sum_{m=1}^{k-1-n_2} \sigma_{1,m}\right| +$$

$$+ \left|\sigma_2^{k-1} - \sum_{m=1}^{n_2} \sigma_{2,m}\right| + \left[\left|\sigma_1^k - (\sigma_1^{k-1} + \sigma_{1,k-n_2})\right| + \left|\sigma_2^{k-1} + \sigma_2^k\right|\right].$$

Thus, the inductive hypothesis $(6.24)_{k-1}$ imply

(6.25)
$$\left|\sigma_1^k - \sum_{m=1}^{k-n_2} \sigma_{1,m}\right| + \left|\sigma_2^k - \sum_{m=1}^{n_2} \sigma_{2,m}\right| \leq \left[Q^1 - Q^{k-1} + E^{k-1} - 2F^{k-1} - 2G^{k-1}\right] +$$

$$+ \left[\left|\sigma_1^k - (\sigma_1^{k-1} + \sigma_{1,k-n_2})\right| + \left|\sigma_2^{k-1} + \sigma_2^k\right|\right].$$

On the other hand, by the definition of E^k, $(6.22)_{k-1}$, $(6.23)_{k-1}$ together with (6.15) yield

(6.26)
$$V^{k-1} \leq \left[V(t_0-) + Q(t_0-)\right] + E^{k-1}$$
$$\leq \left[V(t_0-) + Q(t_0-)\right] + \left(\sum_{i=1}^{2} |\sigma_i^{k-1}|\right)^2$$
$$\leq V(t_0-) + Q(t_0-) + \left(1 + C_5 V(t_0-)\right)^2 (V(t_0-))^2$$
$$\leq \left[V(t_0-) + Q(t_0-)\right] + 4(V(t_0-))^2$$
$$< 5\left[V(t_0-) + Q(t_0-)\right] \leq \min\left\{\frac{1}{8C_5}, \frac{1}{K_1}\right\}.$$

To estimate the second term on the right hand side of (6.25) we use Lemma 5.11. Call v^L the state to the left of $\sigma_{2,1}$, v_k^M the state between $\sigma_{1,k-1-n_2}$ and $\sigma_{1,k-n_2}$, and v_k^R the state to the right of $\sigma_{1,k-n_2}$. Observe that the Riemann problem (v^L, v_k^M) is solved by waves of size $\sigma_1' = \sigma_1^{k-1}$, $\sigma_2' = \sigma_2^{k-1}$ (see figure 6.7), while the Riemann problem (v_k^M, v_k^R) is solved by waves of size $\sigma_1'' = \sigma_{1,k-n_2}$, $\sigma_2'' = 0$. Then, applying Lemma 5.11, and thanks to (6.26), we find

(6.27)
$$\left|\sigma_1^k - (\sigma_1^{k-1} + \sigma_{1,k-n_2})\right| + \left|\sigma_2^k - \sigma_2^{k-1}\right| \leq$$
$$\leq C_5 \Big[|\sigma_1^{k-1} \cdot \sigma_{1,k-n_2}| \max\{|\sigma_1^{k-1}|, |\sigma_{1,k-n_2}|\} +$$
$$+ |\sigma_2^{k-1} \cdot \sigma_{1,k-n_2}| \max\{|\sigma_2^{k-1}|, |\sigma_{1,k-n_2}|\}\Big]$$
$$\leq C_5 \Big[|\sigma_1^{k-1} \cdot \sigma_{1,k-n_2}| + |\sigma_2^{k-1} \cdot \sigma_{1,k-n_2}|\Big] V^{k-1}$$
$$\leq \frac{1}{4}\Big[\hat{K}(\sigma_1^{k-1}, \sigma_{1,k-n_2}) + \hat{K}(\sigma_2^{k-1}, \sigma_{1,k-n_2})\Big].$$

Thus, recalling the definition (3.30) of interaction potential and using (6.26), (6.27) we deduce

(6.28)
$$Q^k - E^k \leq Q^{k-1} - E^{k-1} + K_1\Big[\big|\sigma_1^k - (\sigma_1^{k-1} + \sigma_{1,k-n_2})\big| + \big|\sigma_2^k - \sigma_2^{k-1}\big|\Big]V^{k-1} +$$
$$- \Big[\hat{K}(\sigma_1^{k-1}, \sigma_{1,k-n_2}) + \hat{K}(\sigma_2^{k-1}, \sigma_{1,k-n_2})\Big] - 4\Big[H^k - H^{k-1}\Big] \leq$$
$$\leq Q^{k-1} - E^{k-1} + \Big[\big|\sigma_1^k - (\sigma_1^{k-1} + \sigma_{1,k-n_2})\big| + \big|\sigma_2^k - \sigma_2^{k-1}\big|\Big] +$$
$$- \Big[\hat{K}(\sigma_1^{k-1}, \sigma_{1,k-n_2}) + \hat{K}(\sigma_2^{k-1}, \sigma_{1,k-n_2})\Big] - 4\Big[H^k - H^{k-1}\Big]$$
$$\leq Q^{k-1} - E^{k-1} +$$
$$- \frac{3}{4}\Big[\hat{K}(\sigma_1^{k-1}, \sigma_{1,k-n_2}) + \hat{K}(\sigma_2^{k-1} \cdot \sigma_{1,k-n_2})\Big] - 4\Big[H^k - H^{k-1}\Big]$$
$$= Q^{k-1} - E^{k-1} - \frac{1}{4}\Big[\hat{K}(\sigma_1^{k-1}, \sigma_{1,k-n_2}) + \hat{K}(\sigma_2^{k-1}, \sigma_{1,k-n_2})\Big] +$$
$$- 2\Big[F^k - F^{k-1}\Big] - 2\Big[G^k - G^{k-1}\Big] - 4\Big[H^k - H^{k-1}\Big].$$

From (6.27)-(6.28) it follows

(6.29)
$$V^k - V^{k-1} \leq \Big[\big|\sigma_1^k - (\sigma_1^{k-1} + \sigma_{1,k-n_2})\big| + \big|\sigma_2^k - \sigma_2^{k-1}\big|\Big]$$
$$\leq \Big[Q^{k-1} - Q^k\Big] + \Big[E^k - E^{k-1}\Big] +$$
$$- 2\Big[F^k - F^{k-1}\Big] - 2\Big[G^k - G^{k-1}\Big] - 4\Big[H^k - H^{k-1}\Big].$$

Then, $(6.23)_{k-1}$ together with (6.29) yield $(6.23)_k$, while $(6.24)_k$ follows from (6.25) and (6.29). Finally observe that by the inductive hypothesis $(6.22)_{k-1}$ and using (6.15), (6.26)-(6.27), we deduce

$$\sum_{i=1}^{2}|\sigma_i^k| \leq \sum_{i=1}^{2}|\sigma_i^{k-1}| + |\sigma_{1,k-n_2}| + \Big[\big|\sigma_1^k - (\sigma_1^{k-1} + \sigma_{1,k-n_2})\big| + \big|\sigma_2^k - \sigma_2^{k-1}\big|\Big] \leq$$
$$\leq \sum_{i=1}^{2}|\sigma_i^{k-1}|\Big[1 + C_5|\sigma_{1,k-n_2}|V^{k-1}\Big] + |\sigma_{1,k-n_2}|$$
$$\leq \Bigg[\sum_{m=1}^{k-1-n_2}|\sigma_{1,m}| + \sum_{m=1}^{\min\{k-1,n_2\}}|\sigma_{2,m}|\Bigg] \cdot$$
$$\cdot \Bigg[1 + C_5\sum_{m=1}^{k-2-n_2}|\sigma_{1,m}| + \sum_{m=1}^{\min\{k-2,n_2\}}|\sigma_{2,m}|\Bigg] + |\sigma_{1,k-n_2}| +$$
$$+ C_5|\sigma_{1,k-n_2}|\Bigg[\sum_{m=1}^{k-1-n_2}|\sigma_{1,m}| + \sum_{m=1}^{\min\{k-1,n_2\}}|\sigma_{2,m}|\Bigg]\Big[1 + C_5V(t_0-)\Big]V^{k-1}$$

$$\leq \left[\sum_{m=1}^{k-n_2} |\sigma_{1,m}| + \sum_{m=1}^{\min\{k,n_2\}} |\sigma_{2,m}| \right] \cdot$$

$$\cdot \left[1 + C_5 \sum_{m=1}^{k-1-n_2} |\sigma_{1,m}| + \sum_{m=1}^{\min\{k-1,n_2\}} |\sigma_{2,m}| \right]$$

proving $(6.22)_k$.

Step 2. Using (6.21), $(6.24)_{n_1+n_2}$, recalling the definitions (3.30), (6.10), and by (6.17), we derive
(6.30)
$$\Delta Q \leq -4 \sum_{i=1}^{2} \Big[F_i(\mathcal{J}; t_0, t_0) + G_i(\mathcal{J}; t_0, t_0) + H_i(\mathcal{J}; t_0, t_0) \Big] + \sum_{i=1}^{2} E_i(\mathcal{J}; t_0) +$$
$$+ K_1 V(t_0-) \left[\left| \sigma_1^+ - \sum_{m=1}^{n_1} \sigma_{1,m} \right| + \left| \sigma_2^+ - \sum_{m=1}^{n_2} \sigma_{2,m} \right| \right]$$
$$\leq -4 \sum_{i=1}^{2} \Big[F_i(\mathcal{J}; t_0, t_0) + G_i(\mathcal{J}; t_0, t_0) + H_i(\mathcal{J}; t_0, t_0) \Big] +$$
$$+ K_1 V(t_0-) \Big[-\Delta Q + E^{n_1+n_2} - 2F^{n_1+n_2} - 2G^{n_1+n_2} \Big] + E^{n_1+n_2}$$
$$\leq -4 \sum_{i=1}^{2} \Big[F_i(\mathcal{J}; t_0, t_0) + G_i(\mathcal{J}; t_0, t_0) + H_i(\mathcal{J}; t_0, t_0) \Big] +$$
$$+ \frac{1}{4} \Big[-\Delta Q + 3E^{n_1+n_2} - 3E_0^{n_1+n_2} - 2F^{n_1+n_2} - 2G^{n_1+n_2} \Big],$$

which yields

(6.31)
$$\Delta Q < -3 \sum_{i=1}^{2} \Big[F_i(\mathcal{J}; t_0, t_0) + G_i(\mathcal{J}; t_0, t_0) + H_i(\mathcal{J}; t_0, t_0) \Big] +$$
$$+ E^{n_1+n_2} - \frac{2}{3} \Big[F^{n_1+n_2} + G^{n_1+n_2} \Big].$$

On the other hand, with the same arguments in (6.30), we obtain

(6.32)
$$-\Delta Q + E^{n_1+n_2} \leq 4 \sum_{i=1}^{2} \Big[F_i(\mathcal{J}; t_0, t_0) + G_i(\mathcal{J}; t_0, t_0) \Big] +$$
$$+ \frac{1}{2} \left[\left| \sigma_1^+ - \sum_{m=1}^{n_1} \sigma_{1,m} \right| + \left| \sigma_2^+ - \sum_{m=1}^{n_2} \sigma_{2,m} \right| \right].$$

Step 3. By (6.5), (6.10), (6.15) and (6.21), if $\mathcal{J} \notin \mathcal{B}$, we find

$$E^{n_1+n_2} = \sum_{i=1}^{2} \sum_{\ell} |\widetilde{\sigma}_{i,\ell}^+| \Big[|\sigma_i^+| - |\widetilde{\sigma}_{i,\ell}^+| \Big]$$

$$\leq 4C_{13} \sum_{i=1}^{2} \left[\frac{G_i(\mathcal{J}; t_0, t_0)}{K_1} + \sum_{m=1}^{n_i} |\sigma_{i,m}| \cdot F_i(\mathcal{J}; t_0, t_0) \right]$$

$$\leq 8C_{13} \left(\frac{G^{n_1+n_2}}{K_1} + V(t_0-)F^{n_1+n_2} \right) < \frac{1}{3}(G^{n_1+n_2} + F^{n_1+n_2}),$$

which, using $(6.23)_{n_1+n_2}$, $(6.24)_{n_1+n_2}$ and (6.31)-(6.32), yields (6.18)-(6.20), thus concluding the proof. $\square$

REMARK 6.6. Given a FFB $\mathcal{J}_0 \in \mathcal{B}^k(t_0)$, $k \in \{1,2\} \cap NGNL$, consider a family of fronts $\mathcal{J}$ that interact together at $t \in]t_0, \hat{t}_0 [$, and satisfy the same assumptions of Lemma 6.1. Then, recalling the definition (6.11) of E_k, from the proof of Lemma 6.1 we deduce the further estimates

(6.33)
$$\Delta Q(t) \leq -3 \sum_{i=1}^{2} \left[F_i(\mathcal{J}; t, t) + G_i(\mathcal{J}; t, t) + H_i(\mathcal{J}; t, t) \right] +$$
$$+ \sum_{i \in NGNL} \Xi_i(\mathcal{J}; t) E_i(\mathcal{J}; t) + \left[E_k(\mathcal{J}_0; t+) - E_k(\mathcal{J}_0; t-) \right],$$

(6.34)
$$\Delta(V+Q)(t) \leq -\frac{5}{3} \sum_{i=1}^{2} \left[F_i(\mathcal{J}; t, t) + 2G_i(\mathcal{J}; t, t) + H_i(\mathcal{J}; t, t) \right] +$$
$$+ \sum_{i \in NGNL} \Xi_i(\mathcal{J}; t) E_i(\mathcal{J}; t) + \left[E_k(\mathcal{J}_0; t+) - E_k(\mathcal{J}_0; t-) \right].$$

REMARK 6.7. Consider a family of fronts $\mathcal{J}$ that interact together at t_0, and satisfy the same assumptions of Lemma 6.1, with $C_{14} < (40\,C_1)^{-1}$ in (6.17) (C_1 being the constant of Remark 5.1). Suppose that $\mathcal{J}$ contains two consecutive concordant fronts of a k-th NGNL1 characteristic family: a rarefaction front σ' and a shock σ'', and let v^L be the left state of σ'. Then, following the same arguments of the proof of Lemma 6.1, and relying on (5.3), one easily obtains the sharper estimates

(6.35)
$$\left| \sigma_k^+ - \sum_{m=1}^{n_k} \sigma_{k,m} \right| < 8 \sum_{i=1}^{2} \left[F_i(\mathcal{J}; t_0, t_0) + G_i(\mathcal{J}; t_0, t_0) \right] - 2|\sigma'\sigma''|,$$

(6.36)
$$\Delta Q(t_0) \leq -2 \sum_{i=1}^{2} \left[F_i(\mathcal{J}; t_0, t_0) + G_i(\mathcal{J}; t_0, t_0) + H_i(\mathcal{J}; t_0, t_0) \right] +$$
$$+ \sum_{i \in NGNL} \Xi_i(\mathcal{J}; t_0) E_i(\mathcal{J}; t_0) - \frac{|\sigma'\sigma''|}{2} + \frac{|\sigma'||\sigma' + \sigma'' - \nu_k^\varepsilon(v^L)|}{4(5C_{10})^2},$$

(6.37)
$$\Delta(V+Q)(t_0) \leq -\frac{5}{3}\sum_{i=1}^{2}\Big[F_i(\mathcal{J};t_0,t_0) + 2G_i(\mathcal{J};t_0,t_0) + H_i(\mathcal{J};t_0,t_0)\Big] +$$
$$+ \sum_{i\in NGNL} \Xi_i(\mathcal{J};t_0)\,E_i(\mathcal{J};t_0) - \frac{7}{12}|\sigma'\sigma''| + \frac{|\sigma'||\sigma'+\sigma''-\nu_k^\varepsilon(v^L)|}{4\,(5C_{10})^2},$$

where C_{10} denotes the constant at (6.2). The above estimates will be useful in dealing with the "interactions of good type" (described by Definition 6.1). Clearly, the same estimates hold in the case where $\mathcal{J}$ contains two consecutive concordant fronts of a k-th NGNL2 characteristic family: a shock σ' and a rarefaction σ''.

We next establish some additional interaction estimates that are analogous to those provided by Lemmas 6-7 in [**BC1**]. To this purpose, for any given pair (v,σ) of state v and size σ of a wave-front of a k-th NGNL family, it's useful to introduce the following notation. If σ is a shock, set

(6.38)
$$\vartheta_k(v,\sigma) \doteq \begin{cases} \sigma - \lfloor\delta_k(v)\rfloor^- & \text{if } \sigma > 0, \\ \sigma + \lfloor\delta_k(v)\rfloor^+ & \text{if } \sigma < 0, \end{cases}$$

where $[a]^- = \max\{-a,0\}$, $[a]^+ = \max\{a,0\}$, denote the negative and positive parts of a, while $\delta_k(v)$ is the quantity defined in (2.12). If σ is a rarefaction front, set $\vartheta_k(v,\sigma) \doteq \sigma$. Throughout the following, we shall always refer to an approximate solution v that satisfies the assumption (6.17).

LEMMA 6.2. *For $k \in \{1,2\} \cap NGNL$, $\varepsilon > 0$, $v \in \mathcal{V}$, and any wave-front σ of an ε-approximate solution constructed by the algorithm in Sections 3-4, there holds*

(6.39)
$$\big|\vartheta_k(v,\sigma)\big| \geq \frac{7}{12}\,|\sigma| - C_{15}\cdot\varepsilon$$

where C_{15} denotes some constant independent of ε, v and σ.

PROOF. By definition (6.38), one has
$$\sigma\cdot\delta_k(v) \geq 0 \quad\Longrightarrow\quad \vartheta_k(v,\sigma) = \sigma.$$

Hence, the estimate (6.39) is obviously satisfied whenever σ is a rarefaction front, or a shock such that $\sigma\cdot\delta_k(v) \geq 0$. Next, consider the case where σ is a shock for which $\sigma\cdot\delta_k(v) < 0$, i.e. such that $|\vartheta_k(v,\sigma)| = |\sigma| - |\delta_k(v)|$. Observe that, by possibly taking a restriction of the domain $\mathcal{V}$, because of (4.52) there will be some constant $\bar{c} > 0$ such that

(6.40)
$$\frac{35}{12}|\delta_k(v)| - \bar{c}\cdot\varepsilon \leq \big|\nu_k^\varepsilon(v)\big| \leq \frac{37}{12}|\delta_k(v)| + \bar{c}\cdot\varepsilon \qquad \forall\, v \in \mathcal{V},\, \varepsilon > 0.$$

Then, using (6.40), we obtain
$$\big|\vartheta_k(v,\sigma)\big| \geq |\sigma| - \big|\nu_k^\varepsilon(v)\big| + \frac{23}{12}\big|\delta_k(v)\big| - \bar{c}\cdot\varepsilon$$
$$\geq |\sigma| - \big|\nu_k^\varepsilon(v)\big| + \frac{7}{12}\big|\nu_k^\varepsilon(v)\big| - 2\bar{c}\cdot\varepsilon$$
$$\geq \frac{7}{12}|\sigma| - 2\bar{c}\cdot\varepsilon$$

which proves (6.39), with $C_{15} = 2\bar{c}$. □

LEMMA 6.3. *Consider a family of fronts $\mathcal{J}$ that interact together at t_0. Assume that among the incoming fronts of a k-th NGNL1 characteristic family there is a shock of size σ^-, and that the outgoing wave of the k-th family contains a shock of size σ^+. Call $v^{L,-}$ and $v^{L,+}$ the left states of σ^- and σ^+. Then, there holds*

$$(6.41) \quad F_k(\mathcal{J}; t_0, t_0) + G_K(\mathcal{J}; t_0, t_0) \geq C_{16} |\sigma^-| \Big| \vartheta_k(v^{L,+}, \sigma^+) - \vartheta_k(v^{L,-}, \sigma^-) \Big|,$$

for some constant C_{16} depending only on the system (1.1). Moreover, if $\mathcal{J}$ contains a rarefaction front $\tilde{\sigma}^-$, adjacent to σ^-, located on the left of σ^-, and concordant with σ^-, one has
(6.42)
$$F_k(\mathcal{J}; t_0, t_0) - \frac{|\tilde{\sigma}^- \sigma^-|}{4} + G_k(\mathcal{J}; t_0, t_0) \geq C_{16} |\sigma^-| \Big| \vartheta_k(v^{L,+}, \sigma^+) - \vartheta_k(v^{L,-}, \sigma^-) \Big|.$$

The same results hold in the case where the sets of incoming and outgoing fronts contain a shock of a k-th NGNL2 characteristic family.

PROOF. We will establish only (6.42), the proof of (6.41) being entirely similar. Moreover, to fix the ideas, we shall assume that the first characteristic family is NGNL1, that σ^- is a shock of the first family, and that $\sigma^- > 0$. Let $v^{R,-}$ and $v^{R,+}$ denote the right states of σ^- and σ^+. Observe that, since the map $\tilde{\gamma}_1$ in (2.11) is smooth, because of (2.12) we have

$$\begin{aligned} \Big| \vartheta_1(v^{L,+}, \sigma^+) - \vartheta_1(v^{L,-}, \sigma^-) \Big| &\leq \lfloor v_1^{L,-} - v_1^{L,+} \rfloor^- + |\tilde{\gamma}_1(v_2^{L,+}) - \tilde{\gamma}_1(v_2^{L,-})| \\ &\quad + |v_1^{R,+} - v_1^{R,-}| \\ &\leq \lfloor v_1^{L,-} - v_1^{L,+} \rfloor^- + \mathcal{O}(1) |v_2^{L,+} - v_2^{L,-}| \\ &\quad + |v_1^{R,+} - v_1^{R,-}|. \end{aligned}$$

Therefore, in order to establish (6.42), it will be sufficient to show that there holds

(6.43)
$$F_1(\mathcal{J}; t_0, t_0) - \frac{|\tilde{\sigma}^- \sigma^-|}{4} + G_1(\mathcal{J}; t_0, t_0) \geq$$

$$\geq \frac{1}{6} |\sigma^-| \Big[\lfloor v_1^{L,-} - v_1^{L,+} \rfloor^- + |v_1^{R,+} - v_1^{R,-}| + |v_2^{L,+} - v_2^{L,-}| \Big],$$

under the assumption that $\mathcal{J}$ contains a rarefaction front $\tilde{\sigma}^- > 0$, adjacent to σ^-, and located on the left left of σ^-. Call $\tilde{v}$ the state on the left of $\tilde{\sigma}^-$, and let $\{\sigma^-_{1,1}, \ldots, \sigma^-_{1,p}\}$, $\{\sigma^-_{1,p+3}, \ldots, \sigma^-_{1,n_1}\}$ be the fronts of the first family in $\mathcal{J}$ that are, respectively, on the left of $\tilde{\sigma}^-$, and on the right of σ^-, while let $\sigma^-_{2,1}, \ldots, \sigma^-_{2,n_2}$ be the fronts of the second family in $\mathcal{J}$. Moreover, denote with $\tilde{\sigma}^+$ the (possibly null) rarefaction component of the outgoing wave of the first family. Then, using (6.18),

(6.35), and because of (6.17), we find

(6.44)
$$\lfloor v_1^{L,-} - v_1^{L,+} \rfloor^- \leq |\tilde{v}_1 - v_1^{L,+}|$$
$$\leq \sum_{\ell=1}^{p} |\sigma_{1,\ell}^-| + 2V(t_0-) \cdot \sum_{\ell=1}^{n_2} |\sigma_{2,\ell}^-|$$
$$\leq \sum_{\ell=1}^{p} |\sigma_{1,\ell}^-| + \frac{1}{5} \sum_{\ell=1}^{n_2} |\sigma_{2,\ell}^-|,$$

(6.45)
$$|v_2^{L,+} - v_2^{L,-}| = |v_2^{L,+} - \tilde{v}_2|$$
$$\leq \sum_{\ell=1}^{n_2} |\sigma_{2,\ell}^-| + 2V(t_0-) \cdot \sum_{\ell=1}^{p} |\sigma_{1,\ell}^-|$$
$$\leq \sum_{\ell=1}^{n_2} |\sigma_{2,\ell}^-| + \frac{1}{5} \sum_{\ell=1}^{p} |\sigma_{1,\ell}^-|.$$

(6.46)
$$|v_1^{R,+} - v_1^{R,-}| \leq \left| \tilde{\sigma}^+ + \sigma^+ - \tilde{\sigma}^- - \sigma^- - \sum_{\ell=1}^{p} \sigma_{1,\ell}^- \right| + 2V(t_0-) \cdot \sum_{\ell=1}^{n_2} |\sigma_{2,\ell}^-|$$
$$\leq (1+2V(t_0-)) \cdot \sum_{\ell=p+3}^{n_1} |\sigma_{1,\ell}^-| + 2V(t_0-) \cdot \sum_{\ell=1}^{p} |\sigma_{1,\ell}^-| + 4V(t_0-) \cdot \sum_{\ell=1}^{n_2} |\sigma_{2,\ell}^-|$$
$$\leq \frac{6}{5} \sum_{\ell=p+3}^{n_1} |\sigma_{1,\ell}^-| + \frac{1}{5} \left[\sum_{\ell=1}^{p} |\sigma_{1,\ell}^-| + \sum_{\ell=1}^{n_2} |\sigma_{2,\ell}^-| \right].$$

From (6.44)-(6.46) it follows
$$|\sigma^-| \left[\lfloor v_1^{L,-} - v_1^{L,+} \rfloor^- + |v_1^{R,+} - v_1^{R,-}| + |v_2^{L,+} - v_2^{L,-}| \right] \leq$$
$$\leq \frac{7}{5} |\sigma^-| \left[\sum_{\ell \neq p+1, p+2} |\sigma_{1,\ell}^-| + \sum_{\ell} |\sigma_{2,\ell}^-| \right]$$

which, recalling (6.16), yields (6.43), concluding the proof. □

LEMMA 6.4. *Consider a family of fronts $\mathcal{J}$ that interact together at t_0. Assume that the k-th characteristic family is NGNL1 and let $\sigma_1^-, \ldots, \sigma_{n_k}^-$ be the incoming fronts of the k-th family. Denote their left states by $v^{L,\ell}$, $\ell = 1, \ldots, n_k$. Assume that $|\vartheta_k(v^{L,\ell}, \sigma_\ell^-)| < \sqrt{\varepsilon}/2$ for all $\ell = 1, \ldots, n_k$, and that the outgoing wave of the k-th family contains a shock having size $\sigma^+ > 0$ and left state $v^{L,+}$ such that $\vartheta_k(v^{L,+}, \sigma^+) \geq \sqrt{\varepsilon}/2$. Then, there holds*

(6.47)
$$\sum_{i=1}^{2} \left[F_i(\mathcal{J}; t_0, t_0) + G_i(\mathcal{J}; t_0, t_0) \right] \geq C_{17} \sqrt{\varepsilon} \left| \vartheta_k(v^{L,+}, \sigma^+) - \frac{\sqrt{\varepsilon}}{2} \right|,$$

for some constant C_{17} depending only on the system (1.1). Moreover, if among $\sigma_1^-, \ldots, \sigma_{n_k}^-$ there is a rarefaction front $\sigma_p^- > 0$, and a shock $\sigma_{p+1}^- > 0$ adjacent to σ_p^-, and located on the right of σ_p^-, then the following hold.

a)
$$\sum_{i=1}^{2} \Big[F_i(\mathcal{J}; t_0, t_0) + G_i(\mathcal{J}; t_0, t_0)\Big] - \frac{|\sigma_p^- \sigma_{p+1}^-|}{4} \geq C_{17} \sqrt{\varepsilon} \left|\vartheta_k(v^{L,+}, \sigma^+) - \frac{\sqrt{\varepsilon}}{2}\right|. \quad (6.48)$$

b) Suppose that also the j-th ($j \neq k$) family is NGNL, and assume that among the incoming fronts of the j-th family $\sigma_{n_k+1}^-, \ldots, \sigma_{n_k+n_j}^-$ there is a rarefaction front $\sigma_{q+1}^- > 0$, and a shock $\sigma_q^- > 0$ adjacent to σ_{q+1}^- (located on the left of σ_{q+1}^- if the j-th family is NGNL1, and on the right of σ_{q+1}^- if the j-th family is NGNL2). Then, one has

$$(6.49) \quad \sum_{i=1}^{2}\Big[F_i(\mathcal{J}; t_0, t_0) + G_i(\mathcal{J}; t_0, t_0)\Big] - \frac{|\sigma_p^- \sigma_{p+1}^-|}{4} - \frac{|\sigma_q^- \sigma_{q+1}^-|}{4} \geq$$
$$\geq C_{17} \sqrt{\varepsilon} \left|\vartheta_k(v^{L,+}, \sigma^+) - \frac{\sqrt{\varepsilon}}{2}\right|.$$

The same results hold in the case the outgoing wave of the k-th family contains a shock of size $\sigma^+ < 0$ such that $\vartheta_k(v^{L,+}, \sigma^+) \leq -\sqrt{\varepsilon}/2$, or in the case where the k-th characteristic family is NGNL2 and the same type of assumptions stated above for the outgoing wave of the k-th family are verified.

PROOF. As we have done for the previous lemma, to fix the ideas we assume that the first characteristic family is NGNL1, and let $\sigma_1^-, \ldots, \sigma_{n_1}^-$ be the incoming fronts of the first family. Set

$$\vartheta^+ \doteq \vartheta_1(v^{L,+}, \sigma^+), \qquad \vartheta^\ell \doteq \vartheta_1(v^{L,\ell}, \sigma_\ell^-), \quad \ell = 1, \ldots, n_1.$$

By the regularity properties of the map $\widetilde{\gamma}_1$ at (2.11), let $c > 1$ be some constant such that

$$(6.50) \quad |\widetilde{\gamma}_1(v_2) - \widetilde{\gamma}_1(v_2')| \leq c\,|v_2 - v_2'| \qquad \forall\, v_2, v_2'.$$

We shall consider three cases.

CASE 1: Assume that the total strength of the incoming fronts of the second family is such that

$$(6.51) \quad \sum_{\ell=n_1+1}^{n_1+n_2} |\sigma_\ell^-| > \frac{1}{2c}\sqrt{\varepsilon}.$$

Observe that, using (6.17)-(6.18), we find

$$\vartheta^+ - \frac{\sqrt{\varepsilon}}{2} < \vartheta^+ \leq \sigma^+$$
$$(6.52) \qquad \leq (1 + 2V(t_0-)) \cdot \sum_{\ell=1}^{n_1} |\sigma_\ell^-| + 8\Big[F_2(\mathcal{J}; t_0, t_0) + G_2(\mathcal{J}; t_0, t_0)\Big]$$
$$< \frac{6}{5} \sum_{\ell=1}^{n_1} |\sigma_\ell^-| + 8\Big[F_2(\mathcal{J}; t_0, t_0) + G_2(\mathcal{J}; t_0, t_0)\Big].$$

Thus, since $\vartheta^+ \geq \sqrt{\varepsilon}/2$, and because $\varepsilon < 1$, from (6.17), (6.51)-(6.52) we derive

$$\sqrt{\varepsilon}\left|\vartheta^+ - \frac{\sqrt{\varepsilon}}{2}\right| \leq (10 + 16V(t_0-))\, c\left[F_2(\mathcal{J}; t_0, t_0) + G_2(\mathcal{J}; t_0, t_0)\right]$$

$$\leq 11\, c\left[F_2(\mathcal{J}; t_0, t_0) + G_2(\mathcal{J}; t_0, t_0)\right]$$

which yields (6.47)-(6.48), with $C_{17} = 1/(11\,c)$. Concerning (6.49), suppose that the assumptions in b) are verified by the incoming fronts of the second family $\sigma^-_{n_1+1}, \ldots, \sigma^-_{n_1+n_2}$. Then, if we assume also that

(6.53) $$\left[F_2(\mathcal{J}; t_0, t_0) + G_2(\mathcal{J}; t_0, t_0)\right] - \frac{|\sigma^-_q\, \sigma^-_{q+1}|}{4} \leq \frac{\sqrt{\varepsilon}}{32},$$

by the same computations in (6.52), since $|\sigma^-_{q+1}| < \sqrt{\varepsilon}$ (being a rarefaction) we obtain

$$\vartheta^+ \leq \left(1 + 2V(t_0-)\right) \cdot \sum_{\ell=1}^{n_1} |\sigma^-_\ell| + 2V(t_0-) \cdot \sqrt{\varepsilon} +$$

$$+ 8\left[F_2(\mathcal{J}; t_0, t_0) + G_2(\mathcal{J}; t_0, t_0) - \frac{|\sigma^-_q\, \sigma^-_{q+1}|}{4}\right]$$

$$< \frac{6}{5}\sum_{\ell=1}^{n_1} |\sigma^-_\ell| + \left(\frac{1}{20} + \frac{1}{4}\right)\sqrt{\varepsilon},$$

which implies

(6.54) $$\vartheta^+ - \frac{\sqrt{\varepsilon}}{2} \leq \frac{6}{5}\sum_{\ell=1}^{n_1} |\sigma^-_\ell|.$$

Hence, from (6.51), (6.54) we derive

$$\sqrt{\varepsilon}\left|\vartheta^+ - \frac{\sqrt{\varepsilon}}{2}\right| \leq 10\, c \cdot G_2(\mathcal{J}; t_0, t_0)$$

which yields (6.49), taking $C_{17} = 1/(10\,c)$. On the other hand, if (6.53) is not verified, we recover (6.49) with $C_{17} = 1$ observing that, by the same computations in (6.52), we have

(6.55) $$\left|\vartheta^+ - \frac{\sqrt{\varepsilon}}{2}\right| \leq \left(1 + 2V(t_0-)\right) \cdot V(t_0-) < \frac{1}{32}.$$

CASE 2: Set

(6.56) $$\bar{\ell} \doteq \max\{\ell \leq n_1\ ;\ \sigma^-_\ell \cdot \delta_1(v^{L,\ell}) \leq 0\},$$

and assume that

(6.57) $$\sum_{\ell=n_1+1}^{n_1+n_2} |\sigma^-_\ell| \leq \frac{1}{2c}\sqrt{\varepsilon}$$

and

(6.58) $$\frac{1}{2}\left|\vartheta^+ - \frac{\sqrt{\varepsilon}}{2}\right| \leq \left|\vartheta^+ - \sum_{\ell=\bar{\ell}}^{n_1} \vartheta^\ell + v_1^{L,+} - \widetilde{\gamma}_1(v_2^{L,\bar{\ell}})\right|.$$

Denote with $\tilde\sigma^+$ the (possibly null) rarefaction component of the outgoing wave of the first family and call $\tilde v$ the left state of $\tilde\sigma^+$. Observe that, because of (6.57), the states $v^{L,+}$ $v^{L,1}$ are connected by a rarefaction curve of the second family (see Remark 3.3), and hence one has

(6.59) $$\tilde\sigma^+ = v_1^{L,+} - \tilde v_1, \qquad \tilde v_1 = v_1^{L,1}.$$

Thus, by definitions (2.12), (6.38), (6.56), we have

(6.60)
$$\left|\vartheta^+ - \sum_{\ell=\bar\ell}^{n_1}\vartheta^\ell + \tilde\gamma_1(v_2^{L,+}) - \tilde\gamma_1(v_2^{L,\bar\ell})\right| = \left|\sigma^+ - \sum_{\ell=\bar\ell}^{n_1}\sigma_\ell^- + v_1^{L,+} - v_1^{L,\bar\ell}\right|$$
$$= \left|\tilde\sigma^+ + \sigma^+ - \sum_{\ell=\bar\ell}^{n_1}\sigma_\ell^- + v_1^{L,1} - v_1^{L,\bar\ell}\right|$$
$$= \left|\tilde\sigma^+ + \sigma^+ - \sum_{\ell=1}^{n_1}\sigma_\ell^-\right|,$$

and hence, applying (6.18), we obtain

$$\left|\vartheta^+ - \sum_{\ell=\bar\ell}^{n_1}\vartheta^\ell + \tilde\gamma_1(v_2^{L,+}) - \tilde\gamma_1(v_2^{L,\bar\ell})\right| \le 8\sum_{i=1}^{2}\Big[F_i(\mathcal{J}; t_0, t_0) + G_i(\mathcal{J}; t_0, t_0)\Big]$$

which, together with (6.58), and since $\varepsilon < 1$, yields (6.47) with $C_{17} = 1/16$. On the other hand, if among the incoming fronts of the first family $\sigma_1^-, \ldots, \sigma_{n_1}^-$ there is a rarefaction front $\sigma_p^- > 0$, and a shock $\sigma_{p+1}^- > 0$ adjacent to σ_p^-, applying (6.35) we deduce from (6.60)

$$\left|\vartheta^+ - \sum_{\ell=\bar\ell}^{n_1}\vartheta^\ell + \tilde\gamma_1(v_2^{L,+}) - \tilde\gamma_1(v_2^{L,\bar\ell})\right| \le 8\sum_{i=1}^{2}\Big[F_i(\mathcal{J}; t_0, t_0) + G_i(\mathcal{J}; t_0, t_0)\Big] +$$
$$- 2|\sigma_p^- \, \sigma_{p+1}^-|$$

which, together with (6.58), implies (6.48) with $C_{17} = 1/4$. Concerning (6.49), if we assume also that the assumptions in b) are verified by the incoming fronts of the second family $\sigma_{n_1+1}^-, \ldots, \sigma_{n_1+n_2}^-$, applying (6.35), and thanks to (6.57), we derive from (6.60)

$$\left|\vartheta^+ - \sum_{\ell=\bar\ell}^{n_1}\vartheta^\ell + \tilde\gamma_1(v_2^{L,+}) - \tilde\gamma_1(v_2^{L,\bar\ell})\right| \le 8\sum_{i=1}^{2}\Big[F_i(\mathcal{J}; t_0, t_0) + G_i(\mathcal{J}; t_0, t_0)\Big] +$$
$$- 2|\sigma_p^- \, \sigma_{p+1}^-| - 2|\sigma_q^- \, \sigma_{q+1}^-|$$

which, together with (6.58), implies (6.49) with $C_{17} = 1/4$.

CASE 3: Assume that (6.57) holds and that

(6.61) $$\frac{1}{2}\left|\vartheta^+ - \frac{\sqrt{\varepsilon}}{2}\right| > \left|\vartheta^+ - \sum_{\ell=\bar\ell}^{n_1}\vartheta^\ell + \tilde\gamma_1(v_2^{L,+}) - \tilde\gamma_1(v_2^{L,\bar\ell})\right|.$$

Observe that, since $\vartheta^+ \geq \sqrt{\varepsilon}/2$, relying on (6.61) we deduce

(6.62) $$\sum_{\ell=\bar{\ell}}^{n_1} \vartheta^\ell + \widetilde{\gamma}_1(v_2^{L,+}) - \widetilde{\gamma}_1(v_2^{L,\bar{\ell}}) > \max\left\{\frac{\sqrt{\varepsilon}}{2}, \frac{1}{2}\left(\vartheta^+ - \frac{\sqrt{\varepsilon}}{2}\right)\right\}.$$

On the other hand, with the same notations adopted in Case 2 we have $\widetilde{v}_2 = v_2^{L,+}$, and hence there holds

(6.63) $$|v_2^{L,+} - v_2^{L,\bar{\ell}}| \leq \sum_{\ell=n_1+1}^{n_1+n_2} |\sigma_\ell^-|.$$

Thus, since $|\vartheta^\ell| < \sqrt{\varepsilon}/2$, and because (6.50), (6.57), (6.63) together imply $|\widetilde{\gamma}_1(v_2^{L,+}) - \widetilde{\gamma}_1(v_2^{L,\bar{\ell}})| \leq \sqrt{\varepsilon}/2$, it follows that one of the following three cases occurs:

I) There exists some index $\bar{\ell}' \in \{\bar{\ell}, \ldots n_1\}$ such that

(6.64)
$$\frac{1}{2}\min\left\{\frac{\sqrt{\varepsilon}}{2}, \frac{1}{2}\left(\vartheta^+ - \frac{\sqrt{\varepsilon}}{2}\right)\right\} \leq \sum_{\ell=\bar{\ell}}^{\bar{\ell}'} \vartheta^\ell + \widetilde{\gamma}_1(v_2^{L,+}) - \widetilde{\gamma}_1(v_2^{L,\bar{\ell}})$$
$$\leq \frac{1}{2}\min\left\{\frac{\sqrt{\varepsilon}}{2}, \frac{1}{2}\left(\vartheta^+ - \frac{\sqrt{\varepsilon}}{2}\right)\right\} + \frac{\sqrt{\varepsilon}}{2}.$$

II) There holds

$$\frac{1}{2}\min\left\{\frac{\sqrt{\varepsilon}}{2}, \frac{1}{2}\left(\vartheta^+ - \frac{\sqrt{\varepsilon}}{2}\right)\right\} \leq \widetilde{\gamma}_1(v_2^{L,+}) - \widetilde{\gamma}_1(v_2^{L,\bar{\ell}})$$
$$\leq \frac{1}{2}\min\left\{\frac{\sqrt{\varepsilon}}{2}, \frac{1}{2}\left(\vartheta^+ - \frac{\sqrt{\varepsilon}}{2}\right)\right\} + \frac{\sqrt{\varepsilon}}{2}.$$

III) There exists some index $\bar{\ell}' \in \{\bar{\ell}, \ldots n_1\}$ such that

$$\frac{1}{2}\min\left\{\frac{\sqrt{\varepsilon}}{2}, \frac{1}{2}\left(\vartheta^+ - \frac{\sqrt{\varepsilon}}{2}\right)\right\} \leq \sum_{\ell=\bar{\ell}}^{\bar{\ell}'} \vartheta^\ell$$
$$\leq \frac{1}{2}\min\left\{\frac{\sqrt{\varepsilon}}{2}, \frac{1}{2}\left(\vartheta^+ - \frac{\sqrt{\varepsilon}}{2}\right)\right\} + \frac{\sqrt{\varepsilon}}{2}.$$

To fix the ideas, we shall consider only the case I), the discussion of the other two cases being entirely similar. Observe that, from (6.62), (6.64) it follows
(6.65)
$$m' \doteq \min\left\{\sum_{\ell=\bar{\ell}}^{\bar{\ell}'} \vartheta^\ell + \widetilde{\gamma}_1(v_2^{L,+}) - \widetilde{\gamma}_1(v_2^{L,\bar{\ell}}), \sum_{\ell=\bar{\ell}'}^{n_1} \vartheta^\ell\right\} \geq \frac{1}{2}\min\left\{\frac{\sqrt{\varepsilon}}{2}, \frac{1}{2}\left(\vartheta^+ - \frac{\sqrt{\varepsilon}}{2}\right)\right\},$$

$$m'' \doteq \max\left\{\sum_{\ell=\bar{\ell}}^{\bar{\ell}'} \vartheta^\ell + \widetilde{\gamma}_1(v_2^{L,+}) - \widetilde{\gamma}_1(v_2^{L,\bar{\ell}}), \sum_{\ell=\bar{\ell}'}^{n_1} \vartheta^\ell\right\} \geq \frac{1}{2}\max\left\{\frac{\sqrt{\varepsilon}}{2}, \frac{1}{2}\left(\vartheta^+ - \frac{\sqrt{\varepsilon}}{2}\right)\right\}.$$

On the other hand, using (6.50), (6.63), we obtain

$$(6.66) \quad m' \cdot m'' \leq \left(\sum_{\ell=\bar{\ell}}^{\bar{\ell}'} |\sigma_\ell^-| + c \sum_{\ell=n_1+1}^{n_1+n_2} |\sigma_\ell^-| \right) \sum_{\ell=\bar{\ell}'}^{n_1} |\sigma_\ell^-|$$

$$\leq 4 c \left[F_1(\mathcal{J}; t_0, t_0) + G_1(\mathcal{J}; t_0, t_0) \right].$$

Thus, (6.65)-(6.66) together yield (6.47), with $C_{17} = 1/(32\,c)$. Concerning (6.48)-(6.49), observe that, if among the incoming fronts of the first family there is a rarefaction σ_p^- adjacent to a concordant shock σ_{p+1}^-, it follows that

$$\operatorname{sgn}(\sigma_p^-) \neq \operatorname{sgn}(\delta_1(v^{L,p})), \qquad \operatorname{sgn}(\sigma_{p+1}^-) \neq \operatorname{sgn}(\delta_1(v^{L,p+1})).$$

Therefore, by the definition (6.56) of the index $\bar{\ell}$, it must be $p < \bar{\ell}$, and hence, using (6.50), (6.63), with the same computation in (6.66) we recover also the estimates (6.48)-(6.49) (with the same constant $C_{17} = 1/(32\,c)$). □

6.3. Estimates on the increase of the interaction potential.

We establish here several lemmas providing the basic estimates on the increase E_i of the interaction potential caused by the presence of interactions of bad type.

LEMMA 6.5. *Given a FFB for the k-th family $\mathcal{J}_0 \in \mathcal{B}^k(t_0)$, $k \in \{1,2\} \cap NGNL$, let σ_k^+ be the composed wave emerging from the interaction at t_0, and call σ_k^s the k-shock contained in σ_k^+. Moreover, in the case the k-th family is NGNL1 (NGNL2) let v^s be the left (right) state of σ_k^s. Assume that the total strength of the fronts in $v(t_0)$ satisfy the bound (6.17). Then, there holds*

$$(6.67) \quad \frac{1}{2C_{10}} \varepsilon \leq \left| \sigma_k^s - \nu_k^\varepsilon(v^s) \right| \leq C_{10}\,\varepsilon,$$

where C_{10} is the constant at (6.2).

PROOF. To fix the ideas, we shall consider a FFB for the first family $\mathcal{J}_0 \in \mathcal{B}^1(t_0)$. Moreover, we shall assume that the first characteristic family is NGNL1, the case of a NGNL2 family being entirely similar. Let v^L, v^M be the state to the left of $\mathcal{J}_0$, and the intermediate state between the 2-fronts and the 1-fronts in $\mathcal{J}_0$. Let $\{\sigma_1, \dots, \sigma_p\}$, $\{\sigma_{p+1}, \dots, \sigma_m\}$ and $\{\sigma_{m+1}, \dots, \sigma_n\}$ denote, respectively, the incoming fronts of the second family, the incoming fronts of the first family that are discordant with σ_1^+, and the incoming fronts of the first family that are concordant with σ_1^+. Assume that σ_{m+1} is the incoming 1-front with maximum strength. Observe that by the property of the mixed curves defined in Section 4.4, and by (6.3), (6.4), one has

$$(6.68) \quad \left| v_1^L - v_1^s \right| \leq \varepsilon, \qquad \sum_{h=1}^m |\sigma_h| < C_{11}\,\varepsilon, \qquad \sum_{h=m+2}^n |\sigma_h| < C_{12}\,\varepsilon.$$

Then, using (6.2), (6.3), (6.68), we deduce

$$\left|\nu_1^\varepsilon(v^L) - \nu_1^\varepsilon(v^s) + (v_1^L - v_1^s)\right| \geq \left|\nu_1^\varepsilon(v^L) - \nu_1^\varepsilon(v^s)\right| - \left|v_1^L - v_1^s\right|$$
(6.69)
$$\geq \left(1 + \frac{1}{C_{10}}\right)\varepsilon - \varepsilon \geq \frac{\varepsilon}{C_{10}}.$$

On the other hand, applying (6.18), and thanks to (6.3), (6.15), (6.17), (6.68), we derive
(6.70)
$$\left|\sigma_1^+ - \sum_{h=p+1}^n \sigma_h\right| \leq 2\left[\sum_{h=m+1}^n |\sigma_h|\right] \cdot \left[\sum_{h=m+2}^n |\sigma_h| + K_1 \sum_{h=1}^m |\sigma_h|\right] + 2K_1 \left[\sum_{h=1}^m |\sigma_h|\right]^2$$

$$\leq 2V(t_0-)(C_{12} + 2K_1 C_{11})\varepsilon \leq \frac{\varepsilon}{5C_{10}}.$$

Thus, relying on (6.68)-(6.70), we obtain

$$\left|\sigma_1^s - \nu_1^\varepsilon(v^s)\right| \geq \left|\nu_1^\varepsilon(v^L) - \nu_1^\varepsilon(v^s) + (v_1^L - v_1^s)\right| - \left|\sum_{h=p+1}^m \sigma_h\right| - \left|\sigma_1^+ - \sum_{h=p+1}^n \sigma_h\right|$$

$$\geq \frac{\varepsilon}{C_{10}} - C_{11}\varepsilon - \frac{\varepsilon}{5C_{10}} > \frac{\varepsilon}{2C_{10}}$$

establishing the first inequality in (6.67). Concerning the second inequality, assume that v^L lies on the left side w.r.t. to the curve $\mathcal{V}_1^0$ in (2.11), i.e. that $\sigma_1^s > 0$ (the other case being entirely similar). Then, by the property of the approximate Riemann solver defined in Section 3, one has $\sigma_1^s \geq \nu_1^\varepsilon(v^s)$. On the other hand, using (6.2), (6.68), one finds

$$\sigma_1^s \leq \nu_1^\varepsilon(v^L)$$
$$\leq \nu_1^\varepsilon(v^s) + \left|\nu_1^\varepsilon(v^L) - \nu_1^\varepsilon(v^s)\right|$$
$$\leq \nu_1^\varepsilon(v^s) + C_{10}\varepsilon$$

which thus yields the second inequality in (6.67), completing the proof of the Lemma. □

LEMMA 6.6. *Given a FFB for the k-th family $\mathcal{J}_0 \in \mathcal{B}^k(t_0)$, $k \in \{1,2\} \cap NGNL$, assume that the interaction time of good type $\widehat{t}_0$ for $(\mathcal{J}_0, t_0)$ satisfies either one of the conditions* I)-II) *of Definition 6.1, and let E_k, F_k, G_k be the functionals defined in (6.11)-(6.13). Moreover, assume that $V(t)$, $t < \widehat{t}_0$, satisfies the bound in (6.17). Then, there holds*
(6.71)
$$E_k(\mathcal{J}_0; t_0) \leq \left[F_k(\mathcal{J}_0; t_0, \widehat{t}_0) - F_k(\mathcal{J}_0; t_0, t_0)\right] + \left[G_k(\mathcal{J}_0; t_0, \widehat{t}_0) - G_k(\mathcal{J}_0; t_0, t_0)\right].$$

PROOF. To fix the ideas, we shall consider a FFB for the first family $\mathcal{J}_0 \in \mathcal{B}^1(t_0)$, as we have done in the previous lemma. Moreover, we shall assume that the first family is NGNL1, the case of a NGNL2 family being entirely similar. Throughout the proof we adopt the same notations of Definition 6.1 and thus call σ_h, $1 \leq h \leq H$, the wave-fronts of a polygonal line $x(t)$, $t_0 \leq t \leq \widehat{t}_0$, starting with the shock σ_1 generated by the bad interaction at t_0, and satisfying either one of the conditions I)-II) of Definition 6.1. Let $\{\sigma'_{h,1}, \ldots, \sigma'_{h,n_h}\}$ be the fronts involved in the interaction with σ_h at t_h. Denote $\{\sigma'_{h,1}, \ldots, \sigma'_{h,p_h}\}$, $\{\sigma'_{h,p_h+1}, \ldots, \sigma'_{h,m_h}\}$ and $\{\sigma'_{h,m_h+1}, \ldots, \sigma'_{h,n_h}\}$, respectively, the incoming fronts of the second family, the incoming fronts of the first family that are discordant with σ_h, and the incoming fronts of the first family that are concordant with σ_h, assuming that $\sigma'_{h,m_h+1} = \sigma_h$. Observe that, by Definition 6.1, at every interaction time t_h, $h < H$, the estimates (6.6)-(6.8) are not verified, i.e. there holds

$$(6.72) \quad \sum_{h=1}^{H-1} \sum_{\ell=p_h+1}^{m_h} |\sigma'_{h,\ell}| \leq \frac{\varepsilon}{8C_{10}}, \qquad \sum_{h=1}^{H-1} \sum_{\ell=1}^{p_h} |\sigma'_{h,\ell}| \leq \frac{\varepsilon}{8C_{10}}.$$

Moreover, since the total strength of 2-fronts that hit the polygonal line $x(t)$, $t \leq t_{H-1}$ is certainly $< 2\sqrt{\varepsilon}$, by the properties of the approximate Riemann solver described in Section 3 (see Remark 3.3) it follows that the difference between the strength of the incoming and outgoing 1-fronts at the interaction points along $x(\cdot)$ is independent on the potential interaction between the incoming 2-fronts. Therefore, applying (6.18), and using the bound (6.17) on $V(t_h)$, we find, for any $1 \leq h < H$,

$$|\sigma_{h+1}| \geq \left[|\sigma_h| + \sum_{\ell=m_h+2}^{n_h} |\sigma'_{h,\ell}|\right] \cdot \left[1 - K_1 \sum_{\ell=1}^{n_h} |\sigma'_{h,\ell}|\right] +$$

$$- \left[\sum_{\ell=p_h+1}^{m_h} |\sigma'_{h,\ell}|\right] \cdot \left[1 + K_1 \sum_{\ell=1}^{n_h} |\sigma'_{h,\ell}|\right]$$

$$\geq \left[|\sigma_h| + \sum_{\ell=m_h+2}^{n_h} |\sigma'_{h,\ell}|\right] \cdot \left(1 - K_1 V(t_h-)\right) - \left(1 + K_1 V(t_h-)\right) \cdot \left[\sum_{\ell=p_h+1}^{m_h} |\sigma'_{h,\ell}|\right]$$

$$> |\sigma_h| - \frac{6}{5} \sum_{\ell=p_h+1}^{m_h} |\sigma'_{h,\ell}|,$$

which, in turn, together with (6.72), implies

$$(6.73) \quad |\sigma_{h+1}| > |\sigma_1| - \frac{6}{5} \sum_{h=1}^{H-1} \sum_{\ell=p_h+1}^{m_h} |\sigma'_{h,\ell}| > |\sigma_1| - \frac{\varepsilon}{5C_{10}} \qquad \forall\, 0 \leq h < H.$$

Then, assuming that $\sigma_h > 0$ (the other case being entirely similar), and relying on (6.67), (6.73), we derive

$$(6.74) \qquad \sigma_h > \sigma_1 - \frac{\varepsilon}{5C_{10}}$$
$$> \nu_1^\varepsilon\left(v^{L_{\sigma_1}}\right) + \frac{\varepsilon}{5C_{10}}$$
$$> \frac{\sigma_1}{(5C_{10})^2} \qquad \forall\, 1 < h \leq H.$$

Observe now that, since $\widehat{t}_0$ satisfies either one of the conditions I)-II) of Definition 6.1, it follows that one among the estimates (6.6)-(6.8) is verified at $t = t_H$. Thus, recalling that by (6.15) one has $K_1 > (40)^2 C_{10}^3$, and thanks to (6.74), we

obtain
(6.75)
$$\left[F_1(\mathcal{J}_0; t_0, \widehat{t}_0) - F_1(\mathcal{J}_0; t_0, t_0)\right] + \left[G_1(\mathcal{J}_0; t_0, \widehat{t}_0) - G_1(\mathcal{J}_0; t_0, t_0)\right] \geq$$

$$\geq \frac{K_1}{8} \cdot \sum_{h=1}^{H} \sum_{\ell=1}^{m_h} |\sigma_h \, \sigma'_{h,\ell}| + \frac{1}{4} \cdot \sum_{h=1}^{H} \sum_{\ell=m_h+1}^{n_h} |\sigma_h \, \sigma'_{h,\ell}|$$

$$\geq \frac{1}{4} \cdot \frac{\sigma_1}{(5C_{10})^2} \cdot (10\,C_{10})^2 \cdot \varepsilon \geq \sigma_1 \cdot \varepsilon \,.$$

On the other hand, if we let $\sigma' \in \mathcal{N}_R^1(\mathcal{J}_0, t_0+)$ be the rarefaction front generated by the bad interaction at t_0, one clearly has

$$E_1(\mathcal{J}_0; t_0) = |\sigma_1 \cdot \sigma'| \leq |\sigma_1| \cdot \varepsilon \,,$$

which, together with (6.75), yields (6.71). □

LEMMA 6.7. *Given a FFB for the k-th family $\mathcal{J}_0 \in \mathcal{B}^k(t_0)$, $k \in \{1,2\} \cap NGNL$, assume that the interaction time of good type $\widehat{t}_0$ for $(\mathcal{J}_0, t_0)$ satisfies either one of the conditions* III)-IV) *of Definition 6.1. Moreover, assume that $V(t)$, $t < \widehat{t}_0$, satisfies the bound in (6.17). Then, letting E_k, H_k be the functionals defined in (6.11), (6.14), there holds*

(6.76) $$E_k(\mathcal{J}_0; t_0) \leq \left[H_k(\mathcal{J}_0; t_0, \widehat{t}_0) - H_k(\mathcal{J}_0; t_0, t_0)\right].$$

PROOF. As in the previous lemmas, we shall consider only a FFB for the first family $\mathcal{J}_0 \in \mathcal{B}^1(t_0)$, and we shall assume that the first family is NGNL1, the other cases being entirely similar. Let σ_1, σ'_1 be, respectively, the rarefaction front and the shock generated by the bad interaction for the first family at t_0, so that one has

(6.77) $$E_1(\mathcal{J}_0; t_0) = |\sigma_1 \cdot \sigma'_1| \,.$$

We shall distinguish two cases.

CASE 1: Let $\widehat{t}_0$ be an interaction time of good type satisfying condition III) of Definition 6.1. Then, according with Definition 6.1, there will be a polygonal line $x(t)$, $t_0 \leq t \leq \widehat{t}_0$ satisfying conditions I.I), III.II), that starts with the front σ_1 and terminates with a front $\sigma_H = \sigma^r \in \mathcal{N}_R^1(\mathcal{J}_0, \widehat{t}_0) \cap \mathcal{C}^1(\widehat{t}_0)$. We first observe that the total amount of 1-fronts canceled in the time interval $]t_0, \widehat{t}_0]$ by interactions of the fronts in $x(\cdot)$ with other 1-fronts (coming from the left) is $\geq \sigma_1/2$, i.e. there holds

(6.78) $$|\sigma^r| + \sum_{\substack{\sigma''_h \in \mathcal{I}^1(\sigma_h, t_h) \cap \mathcal{C}^1(t_h) \\ 0 < h < H}} |\sigma''_h| \geq \frac{\sigma_1}{2} \,,$$

where σ_h, $0 < h \leq H$, are the fronts of $x(\cdot)$. Indeed, (6.78) is clearly verified if

(6.79) $$\sum_{\substack{\sigma''_h \in \mathcal{I}^1(\sigma_h, t_h) \cap \mathcal{C}^1(t_h) \\ 0 < h < H}} |\sigma''_h| \geq \frac{\sigma_1}{2} \,,$$

while, in the case (6.79) does not hold, recalling that by condition III) of Definition 6.1 the total strength of 2-fronts that hit $x(\cdot)$ is $\leq \sqrt{\varepsilon}$, we deduce, by the properties of the approximate Riemann solver described in Section 3 (see Remark 3.3), that all the fronts in $x(\cdot)$ have strength $\geq \sigma_1/2$. Thus, in particular, one has $|\sigma^r| \geq \sigma_1/2$, which implies (6.78).

Next, we observe that, if we let σ'_h denote the shock in the set $\mathcal{N}^1_S(\mathcal{J}_0, t_h)$ (consisting of a single front: see Remark 6.3), since σ'_h belongs to a polygonal line satisfying conditions I.I)-I.II) of Definition 6.1, we deduce a bound as in (6.74), i.e. one has

$$(6.80) \qquad |\sigma'_h| > \frac{|\sigma'_1|}{(5C_{10})^2} \qquad \forall\, 1 < h \leq H.$$

Thus, recalling that by (6.15) one has $K_1 > 200\, C_{10}^2$, and relying on (6.78), (6.80) we derive

$$\left[H_1(\mathcal{J}_0; t_0, \widehat{t_0}) - H_1(\mathcal{J}_0; t_0, t_0)\right] \geq \frac{K_1}{4} \cdot \frac{|\sigma'_1|}{(5C_{10})^2} \cdot \left[|\sigma^r| + \sum_{\substack{\sigma''_h \in \mathcal{I}^1(\sigma_h, t_h) \cap \mathcal{C}^1(t_h) \\ 0 < h < H}} |\sigma''_h|\right]$$

$$\geq \frac{K_1}{200} \cdot \frac{|\sigma'_1 \cdot \sigma_1|}{C_{10}^2} > |\sigma'_1 \cdot \sigma_1|$$

which, together with (6.77), yields

$$(6.81) \qquad E_1(\mathcal{J}_0; t_0) \leq \left[H_1(\mathcal{J}_0; t_0, \widehat{t_0}) - H_1(\mathcal{J}_0; t_0, t_0)\right]$$

proving (6.76).

CASE 2: Let $\widehat{t_0}$ be an interaction time of good type satisfying condition IV) of Definition 6.1. Then, according with Definition 6.1, there will be a polygonal line $x(t)$, $t_0 \leq t \leq \overline{t} \leq \widehat{t_0}$ satisfying conditions I.I), III.II), that starts with the rarefaction front σ_1 and has the following property. The total strength of 2-fronts originated by interactions of the fronts in $x(t)$, $t \in]t_0, \overline{t}]$ that are canceled in the time interval $]t_0, \widehat{t_0}]$, and were approaching the fronts in $\mathcal{N}^1_S(\mathcal{J}_0, t)$, $t \in]t_0, \widehat{t_0}]$, is $> \sqrt{\varepsilon}/2$.

Thus, if we let $t_0 < t_1 < \cdots < t_H = \overline{t}$ be the interaction times for the fronts of $x(\cdot)$, denote $\mathcal{J}_h$, $0 < h \leq H$ the set of fronts interacting at t_h with the front in $x(\cdot)$, and let t'_k, $0 < k \leq K$ be the times at which take place some cancellation of a front in $\mathcal{N}^2(\mathcal{J}_h)$, $t_h < t'_k$ (that was approaching the fronts in $\mathcal{N}^1_S(\mathcal{J}_0, t)$, $t \in]t_0, \widehat{t_0}]$), we have

$$(6.82) \qquad \sum_{k=1}^{K} \sum_{\substack{\sigma'' \in \mathcal{C}^2(t'_k) \cap \mathcal{N}^2(\mathcal{J}_h, t'_k) \\ t_0 < t_h < t'_k \leq \widehat{t_0}}} |\sigma''| > \frac{\sqrt{\varepsilon}}{2} > |\sigma_1|.$$

Observe that, if we let σ'_k denote the shock in the set $\mathcal{N}^1_S(\mathcal{J}_0, t'_k)$ (consisting of a single front: see Remark 6.3), as in Case 1 we deduce the bound

$$(6.83) \qquad |\sigma'_k| > \frac{|\sigma'_1|}{(5C_{10})^2} \qquad \forall\, 1 < k \leq K.$$

Then, relying on (6.82)-(6.83), observing that $\mathcal{N}^2(\mathcal{J}_h, t'_k) \subset \mathcal{N}^2(\mathcal{N}_R^1(\mathcal{J}_0, t_h), t'_k)$, and recalling that by (6.15) one has $K_1 > 100\, C_{10}^2$, we deduce

(6.84)
$$\left[H_1(\mathcal{J}_0; t_0, \widehat{t}_0) - H_1(\mathcal{J}_0; t_0, t_0)\right] \geq \frac{K_1}{4} \cdot \frac{|\sigma'_1|}{(5C_{10})^2} \cdot \left[\sum_{k=1}^{K} \sum_{\substack{\sigma'' \in \mathcal{C}^2(t'_k) \cap \mathcal{N}^2(\mathcal{J}_h, t'_k) \\ t_0 < t_h < t'_k \leq \widehat{t}_0}} |\sigma''|\right]$$
$$> |\sigma'_1 \cdot \sigma_1|,$$

which, together with (6.77), yields (6.80) completing the proof of the Lemma. □

LEMMA 6.8. *Given a FFB for the k-th family $\mathcal{J}_0 \in \mathcal{B}^k(t_0)$, $k \in \{1,2\} \cap$ NGNL1, assume that the interaction time of good type $\widehat{t}_0$ for $(\mathcal{J}_0, t_0)$ satisfies condition* V) *of Definition 6.1 (and none of the other conditions* I)-IV)). *Let σ^s, σ^r be, respectively, the shock in $\mathcal{N}_S^k(\mathcal{J}_0, \widehat{t}_0)$ and the wave-front in $\mathcal{N}_R^k(\mathcal{J}_0, \widehat{t}_0)$ that interact together at $\widehat{t}_0$ (σ^s located on the right of σ^r) according with condition* V), *and call v^L the left state of σ^r. Assume that $V(t)$, $t < \widehat{t}_0$, satisfies the bound in (6.17). Then, letting F_k, G_k be the functionals defined in (6.12)-(6.13), there holds*

(6.85)
$$\frac{|\sigma^r| |\sigma^s + \sigma^r - \nu_k^\varepsilon(v^L)|}{4\,(5C_{10})^2} \leq \left[F_k(\mathcal{J}_0; t_0, \widehat{t}_0-) - F_k(\mathcal{J}_0; t_0, t_0\,)\right] +$$
$$+ \left[G_k(\mathcal{J}_0; t_0, \widehat{t}_0\,) - G_k(\mathcal{J}_0; t_0, t_0\,)\right]$$

where C_{10} denotes the constant at (6.2). The same type of estimate holds if we are considering a FFB for a k-th NGNL2 family $\mathcal{J}_0 \in \mathcal{B}^k(t_0)$, and the interaction time of good type for $(\mathcal{J}_0, t_0)$ satisfies condition V) *of Definition 6.1.*

PROOF. As in the previous lemmas, we shall consider only a FFB for the first family $\mathcal{J}_0 \in \mathcal{B}^1(t_0)$, and we shall assume that the first family is NGNL1, the other cases being entirely similar. Let $\sigma_1, \sigma'_1 > 0$ be, respectively, the shock and the rarefaction front generated by the bad interaction for the first family at t_0. According with condition V) of Definition 6.1, there is a polygonal line $x(t)$, $t_0 \leq t \leq \widehat{t}_0$ having the properties I.I)-I.II), which starts with σ_1 and terminates with the shock $\sigma^s \in \mathcal{N}_S^1(\mathcal{J}_0, \widehat{t}_0)$ that interacts at $\widehat{t}_0$ with the front $\sigma^r \in \mathcal{N}_R^1(\mathcal{J}_0, \widehat{t}_0)$. Let $\sigma_h > 0$, $0 < h \leq H$, be the fronts of $x(\cdot)$, where $\sigma_H = \sigma^s$. Recall that, by Remark 6.3, the left states $v^{L_{\sigma_h}}$, $h \geq 1$ of all the fronts σ_h, $h \geq 1$ stay on the same (left) side w.r.t the curve $\mathcal{V}_1^0$ in (2.11), and lie on the same vertical line of the grid of step ε. Thus, in particular, one has

(6.86)
$$\lfloor v^{L_{\sigma^s}} \rfloor_1 \varepsilon = \lfloor v^{L_{\sigma_1}} \rfloor_1 \varepsilon.$$

Since the shock σ^s interacts at $\widehat{t}_0$ (on its left) with the rarefaction front σ^r, it follows that σ^s travels with smaller speed than $\lambda_1^{\varepsilon-}(v^{L_{\sigma^s}})$, and hence, by construction, there holds

(6.87)
$$\sigma^s > \nu_1^{\varepsilon-}(v^{L_{\sigma^s}}) \doteq \lim_{s \to 0-} \nu_1^\varepsilon(R_1(v^{L_{\sigma^s}}, s)).$$

Conversely, observing that the front σ_1 has larger speed than $\lambda_1^{\varepsilon-}(v^{L_{\sigma_1}})$ (being σ_1 the shock component of the composed wave generated by the bad interaction at

t_0), we deduce that

(6.88) $$\sigma_1 < \nu_1^{\varepsilon-}(v^{L_{\sigma_1}}) \doteq \lim_{s\to 0-} \nu_1^\varepsilon\bigl(R_1(v^{L_{\sigma_1}}, s)\bigr).$$

Therefore, using (6.2), and relying on (6.86)-(6.88), we derive

(6.89) $$\bigl|\sigma^s - \nu_1^{\varepsilon-}(v^{L_{\sigma^s}})\bigr| \le \bigl|\nu_1^{\varepsilon-}(v^{L_{\sigma^s}}) - \nu_1^{\varepsilon-}(v^{L_{\sigma_1}})\bigr| + [\sigma^s - \sigma_1]^+$$
$$\le C_{10} \sum_{\substack{\sigma_h'' \in \mathcal{I}^2(\sigma_h, t_h) \\ 0 < h < H}} |\sigma_h''| + \sum_{\substack{\sigma_h''' \in \mathcal{I}^1(\sigma_h, t_h),\ \sigma_h''' > 0 \\ 0 < h < H}} |\sigma_h'''|.$$

The other fronts of $x(\cdot)$, belonging to a polygonal line with the properties I.I)-I.II) of Definition 6.1, clearly satisfy a bound as in (6.74), and hence, since $\sigma_1 \ge \varepsilon \ge \sigma^r$ (cfr. Remark 6.1), we derive

(6.90) $$\sigma_h > \frac{\sigma^r}{(5C_{10})^2} \qquad \forall\ 1 \le h \le H.$$

Then, relying on (6.89)-(6.90), and observing that by (6.15) one has $K_1 > 10^2\, C_{10}^3$, we find

(6.91)
$$|\sigma^r||\sigma^s + \sigma^r - \nu_1^\varepsilon(v^{L_{\sigma^r}})| =$$
$$= |\sigma^r||\sigma^s - \nu_1^{\varepsilon-}(v^{L_{\sigma^s}})|$$
$$\le 25\, C_{10}^3 \sum_{\substack{\sigma_h'' \in \mathcal{I}^2(\sigma_h, t_h) \\ 0 < h < H}} |\sigma_h\, \sigma_h''| + (5\, C_{10})^2 \sum_{\substack{\sigma_h''' \in \mathcal{I}^1(\sigma_h, t_h),\ \sigma_h''' > 0 \\ 0 < h < H}} |\sigma_h\, \sigma_h'''|$$
$$\le 4\,(5C_{10})^2 \Bigl(\bigl[F_1(\mathcal{J}_0; t_0, \widehat{t}_0) - F_1(\mathcal{J}_0; t_0, t_0)\bigr] + \bigl[G_1(\mathcal{J}_0; t_0, \widehat{t}_0) - G_1(\mathcal{J}_0; t_0, t_0)\bigr]\Bigr)$$

which yields (6.85). □

Before stating the next results, it's useful to introduce the following notation in connection with a k-th NGNL1 family and with a given FFB $\mathcal{J}_0 \in \mathcal{B}^k(t_0)$. Let $\widehat{t}_0$ be the interaction time of good type for $(\mathcal{J}_0, t_0)$, and assume that $\widehat{t}_0$ satisfies condition V) of Definition 6.1. Let $\sigma' > 0$, $\sigma'' > 0$ be, respectively, the rarefaction front in $\mathcal{N}_R^k(\mathcal{J}_0, \widehat{t}_0)$ and the shock in $\mathcal{N}_S^k(\mathcal{J}_0, \widehat{t}_0)$ that interact together at $\widehat{t}_0$ according with condition V). Then, letting $\widehat{\mathcal{J}}_0$ denote the set of fronts interacting at $\widehat{t}_0$, we define

(6.92) $$\widehat{F}_k(\widehat{\mathcal{J}}_0; \widehat{t}_0) \doteq \frac{|\sigma'\, \sigma''|}{4}.$$

An entirely similar definition is given in connection with a k-th NGNL2 family and with a given FFB $\mathcal{J}_0 \in \mathcal{B}^k(t_0)$. On the other hand, for any k-th characteristic family, and for any given set of fronts $\widehat{\mathcal{J}}_0$ involved in an interaction at $\widehat{t}_0$ (not necessarily related to an FFB $\mathcal{J}_0$), we will set $\widehat{F}_k(\widehat{\mathcal{J}}_0; \widehat{t}_0) \doteq 0$ if either one of the following two cases occurs:

- $\widehat{t}_0$ is an interaction time of good type (for a FFB $\mathcal{J}_0 \in \mathcal{B}^k(t_0)$) that does not satisfy condition V) of Definition 6.1;
- the interaction of the fronts in $\widehat{\mathcal{J}}_0$ is not of good type for the k-th family (according with Definition 6.1).

This clearly means, in particular, that if the k-th family is GNL or LD, then one has $\widehat{F}_k(\widehat{\mathcal{J}}_0; \widehat{t}_0) = 0$ for any set of fronts $\widehat{\mathcal{J}}_0$, and any time $\widehat{t}_0$.

LEMMA 6.9. *Given a FFB for the k-th family $\mathcal{J}_0 \in \mathcal{B}^k(t_0)$, $k \in \{1,2\} \cap NGNL$, assume that the interaction time of good type $\widehat{t}_0$ for $(\mathcal{J}_0, t_0)$ either satisfies one of the conditions* I)-II) *of Definition 6.1 and none of the other conditions* III)-V), *or satisfies condition* V) *of Definition 6.1. Moreover, assume that $V(t)$, $t < \widehat{t}_0$, satisfies the bound in (6.17). Then, letting $\widehat{\mathcal{J}}_0$ be the set of fronts interacting at $\widehat{t}_0$, and letting E_i, F_i, G_i, Ξ_i be the functionals defined in Section 6.1, there holds*

(6.93)
$$\Delta Q(\widehat{t}_0) \leq -2 \sum_{i=1}^{2} \Big[F_i(\widehat{\mathcal{J}}_0; \widehat{t}_0, \widehat{t}_0) + G_i(\widehat{\mathcal{J}}_0; \widehat{t}_0, \widehat{t}_0) \Big] + \widehat{F}_k(\widehat{\mathcal{J}}_0; \widehat{t}_0) +$$
$$+ \sum_{i \in NGNL} \Xi_i(\widehat{\mathcal{J}}_0; \widehat{t}_0) E_i(\widehat{\mathcal{J}}_0; \widehat{t}_0) - E_k(\mathcal{J}_0; \widehat{t}_0-) +$$
$$+ \Big[F_k(\mathcal{J}_0; t_0, \widehat{t}_0) - F_k(\mathcal{J}_0; t_0, t_0) \Big] + \Big[G_k(\mathcal{J}_0; t_0, \widehat{t}_0) - G_k(\mathcal{J}_0; t_0, t_0) \Big],$$

(6.94)
$$\Delta(V+Q)(\widehat{t}_0) \leq -\sum_{i=1}^{2} \Big[F_i(\widehat{\mathcal{J}}_0; \widehat{t}_0, \widehat{t}_0) + G_i(\widehat{\mathcal{J}}_0; \widehat{t}_0, \widehat{t}_0) \Big] +$$
$$+ \sum_{i \in NGNL} \Xi_i(\widehat{\mathcal{J}}_0; \widehat{t}_0) E_i(\widehat{\mathcal{J}}_0; \widehat{t}_0) - E_k(\mathcal{J}_0; \widehat{t}_0-) +$$
$$+ \Big[F_k(\mathcal{J}_0; t_0, \widehat{t}_0) - F_k(\mathcal{J}_0; t_0, t_0) \Big] + \Big[G_k(\mathcal{J}_0; t_0, \widehat{t}_0) - G_k(\mathcal{J}_0; t_0, t_0) \Big].$$

PROOF. As in the previous lemmas, we shall consider only a FFB for the first family $\mathcal{J}_0 \in \mathcal{B}^1(t_0)$, and we shall assume that the first family is NGNL1, the other cases being entirely similar. Observe first that, if the ITG $\widehat{t}_0$ for $(\mathcal{J}_0, t_0)$ satisfies either one of conditions I)-II) of Definition 6.1 and does not satisfy condition V), then, by (6.11), one has $E_1(\mathcal{J}_0; \widehat{t}_0-) = E_1(\mathcal{J}_0; t_0)$, and hence the estimates (6.93)-(6.94) are obtained applying Lemma 6.1 and Lemma 6.6. Next, consider the case that $\widehat{t}_0$ satisfies condition V) of Definition 6.1. Let σ^s, σ^r be, respectively, the shock in $\mathcal{N}_S^1(\mathcal{J}_0, \widehat{t}_0)$ and the front in $\mathcal{N}_R^1(\mathcal{J}_0, \widehat{t}_0)$ that interact together at $\widehat{t}_0$, and call v^L the left state of σ^s. Observe that, by (6.11), (6.92), one has $E_1(\mathcal{J}_0; \widehat{t}_0-) = 4\widehat{F}_1(\widehat{\mathcal{J}}_0; \widehat{t}_0) = |\sigma^r \sigma^s|$, and that (6.12) implies $\widehat{F}_1(\widehat{\mathcal{J}}_0; \widehat{t}_0) \leq F_1(\mathcal{J}_0; t_0, \widehat{t}_0) - F_1(\mathcal{J}_0; t_0, \widehat{t}_0-)$. Then, using (6.36), and applying Lemma 6.8, we find

$$\Delta Q(\widehat{t}_0) \leq -2 \sum_{i=1}^{2} \Big[F_i(\widehat{\mathcal{J}}_0; \widehat{t}_0, \widehat{t}_0) + G_i(\widehat{\mathcal{J}}_0; \widehat{t}_0, \widehat{t}_0) \Big] +$$

$$\sum_{i \in NGNL} \Xi_i(\widehat{\mathcal{J}}_0; \widehat{t}_0)\, E_i(\widehat{\mathcal{J}}_0; \widehat{t}_0) - \frac{|\sigma^s \sigma^r|}{2} + \frac{|\sigma^r| \cdot |\sigma^s + \sigma^r - \nu_1^\varepsilon(v^L)|}{4(5C_{10})^2}$$

$$< -2 \sum_{i=1}^{2} \Big[F_i(\widehat{\mathcal{J}}_0; \widehat{t}_0, \widehat{t}_0) + G_i(\widehat{\mathcal{J}}_0; \widehat{t}_0, \widehat{t}_0) \Big] + \widehat{F}_1(\widehat{\mathcal{J}}_0; \widehat{t}_0) +$$

$$+ \sum_{i \in NGNL} \Xi_i(\widehat{\mathcal{J}}_0; \widehat{t}_0)\, E_i(\widehat{\mathcal{J}}_0; \widehat{t}_0) - E_1(\mathcal{J}_0; \widehat{t}_0-) +$$

$$+ \Big[F_1(\mathcal{J}_0; t_0, \widehat{t}_0) - F_1(\mathcal{J}_0; t_0, t_0) \Big] + \Big[G_1(\mathcal{J}_0; t_0, \widehat{t}_0) - G_1(\mathcal{J}_0; t_0, t_0) \Big]$$

proving (6.93). Clearly, using (6.37), and applying Lemma 6.8, we obtain the corresponding estimate (6.94) for $\Delta(V+Q)(\widehat{t}_0)$. □

LEMMA 6.10. *Given a FFB for the k-th family $\mathcal{J}_0 \in \mathcal{B}^k(t_0)$, $k \in \{1,2\} \cap NGNL$, assume that the interaction time of good type $\widehat{t}_0$ for $(\mathcal{J}_0, t_0)$ satisfies either one of the conditions* III)-IV) *of Definition 6.1 and does not satisfy condition* V). *Moreover, assume that $V(t)$, $t < \widehat{t}_0$, satisfies the bound in (6.17). Then, letting $\widehat{\mathcal{J}}_0$ be the set of fronts interacting at $\widehat{t}_0$, and letting $E_i, F_i, G_i, H_i, \Xi_i$ be the functionals defined in Section 6.1, there holds*

(6.95)
$$\Delta Q(\widehat{t}_0) \leq -2 \sum_{i=1}^{2} \Big[F_i(\widehat{\mathcal{J}}_0; \widehat{t}_0, \widehat{t}_0) + G_i(\widehat{\mathcal{J}}_0; \widehat{t}_0, \widehat{t}_0) + H_i(\widehat{\mathcal{J}}_0; \widehat{t}_0, \widehat{t}_0) \Big]$$
$$+ \sum_{i \in NGNL} \Xi_i(\widehat{\mathcal{J}}_0; \widehat{t}_0)\, E_i(\widehat{\mathcal{J}}_0; \widehat{t}_0) - E_k(\mathcal{J}_0; \widehat{t}_0-) +$$
$$+ \Big[H_k(\mathcal{J}_0; t_0, \widehat{t}_0) - H_k(\mathcal{J}_0; t_0, t_0) \Big],$$

(6.96)
$$\Delta(V+Q)(\widehat{t}_0) \leq -\sum_{i=1}^{2} \Big[F_i(\widehat{\mathcal{J}}_0; \widehat{t}_0, \widehat{t}_0) + G_i(\widehat{\mathcal{J}}_0; \widehat{t}_0, \widehat{t}_0) + H_i(\widehat{\mathcal{J}}_0; \widehat{t}_0, \widehat{t}_0) \Big] +$$
$$+ \sum_{i \in NGNL} \Xi_i(\widehat{\mathcal{J}}_0; \widehat{t}_0)\, E_i(\widehat{\mathcal{J}}_0; \widehat{t}_0) - E_k(\mathcal{J}_0; \widehat{t}_0-) +$$
$$+ \Big[H_k(\mathcal{J}_0; t_0, \widehat{t}_0) - H_k(\mathcal{J}_0; t_0, t_0) \Big].$$

PROOF. Since the ITG $\widehat{t}_0$ for $(\mathcal{J}_0, t_0)$ does not satisfy condition V) of Definition 6.1, then, by (6.11), one has $E_k(\mathcal{J}_0; \widehat{t}_0-) = E_k(\mathcal{J}_0; t_0)$. Hence, the estimates (6.95)-(6.96) are obtained applying Lemma 6.1 and Lemma 6.7. □

LEMMA 6.11. *For a 2×2 system satisfying the assumption* (**A**), *there exists some constant C_{18} independent on ε so that, letting E_k, Ξ_k be the functionals* *defined in Section 6.1, there holds*

(6.97) $$\sum_{k \in NGNL} \sum_{\substack{\mathcal{J} \in \mathcal{B}^k(s) \\ 0 < s \leq t}} \Xi_k(\mathcal{J}; t) \, E_k(\mathcal{J}; t+) \leq C_{18} \cdot V(t+) \cdot \varepsilon \qquad \forall \, t > 0.$$

PROOF. Given any $\mathcal{J} \in \mathcal{B}^k(t_0)$, $k \in \{1, 2\} \cap NGNL$, from the definition (6.9) of the map $t \mapsto \Xi_k(\mathcal{J}; t)$, $t \geq t_0$, and by Remark 6.3, it follows that, as long as $\Xi_k(\mathcal{J}; t) = 1$, the set $\mathcal{N}_S^k(\mathcal{J}; t)$ consists of a single wave-front, say σ_t, that cannot be involved in another interaction of bad type for the k-th family. Thus, for $k \in \{1, 2\} \cap NGNL$, one has

(6.98)
$$\mathcal{J}', \mathcal{J}'' \in \bigcup_{s \leq t} \mathcal{B}^k(s), \quad \mathcal{J}' \neq \mathcal{J}''$$

$$\Xi_k(\mathcal{J}'; t) = \Xi_k(\mathcal{J}''; t) = 1$$

$$\mathcal{N}_S^k(\mathcal{J}', t) = \{\sigma_t'\}, \qquad \mathcal{N}_S^k(\mathcal{J}'', t) = \{\sigma_t''\}$$

$$\Downarrow$$

$$\sigma_t' \neq \sigma_t''.$$

On the other hand, by the arguments in the proof of Lemma 6.6, since $\sigma_t \in \mathcal{N}_S^k(\mathcal{J}; t)$ belongs to a polygonal line with the properties I)-II) of Definition 6.1, we deduce a bound as in (6.74). Hence, letting σ_1 be the shock generated at t_0 by the bad interaction among the fronts in $\mathcal{J} \in \mathcal{B}^k(t_0)$, and recalling (6.11), we get

(6.99) $$E_k(\mathcal{J}; t) \leq \varepsilon \cdot \max\{|\sigma_1|, |\sigma_t|\} \leq \varepsilon \cdot (5C_{10})^2 \, |\sigma_t|.$$

Then, the estimate (6.97) clearly follows from (6.98)-(6.99) taking $C_{18} = (5C_{10})^2$. □

6.4. Proof of Proposition 3.1.

We are now ready to prove the main lemma which allows to establish Proposition 3.1. In the following, we will denote by $\mathrm{Int}(t)$ the collection of all families of fronts that interact together at time t.

LEMMA 6.12. *For a 2×2 system satisfying the assumption* (**A**), *let C_{18} be the constant of Lemma 6.11, fix $0 < \varepsilon < 1/(2C_{18})$, and consider an ε-approximate solution v constructed by the algorithm in Sections 3-4 on some initial time interval $[0, T[$. Given $0 \leq \tau_1 < \tau_2 < T$, assume that*

(6.100) $$V(\tau_1+) + Q(\tau_1+) < \min\left\{\frac{1}{80\, C_1}, \frac{C_{14}}{2}\right\},$$

where C_1 denotes the constant of Remark 5.1, and C_{14} is the constant at (6.17). Then, letting $E_i, F_i, G_i, H_i, \Xi_i$ be the functionals defined in Section 6.1, and $\widehat{F}_i$ the functional introduced before Lemma 6.9, the following estimates hold:

(6.101)
$$Q(\tau_2+) \leq Q(\tau_1+) - \sum_{i=1}^{2} \sum_{\substack{\mathcal{J}_\ell \in \text{Int}(t_\ell) \\ \tau_1 < t_\ell \leq \tau_2}} \left[F_i(\mathcal{J}_\ell; t_\ell, t_\ell) - \widehat{F}_i(\mathcal{J}_\ell, t_\ell) + G_i(\mathcal{J}_\ell; t_\ell, t_\ell) \right] +$$
$$+ \sum_{i \in NGNL} \sum_{\substack{\mathcal{J}_\ell \in \mathcal{B}^i(t_\ell) \\ \tau_1 < t_\ell \leq \tau_2}} \Xi_i(\mathcal{J}'_\ell; \tau_2) \left[E_i(\mathcal{J}_\ell; \tau_2) - F_i(\mathcal{J}_\ell; t_\ell, \tau_2) - G_i(\mathcal{J}_\ell; t_\ell, \tau_2) - H_i(\mathcal{J}_\ell; t_\ell, \tau_2) \right],$$

(6.102)
$$\left[V(\tau_2+) + Q(\tau_2+) \right] \leq \left[V(\tau_1+) + Q(\tau_1+) \right] + \sum_{i \in NGNL} \sum_{\substack{\mathcal{J}_\ell \in \mathcal{B}^i(t_\ell) \\ \tau_1 < t_\ell \leq \tau_2}} \Xi_i(\mathcal{J}_\ell; \tau_2) E_i(\mathcal{J}_\ell; \tau_2),$$

(6.103) $\qquad Q(\tau_2+) \leq Q(\tau_1+) + C_{18} \cdot \varepsilon \cdot V(\tau_2+),$

(6.104) $\qquad \left[V(\tau_2+) + Q(\tau_2+) \right] \leq (1 + 2C_{18} \cdot \varepsilon) \left[V(\tau_1+) + Q(\tau_1+) \right],$

(t_ℓ being the interaction times of the fronts in v^θ, and $\mathcal{J}_\ell$ denoting the pair of fronts interacting at t_ℓ).

PROOF. To establish (6.101)-(6.104) we will argue by induction on the number of interactions occurring in the time interval $]\tau_1, \tau_2]$. Namely, letting $\tau_1 < t_1 < t_2 < \cdots < t_\ell < t_{\ell+1}, \cdots < t_N \leq \tau_2$, be the times when the interactions take place, and denoting with $\mathcal{J}_\ell, 1 \leq \ell \leq N$, the set of fronts interacting at t_ℓ, we will show by induction on $\ell \geq 1$ that the following estimates hold:

(6.105)$_\ell$ $\qquad \left[V(t) + Q(t) \right] < C_{14} \qquad \forall\, t < t_\ell,$

(6.106)$_\ell$
$$Q(t_\ell+) \leq Q(\tau_1+) - \sum_{i=1}^{2} \sum_{h=1}^{\ell} \left[F_i(\mathcal{J}_h; t_h, t_h) - \widehat{F}_i(\mathcal{J}_h, t_h) + G_i(\mathcal{J}_h; t_h, t_h) \right] +$$
$$+ \sum_{i \in NGNL} \sum_{\substack{\mathcal{J}_h \in \mathcal{B}^i(t_h) \\ \tau_1 < t_h \leq t_\ell}} \Xi_i(\mathcal{J}_h; t_\ell) E_i(\mathcal{J}_h; t_\ell) +$$
$$- \sum_{i \in NGNL} \sum_{\substack{\mathcal{J}_h \in \mathcal{B}^i(t_h) \\ \tau_1 < t_h \leq t_\ell}} \Xi_i(\mathcal{J}_h; t_\ell) \left[F_i(\mathcal{J}_h; t_h, t_\ell) + G_i(\mathcal{J}_h; t_h, t_\ell) + H_i(\mathcal{J}_h; t_h, t_\ell) \right],$$

(6.107)$_\ell$
$$\left[V(t_\ell+) + Q(t_\ell+) \right] \leq \left[V(\tau_1+) + Q(\tau_1+) \right] + \sum_{i \in NGNL} \sum_{\substack{\mathcal{J}_h \in \mathcal{B}^i(t_h) \\ \tau_1 < t_h \leq t_\ell}} \Xi_i(\mathcal{J}_h; t_\ell) E_i(\mathcal{J}_h; t_\ell).$$

Observe that, under the assumption (6.105)$_\ell$, one can apply Lemma 6.1, Lemma 6.9, Lemma 6.10 and use the estimates (6.33)-(6.34) of Remark 6.6 which, all together, yield (6.106)$_\ell$ − (6.107)$_\ell$. Therefore, since the bound (6.105)$_1$ certainly holds because of (6.100), in order to be able to carry out the inductive argument we only need to show that, for any $\ell > 1$, the inductive hypotheses $(6.107)_{\ell-1}$ implies $(6.105)_\ell$. But, since $\varepsilon < 1/(2C_{18})$, using (6.97), $(6.107)_{\ell-1}$, we derive

$$(6.108)_\ell \quad \begin{aligned} \left[V(t_\ell-)+Q(t_\ell-)\right] &= \left[V(t_{\ell-1}+)+Q(t_{\ell-1}+)\right] \\ &\leq \frac{1}{1-C_{18}\cdot\varepsilon}\left[V(\tau_1+)+Q(\tau_1+)\right] \\ &\leq (1+2C_{18}\cdot\varepsilon)\left[V(\tau_1+)+Q(\tau_1+)\right] \end{aligned}$$

which, together with (6.100), implies $(6.105)_\ell$. By induction, we thus obtain $(6.106)_N - (6.107)_N$, and hence (6.101)-(6.102), since $V(\tau_2+) = V(t_N+)$, $Q(\tau_2+) = Q(t_N+)$. The estimates (6.103)-(6.104) then follow from (6.97) and (6.101)-(6.102). □

PROOF OF PROPOSITION 3.1.
Let C_1, C_{14}, C_{18} be the constants at (5.3), (6.17), and (6.97). Set

$$(6.109) \quad \delta_1 \doteq \min\left\{\frac{1}{80\,C_1}, \frac{C_{14}}{2}\right\}, \qquad c_0 \doteq 2C_{18},$$

and consider an ε-approximate solution $v = v(t,x)$ constructed on the time interval $[0, T[$, with $\varepsilon \leq c_0^{-1}$, and satisfying the bound

$$V\bigl(v(0,\cdot)\bigr) + Q\bigl(v(0,\cdot)\bigr) < \delta.$$

Then, applying Lemma 6.12 with $\tau_1 = 0$, $\tau_2 = t < T$, we recover (3.32) from (6.104). □

7. ESTIMATES ON THE NUMBER OF DISCONTINUITIES

In this section, following a procedure similar to the one adopted in [**BC1**], we will show that the number of wave-fronts and of interaction points in any approximate solution $v = v(t, x)$ constructed by our algorithm in Sections 3-4 remains finite. According with Proposition 3.1, we assume that v is an ε-approximate solution, with $\varepsilon \leq \varepsilon_0$, defined on some initial interval $[0, T[$, and satisfying the bound

$$(7.1) \quad V^\varepsilon\bigl(v(t,\cdot)\bigr) + Q^\varepsilon\bigl(v(t,\cdot)\bigr) < 2\delta \qquad \forall\, t \in [0,T[$$

for some $\delta \leq \delta_1$. Moreover, we shall assume that the set of interaction points has no limit point in the strip $[0, T[\times\mathbb{R}$. We will adopt the same notation used in Section 6. In particular, a wave σ connecting the left state v^L with the right state v^R will be often equivalently denoted by $\bigl(v^L, v^R\bigr)$ or by $\bigl(v^{L_\sigma}, v^{R_\sigma}\bigr)$, while two waves of the i-th family σ', σ'' will be called *"concordant"* if $\mathrm{sgn}(\sigma') = \mathrm{sgn}(\sigma'')$, *"discordant"* otherwise. We next recall the definitions of *Big Shock Front* (BSF) and *Maximal Shock Front* (MSF) introduced in [**BC1**] for each characteristic family.

A BSF of the $i \in \{1,2\}$-th family is a polygonal line in the (t,x)-plane, with a finite or countable number of nodes (t_0, x_0), $(t_1, x_1), \ldots$, having the properties:

I) The points (t_h, x_h) are all interactions points, with $0 \leq t_0 < \cdots < t_{h-1} < t_h < \cdots < T$.

II) On each segment joining (t_{h-1}, x_{h-1}) and (t_h, x_h), the solution v has a shock of the i-th family with strength $|\sigma_h| \geq \sqrt{\varepsilon}$. Moreover, there exists some integer $H \geq 1$ such that $|\sigma_H| \geq 2\sqrt{\varepsilon}$.

III) If there are two or more shocks all with strength $\geq \sqrt{\varepsilon}$, belonging to the i-th family, and entering the point (t_h, x_h), then the shock coming from (t_{h-1}, x_{h-1}) is the one traveling with the largest speed, i.e. the one coming farther from the left.

A BSF which is maximal w.r.t. set-theoretic inclusion will be called a MSF. Thus each BSF is contained in some MSF and, because of III), any two MSFs of the same family either coincide or are disjoint. We will first show that the total number of MSFs is finite as stated in Proposition 3.2.

7.1. Proof of Proposition 3.2.

Let C_{15} be the constant of Lemma 6.2, and take $\varepsilon \leq \varepsilon_0' \doteq (12C_{15})^{-2}$. First observe that, if the k-th family is GNL or LD, relying on analogous interaction estimates to those provided by Lemmas 6-7 in [**BC1**], one can show that the potential interaction Q^ε must decrease along an MSF by some amount, uniform w.r.t. ε. Hence, using the a-priori bound (3.32) on Q^ε, one can prove that the number of MSFs of a k-th GNL or LD family is finite with the same arguments in [**BC1**, Section 5]. Next, consider the case of MSF of a k-th NGNL family. By Lemma 6.2, if the k-th family is NGNL1 any BSF of the k-th family enjoy also the properties:

II)′ On each segment joining (t_{h-1}, x_{h-1}) and (t_h, x_h), the solution v has a shock of the k-th family with size σ_h, and left state v^h, such that $|\vartheta_k(v^h, \sigma_h)| \geq \sqrt{\varepsilon}/2$, where ϑ_k is defined at (6.38). Moreover, there exists some integer $H \geq 1$ such that $|\vartheta_k(v^H, \sigma_H)| \geq \sqrt{\varepsilon}$.

III)′ If there are two or more shocks $\sigma_{h_1}, \sigma_{h_2}, \ldots$, with left states $v^{h_1}, v^{h_2}, \ldots$, belonging to the k-th family and entering the point (t_h, x_h), such that $|\vartheta_k(v^{h_\ell}, \sigma_{h_\ell})| \geq \sqrt{\varepsilon}/2$, for all $\ell \geq 1$, then the shock coming from (t_{h-1}, x_{h-1}) is the one traveling with the largest speed, i.e. the one coming farther from the left.

Similar properties are satisfied by an MSF of a k-th NGNL2 characteristic family. Therefore, in order to prove that the number of MSFs of a k-th NGNL family is finite, it suffices to show that it is finite the total number of maximal (w.r.t. set-theoretic inclusion) polygonal lines, with an at most countable number of nodes, that enjoy the properties I), II)′, III)′ (or the corresponding ones for a k-th NGNL2 family). We shall denote any such polygonal line as MSFT.

Observe first that, by (7.1) and because of II)′, the number of MSFT of a k-th NGNL family starting at $t_0 = 0$ is at most $4\delta/\sqrt{\varepsilon}$. On the other hand, if a MSFT of of a k-th NGNL1 family starts at a point (t_0, x_0), $t_0 > 0$, then its size must grow from a value σ such that $|\vartheta_k(v^{L_\sigma}, \sigma)| < \sqrt{\varepsilon}/2$, before t_0, to a value σ' such that $|\vartheta_k(v^{L_{\sigma'}}, \sigma')| \geq \sqrt{\varepsilon}$ at some time t_H. This implies that some decrease on the interaction potential must take place along the shock front. Clearly, the same happens for a MSFT of a k-th NGNL2 family. The a priori estimates on the decrease of the interaction potential over any interval $]\tau_1, \tau_2]$ given by Lemma 6.11, yield an estimate on the number of MSFT.

Indeed, first observe that, by (7.1), and because of (6.109), one has

$$V^\varepsilon\big(v(t_h\pm, \cdot)\big) + Q^\varepsilon\big(v(t_h\pm, \cdot)\big) < \frac{C_{14}}{2} \qquad \forall\, 0 \leq h \leq H,$$

which guarantees that the assumptions of Lemmas 6.3, 6.4 and 6.11 are satisfied. Next, consider an MSFT of a k-th NGNL1 family (the same arguments will apply to MSFT of a a k-th NGNL2 family) with vertices at (t_h, x_h), $h \geq 0$, starting at t_0, and set $\vartheta^h \doteq \vartheta_k(v^{L_{\sigma_h}}, \sigma_h)$, $h \geq 0$. Let $\mathcal{J}_0$ be the set of fronts involved in the interaction at t_0, and denote by $\sigma_{0,1}, \ldots \sigma_{0,n_1}$ the fronts in $\mathcal{J}_0$ belonging to the k-th characteristic family. The maximality assumption implies that $|\vartheta_k(v^{L_{\sigma_{0,\ell}}}, \sigma_{0,\ell})| < \sqrt{\varepsilon}/2$ for all $\ell = 1, \ldots, n_1$. Then, recalling the definition of the functional $\widehat{F}_i$ in (6.94), an application of Lemma 6.4 yields the estimate

$$(7.2) \quad \sum_{i=1}^{2} \Big[F_i(\mathcal{J}_0; t_0, t_0) - \widehat{F}_i(\mathcal{J}_0, t_0) + G_i(\mathcal{J}_0; t_0, t_0) \Big] \geq C_{17}\sqrt{\varepsilon} \left| \vartheta^1 - \frac{\sqrt{\varepsilon}}{2} \right|.$$

On the other hand, at every interaction time $t_h > t_0$ the shock σ_h satisfies $|\vartheta^h| \geq \sqrt{\varepsilon}/2$. Thus, if we let $\mathcal{J}_h$ denote the set of fronts interacting together at t_h, applying Lemma 6.3 we obtain

$$(7.3)_h \quad \sum_{i=1}^{2} \Big[F_i(\mathcal{J}_h; t_h, t_h) - \widehat{F}_i(\mathcal{J}_h, t_h) + G_i(\mathcal{J}_h; t_h, t_h) \Big] \geq C_{16} \frac{\sqrt{\varepsilon}}{2} \left| \vartheta^{h+1} - \vartheta^h \right|.$$

Then, relying on (7.1)-(7.2), $(7.3)_h, 1 < h \leq H$, and using (6.97) and (6.101), with $\tau_1 = t_0$, $\tau_2 = t_H$, we derive
(7.4)
$$Q^\varepsilon\big(v(t_H+,\cdot)\big) - Q^\varepsilon\big(v(t_0-,\cdot)\big) \leq$$

$$\leq -\min\{C_{16}/2,\ C_{17}\} \cdot \sqrt{\varepsilon} \left[\left| \vartheta^1 - \frac{\sqrt{\varepsilon}}{2} \right| + \sum_{h=1}^{H} |\vartheta^{h+1} - \vartheta^h| \right] +$$

$$+ \sum_{i \in NGNL} \sum_{\substack{\mathcal{J}' \in \mathcal{B}^i(t) \\ t_0 \leq t \leq t_H}} \Xi_i(\mathcal{J}'; \tau_2) E_i(\mathcal{J}')$$

$$\leq -\min\{C_{16}/2,\ C_{17}\} \cdot \sqrt{\varepsilon} \left| \vartheta^{H+1} - \frac{\sqrt{\varepsilon}}{2} \right| + C_{18} \cdot V(t_H+) \cdot \varepsilon$$

$$\leq -\big(\min\{C_{16}/4,\ C_{17}/2\} - C_{18} \cdot 2\delta\big)\varepsilon.$$

Choose a positive constant $\delta_2 \leq \delta_1$ such that

$$(7.5) \qquad \delta_2 \leq \frac{C_{19}}{4C_{18}}, \qquad C_{19} \doteq \min\{C_{16}/4,\ C_{17}/2\}.$$

If the ε-approximate solution $v = v(t,x)$ satisfies the bound

$$(7.6) \qquad V^\varepsilon\big(v(0,\cdot)\big) + Q^\varepsilon\big(v(0,\cdot)\big) < \delta_2,$$

it follows by Proposition 3.1 that (7.1) holds with $\delta = \delta_2$, and hence we deduce from (7.4) that each MSF forces the interaction potential Q to decrease by an amount $\geq \dfrac{C_{19}}{2} \cdot \varepsilon$. Therefore, since $Q\big(v(0,\cdot)\big) < \delta_2$, the total number of MSFs of the k-th family cannot be greater than $\dfrac{2\delta_2}{C_{19}\varepsilon}$, completing the proof of Proposition 3.2.

□

Throughout the next part of the section we shall assume that the ε-approximate solution v satisfies the initial bound (7.6) and hence, by Proposition 3.2, has a finite number of MSFs. For convenience, we will work out all the proofs for a system satisfying the assumption $(\mathbf{A'})$ (in which the first family is NGNL1 and the second family is NGNL2). The case of a general 2×2 NGNL system satisfying the assumption $(\mathbf{A})$ can be covered with entirely similar arguments and relying on the analysis in [**BC1**, Section 5].

7.2. Proof of Proposition 3.3.

From now on, we will call the wave-fronts *strong* or *weak*, respectively, if their strength $|\sigma|$ is greater or smaller than $2\sqrt{\varepsilon}$. Hence rarefaction fronts are always weak. There are six types of possible interactions involving a strong wave of the i-th family (iS-wave) and a weak wave of the j-th family (jW-wave), depending on which family the waves belong to and on their respective locations (see figures 7.1-7.2):

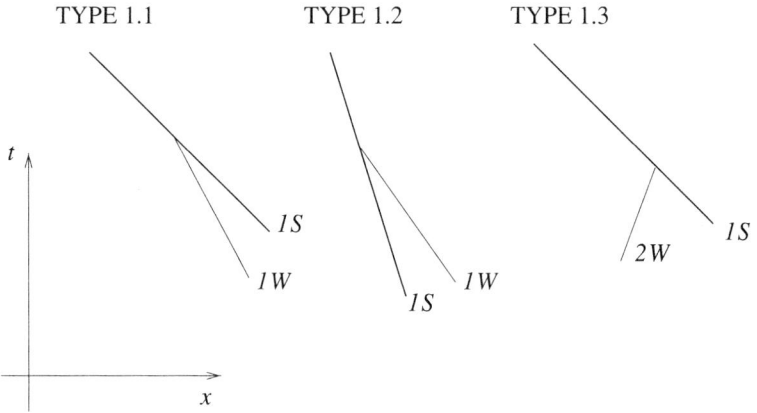

FIGURE 7.1

TYPE 1.1: the interaction involves a $1S$-wave and a $1W$-wave located on the left of the $1S$-wave;

TYPE 1.2: the interaction involves a $1S$-wave and a $1W$-wave located on the right of the $1S$-wave;

TYPE 1.3: the interaction involves a $2W$-wave and a $1S$-wave;

TYPE 2.1: the interaction involves a $2S$-wave and a $2W$-wave located on the left of the $2S$-wave;

TYPE 2.2: the interaction involves a $2S$-wave and a $2W$-wave located on the right of the $2S$-wave;

TYPE 2.3: the interaction involves a $1W$-wave and a $2S$-wave.

Notice that, by the properties of the approximate Riemann solver described in Section 3, any interaction involving a strong wave that produces a composed wave, consisting of a piecewise constant rarefaction fan and a of a strong shock, is always an interaction of type 1.2, 1.3, 2.2, 2.3.

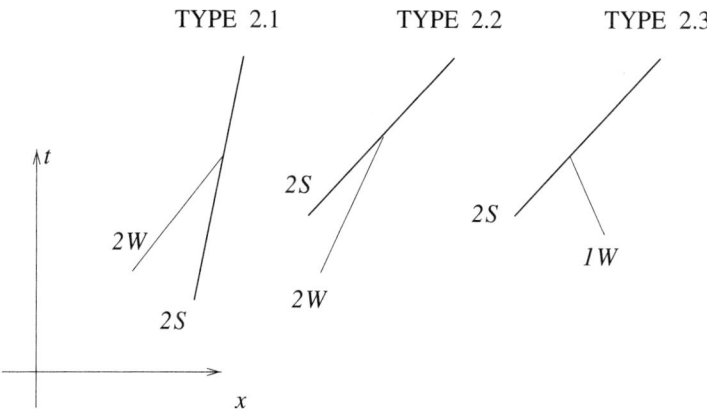

FIGURE 7.2

Any interaction between weak waves involving discordant waves of the i-th family will be called an *interaction of type* $i.4$.

PROOF OF PROPOSITION 3.3. Assume by contradiction that, as $t \to T-$, the set of interaction points has a limit point, say $(T, \bar{x})$, in the (t, x)-plane (and no limit point in the strip $[0, T[\times \mathbb{R})$. By Proposition 3.2 there are finitely many MSF which approach $\bar{x}$ as $t \to T-$. Let these MSF be located at (see figure 7.3):

$$x_{2,1}(t) < \ldots < x_{2,n_2}(t) < x_{1,1}(t) < \ldots < x_{1,n_1}(t)$$

($x_{i,\ell}$ being the locations of the MSF of the i-th family), and satisfy

$$\lim_{t \to T-} x_{1,\ell}(t) = \lim_{t \to T-} x_{2,\ell}(t) = \bar{x}.$$

Choose $\delta, \rho > 0$ so small that all of the above shock fronts are defined on the common interval $[T - \delta, T[$, and such that the trapezoid

$$(7.7) \qquad \Delta \doteq \left\{ (t, x) : \quad t \in [T - \delta, T[, \ |x - \bar{x}| \leq \rho + (T - t)\lambda^{\max} \right\}$$

does not intersect any other MSF (see Fig. 7.3). Here $\lambda^{\max}$ is the upper bound for all characteristic speeds introduced at (2.1).
In order to establish Proposition 3.3 we need the following lemma whose proof is given below.

LEMMA 7.1. *The total number of interaction points of type* $1.1 - 1.4$, $2.1 - 2.4$, *inside Δ is finite.*

Relying on the above lemma one can reach the conclusion of Proposition 3.3 arguing as follows. By Lemma 7.1 we deduce that, after some time $\bar{t} < T$ sufficiently large, no further interaction of type 1.1-1.4, 2.1-2.4 occurs inside Δ. Thus, every interaction point (t, x), $t \geq \bar{t}$, in Δ satisfies one of the following two conditions:

I) all waves interacting at (t, x) are weak waves of different families;

II) all waves interacting at (t, x) are concordant weak waves of the same family.

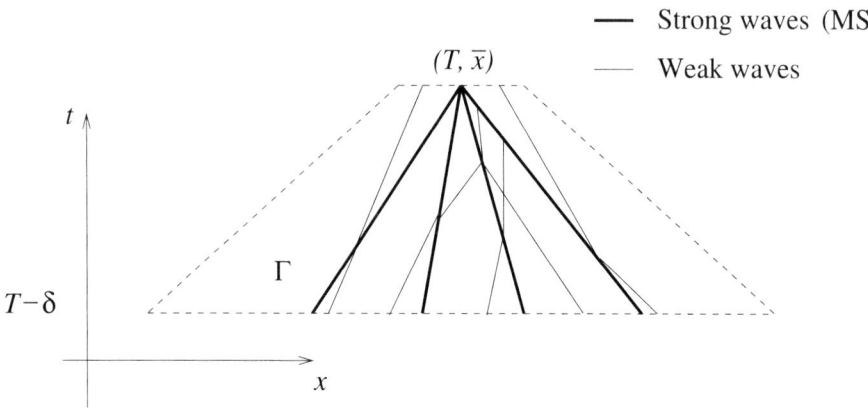

FIGURE 7.3

In the first case the total number of wave-fronts in $v(t,x)$, $(t,x) \in \Delta$ may increase by at most $2(\lfloor 1/\sqrt{\varepsilon} \rfloor + 1)$ times the number of wave-fronts in $v(\bar{t})$, within the time interval $[\bar{t}, T]$. On the other hand, the outgoing wave-fronts of different families produced by this interaction cannot interact together at any later time. In the second case, recalling that by construction every strong wave inside Δ is contained in one of the MSFs considered above, it follows that the interaction generates only a single weak wave-front. Thus the number of wave-fronts decreases. Since no wave-front may enter Δ across the oblique sides due to the choice of $\lambda^{\max}$, we deduce that the total number of wave-fronts inside Δ must be finite. Hence the total number of interaction points inside Δ is also finite. This contradicts the initial assumption that $(T, \bar{x})$ is a limit of interaction points, proving Proposition 3.3. □

PROOF OF LEMMA 7.1. The proof is given in two steps:

Step 1: For each $j = 1, 2$, in order to estimate the number of interaction points of type $j.2$, $j.4$, we will introduce a function $M_j : [T-\delta, T[\to \mathbb{N}$ with the following properties:

I) M_j is non increasing at any time t;
II)$_j$ M_j is strictly decreasing at any time t where an interaction of type $j.2$, $j.4$ occurs inside Δ.

Step 2: In order to estimate the number of interaction points of type $j.1$, $j.3$, $j = 1, 2$, we will introduce a function $N : [T - \delta, T[\to \mathbb{N}$ with the following properties:

III) N is non increasing at any time t where no interaction of type $j.2$, $j = 1, 2$ occurs inside Δ;
IV) N is strictly decreasing at any time t where an interaction of type $j.1$, $j.3$, $j = 1, 2$ occurs inside Δ, and no interaction of type $j.2$, $j = 1, 2$ takes place.

The conclusion of the lemma follows easily from the existence of the maps M_1, M_2, N with the above properties. Indeed, by the properties of M_1, M_2, one deduces that the total number of interactions of type 1.2, 1.4 and 2.2, 2.4 inside Δ is finite. In particular, this means that after some time $\bar{\tau} < T$ sufficiently large, no further interaction of type 1.2, or 2.2 occurs inside Δ. In turn, the properties of N imply that the total number of interaction points of type 1.1, 1.3, and 2.1, 2.3 inside the region $\Delta \cap (]\bar{\tau}, T[\times \mathbb{R})$ is finite as well. On the other hand, since we are assuming that the set of interaction points has no limit point in the strip $[0, T[\times \mathbb{R}$, it follows that the total number of interactions of any type in the strip $[0, \bar{\tau}] \times \mathbb{R}$ is certainly finite, concluding the proof.

Step 1: Here we define a function $M_1 : [T-\delta, T[\to \mathbb{N}$ that enjoys the properties I), II)$_1$. The construction of a function M_2 enjoying the properties I), II)$_2$ is entirely similar. Let $\tilde{\gamma}_i$, $i = 1, 2$ be the maps at (2.11) defining the parameterization in Riemann coordinates of the curves Γ_i, $i = 1, 2$ at (1.4). Call

(7.8)
$$\widehat{a}_i \doteq \inf_{v \in \mathcal{V}} \tilde{\gamma}_i(v_j),$$
$$\widehat{b}_i \doteq \sup_{v \in \mathcal{V}} \tilde{\gamma}_i(v_j), \qquad i, j = 1, 2, \quad i \neq j,$$

and, for any given wave (v^L, v^R), define

(7.9)
$$W_1(v^L, v^R) \doteq \begin{cases} 1 + \#\left\{k \in \mathbb{Z}: \ v_1^L < k\varepsilon < \min\{v_1^R, \widehat{b}_1\}\right\} & \text{if } v_1^L < \min\{v_1^R, \widehat{b}_1\}, \\ 1 + \#\left\{k \in \mathbb{Z}: \ \max\{v_1^R, \widehat{a}_1\} < k\varepsilon < v_1^L\right\} & \text{if } v_1^L > \max\{v_1^R, \widehat{a}_1\}, \\ 1 & \text{otherwise}, \end{cases}$$

(7.10)
$$W_2(v^L, v^R) \doteq \begin{cases} 1 + \#\left\{k \in \mathbb{Z}: \ \max\{v_2^L, \widehat{a}_2\} < k\varepsilon < v_2^R\right\} & \text{if } v_2^R > \max\{v_2^L, \widehat{a}_2\}, \\ 1 + \#\left\{k \in \mathbb{Z}: \ v_2^R < k\varepsilon < \min\{v_2^L, \widehat{b}_2\}\right\} & \text{if } v_2^R < \min\{v_2^L, \widehat{b}_2\}, \\ 1 & \text{otherwise}. \end{cases}$$

Notice that, by construction, $W_i(v^L, v^R) = 1$ whenever (v^L, v^R) is a rarefaction front of the i-th family. Call $v^{L_\ell^i} = v^{L_\ell^i}(t)$ and $v^{R_\ell^i} = v^{R_\ell^i}(t)$, respectively, the left and right state of the strong wave contained in the MSF located at $x = x_{i,\ell}(t)$, $\ell = 1, \ldots, n_1$. Besides, call $w^{L_\alpha^i} = w^{L_\alpha^i}(t)$ and $w^{R_\alpha^i} = w^{R_\alpha^i}(t)$ the left and right state of a weak wave of the i-th family located at $x = y_{i,\alpha}(t)$. Then, define

(7.11) $$M_1(t) \doteq \sum_{y_{1,\alpha}(t) \geq \bar{x}} W_1(w^{L_\alpha^1}, w^{R_\alpha^1}) + \sum_{\ell=1}^{n_1} W_1(v^{L_\ell^1}, v^{R_\ell^1}) \qquad t \in [T - \delta, T[,$$

where the first sum ranges over the set of all weak waves of the first family $(w^{L_\alpha^1}, w^{R_\alpha^1})$ inside Δ, located at time t on the right of $\bar{x}$. Notice that, because of (7.1), the number of wave-fronts at any time $t < T$ is finite. Hence, the map M_1 defined in (7.11) takes values in the set $\mathbb{N}$ for all $t \in [T - \delta, T[$. Concerning the properties I)-II)$_1$, observe that any interaction occurring between fronts located on the left of $\bar{x}$, produces fronts of the first family that remain on the left of $\bar{x}$ for all times. Therefore it will be sufficient to check that the function M_1 satisfies the properties I)-II)$_1$ at every time $t > T - \delta$ where it occurs an interaction between waves located on the right of $\bar{x}$. We shall consider three cases:

CASE 1. The interaction involves only weak waves.

CASE 2. The interaction is of type 1.1 or 1.3 and not of type 1.2.

CASE 3. The interaction is of type 1.2.

CASE 1. Assume that ν_1 weak fronts of the first characteristic family and ν_2 weak fronts of the second characteristic family interact together at a point (t,x) inside Δ. Since by construction no strong wave can be generated inside Δ, this interaction produces only weak waves. Call w^L, w^R, respectively, the states to the left and to the right of (t, x) and let w^{M-}, w^{M+}, be the middle states between the two families of interacting waves before and after the interaction (see Fig. 7.4). We shall first consider the cases where at most two fronts of the first family are involved in the interaction, and then we will argue by induction on the number ν_1 of interacting fronts of the first family.

Let $\nu_1 = 1$. Then, by construction one has $W_1(w^{M-}, w^R) = W_1(w^L, w^R) = W_1(w^L, w^{M+})$. Hence $M_1(t+) = M_1(t-)$.

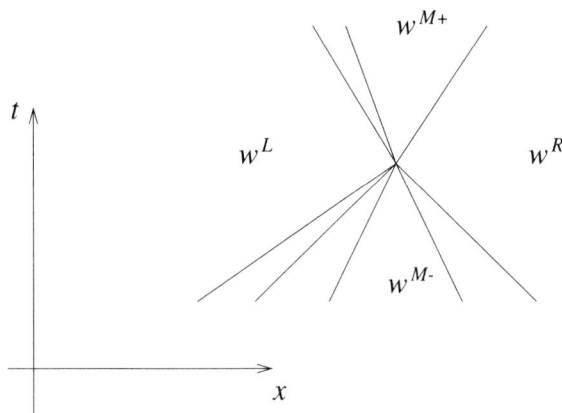

FIGURE 7.4

Let $\nu_1 = 2$, and call σ', σ'', the sizes of the incoming fronts of the first family. Set $w' = R_1(w^{M-}, \sigma') = R_1(w^R, -\sigma'')$. Six cases can occur:

a) $w_1^{M-} < w_1' < w_1^R$: σ', σ'', are concordant. Then there holds

(7.12) $\quad W_1(w^{M-}, w') + W_1(w', w^R) \geq W_1(w^{M-}, w^R) \geq W_1(w^L, w^{M+}).$

It follows that $M_1(t+) \leq M_1(t-)$.

b) $w_1^R < w_1^{M-} < w_1'$: σ', σ'', are discordant. If $\widehat{a}_1 < w_1^{M-}$, then one has

(7.13) $\quad W_1(w', w^R) \geq W_1(w^{M-}, w^R) \geq W_1(w^L, w^{M+}),$

and hence $M_1(t+) \leq M_1(t-) - 1$. Otherwise only one wave-front emerges from the interaction and $W_1(w^L, w^{M+}) = 1$. Again $M_1(t+) \leq M_1(t-) - 1$.

c) $w_1^{M-} < w_1^R < w_1'$: σ', σ'', are discordant. We have $W_1(w', w^R) = 1$. It follows that $W_1(w^L, w^{M+}) \leq W_1(w^{M-}, w')$, from which we obtain $M_1(t+) \leq M_1(t-) - 1$.

d) $w_1^R < w_1' < w_1^{M-}$: similar to case a).

e) $w_1' < w_1^{M-} < w_1^R$: similar to case b).

f) $w'_1 < w_1^R < w_1^{M-}$: similar to case c).

Now assume that M_1 enjoys the properties I)-II)$_1$ any time that it occurs an interaction involving $\nu_1 - 1$ weak fronts of the first family, and consider an interaction involving ν_1 weak fronts of the first family. Since we are dealing with an interaction that produces only weak waves, we can let the waves interact in the following order. First we let interact together all the waves of the second characteristic family and the first $\nu_1 - 1$ waves of the first family. Next, we let the waves generated by this preliminary interaction interact with the last weak wave of the first family. Thus, using the inductive hypotheses and similar arguments as above, one concludes that M_1 enjoys the properties I)-II)$_1$ at any interaction time falling in Case 1.

CASE 2. Assume that at time t it occurs an interaction of type 1.1 or 1.3 (and not of type 1.2) inside Δ involving:

- a strong shock of the first family contained in the MSF located at $x_{1,\ell}$, $\ell \in \{1, \ldots, n_1\}$;
- ν_1 weak fronts of the first characteristic family, located on the left of $x_{1,\ell}$, $\ell \in \{1, \ldots, n_1\}$, having sizes $\sigma^-_{1,1}, \ldots, \sigma^-_{1,\nu_1}$;
- ν_2 weak fronts of the second family, located on the right of $x_{1,\ell}$, $\ell \in \{1, \ldots, n_1\}$, having sizes $\sigma^-_{2,1}, \ldots, \sigma^-_{2,\nu_2}$.

Call $w^{L_m^-}$, $w^{R_m^-}$, the left and right states of $\sigma^-_{1,m}$, $1 \leq m \leq \nu_1$, and set $\sigma_2^- \doteq \sum_{m=1}^{\nu_2} \sigma^-_{2,m}$. If the wave of the first family generated by this interaction is a composed wave that contains a piecewise constant rarefaction fan $\widetilde{\sigma}_1^+$, let $\widetilde{\sigma}^+_{1,1}, \ldots \widetilde{\sigma}^+_{1,p}$ denote the wave-fronts of $\widetilde{\sigma}_1^+$, and call $w^{L_\alpha^+}$, $w^{R_\alpha^+}$, the left and right states of $\widetilde{\sigma}^+_{1,\alpha}$, $1 \leq \alpha \leq p$. Otherwise, set $\widetilde{\sigma}_1^+ \doteq 0$. As usual, let $v^{L_\ell^1}(t)$, $v^{R_\ell^1}(t)$, denote the left and right states of the strong shock located at $x_{1,\ell}(t)$, $\ell \in \{1, \ldots, n_1\}$. Then, one has

(7.14)
$$\sum_{\alpha=1}^p W_1\big(w^{L_\alpha^+}, w^{R_\alpha^+}\big) + W_1\big(v^{L_\ell^1}(t+), v^{R_\ell^1}(t+)\big) =$$
$$= W_1\Big[R_1\big(v^{L_\ell^1}(t+), -\widetilde{\sigma}_1^+\big), v^{R_\ell^1}(t+)\Big]$$
$$= W_1\Big[R_2\Big(R_1\big(v^{L_\ell^1}(t+), -\widetilde{\sigma}_1^+\big), \sigma_2^-\Big), v^{R_\ell^1}(t-)\Big]$$
$$\leq \sum_{m=1}^{\nu_1} W_1\big(w^{L_m^-}, w^{R_m^-}\big) + W_1\big(v^{L_\ell^1}(t-), v^{R_\ell^1}(t-)\big).$$

Hence $M_1(t+) \leq M_1(t-)$.

CASE 3. Assume that at time t it occurs inside Δ an interaction involving the same type of fronts as in Case 2, plus at least one weak wave of the first family located on the right of the MSF in $x_{1,\ell}$, $\ell \in \{1, \ldots, n_1\}$. Denote it by σ^-_{1,ν_1+1}. Then, one can check that a similar estimate as in (7.14) holds

$$\sum_{\alpha=1}^p W_1\big(w^{L_\alpha^+}, w^{R_\alpha^+}\big) + W_1\big(v^{L_\ell^1}(t+), v^{R_\ell^1}(t+)\big) =$$

$$= W_1\Big[R_1\big(v^{L_\ell^1}(t+),\,-\widetilde{\sigma}_1^+\big),\,v^{R_\ell^1}(t+)\Big]$$

$$\leq \sum_{m=1}^{\nu_1} W_1\big(w^{L_m^1},w^{R_m^1}\big) + W_1\big(v^{L_\ell^1}(t-),v^{R_\ell^1}(t-)\big)$$

(7.15)
$$\leq \bigg[\sum_{m=1}^{\nu_1+1} W_1\big(w^{L_m^1},w^{R_m^1}\big) + W_1\big(v^{L_\ell^1}(t-),v^{R_\ell^1}(t-)\big)\bigg] - 1$$

and hence $M_1(t+) \leq M_1(t-) - 1$, concluding the proof that M_1 enjoys the properties I)-II)$_1$.

Step 2: Here we define a function $N : [T-\delta, T[\to \mathbb{N}$ that enjoys the properties III)-IV). To keep track of the number of weak waves, to each weak wave of the i-th family $(w^{L_\alpha^i}, w^{R_\alpha^i})$ located at $x = y_{i,\alpha}(t)$ we associate a weight $P_i(w^{L_\alpha^i}, w^{R_\alpha^i})$ defined as follows:

(7.16)$_1$

position $y_{1,\alpha}(t) \in$		weight $P_1(w^{L_\alpha^1}, w^{R_\alpha^1})$
$]-\infty, x_{2,1}(t)[$		$W_1(w^{L_\alpha^1}, w^{R_\alpha^1})(N_\varepsilon+1)^3$
$]x_{2,\ell}(t), x_{2,\ell+1}(t)[$,	$1 \leq \ell < n_2$	$W_1(w^{L_\alpha^1}, w^{R_\alpha^1})(N_\varepsilon+1)^{3(\ell+1)}$
$]x_{2,n_2}(t), x_{1,1}(t)[$		$W_1(w^{L_\alpha^1}, w^{R_\alpha^1})(N_\varepsilon+1)^{3(n_1+n_2)+4}$
$]x_{1,\ell}(t), x_{1,\ell+1}(t)[$,	$1 \leq \ell < n_1$	$W_1(w^{L_\alpha^1}, w^{R_\alpha^1})(N_\varepsilon+1)^{3(n_1-\ell)+1}$
$]x_{1,n_1}(t), +\infty[$		$W_1(w^{L_\alpha^1}, w^{R_\alpha^1})(N_\varepsilon+1)$

(7.16)$_2$

position $y_{2,\alpha}(t) \in$		weight $P_2(w^{L_\alpha^2}, w^{R_\alpha^2})$
$]-\infty, x_{2,1}(t)[$		$W_2(w^{L_\alpha^2}, w^{R_\alpha^2})(N_\varepsilon+1)$
$]x_{2,\ell}(t), x_{2,\ell+1}(t)[$,	$1 \leq \ell < n_2$	$W_2(w^{L_\alpha^2}, w^{R_\alpha^2})(N_\varepsilon+1)^{3\ell+1}$
$]x_{2,n_2}(t), x_{1,1}(t)[$		$W_2(w^{L_\alpha^2}, w^{R_\alpha^2})(N_\varepsilon+1)^{3(n_1+n_2)+4}$
$]x_{1,\ell}(t), x_{1,\ell+1}(t)[$,	$1 \leq \ell < n_1$	$W_2(w^{L_\alpha^2}, w^{R_\alpha^2})(N_\varepsilon+1)^{3(n_1-\ell+1)}$
$]x_{1,n_1}(t), +\infty[$		$W_2(w^{L_\alpha^2}, w^{R_\alpha^2})(N_\varepsilon+1)^3$,

with $N_\varepsilon \doteq \lfloor \delta/\varepsilon \rfloor + 1$. Then, for any $t \in [T-\delta, T[$, define
(7.17)
$$N(t) \doteq \sum_{i=1,2}\sum_\alpha P_i\big(w^{L_\alpha^i}, w^{R_\alpha^i}\big) +$$
$$+ (N_\varepsilon+1)^{3(n_1+n_2)+4}\big[W_1\big(v^{L_1^1}(t), v^{R_1^1}(t)\big) + W_2\big(v^{L_{n_2}^2}(t), v^{R_{n_2}^2}(t)\big)\big],$$

where the first sum ranges over the set of all weak waves located at time t inside Δ. Call $N(t)$ the *weighted number of weak waves present at time t inside Δ*. Since, because of (7.1), the number of wave-fronts at any time $t < T$ is finite, it follows that the map N defined in (7.17) takes values in the set $\mathbb{N}$ for all $t \in [T - \delta, T[$. Concerning the properties III)-IV), with the same arguments used in Case 1 of Step 1, one can easily verify that N cannot increase after any interaction involving only weak waves. Thus, if we show that property IV) holds, we obtain III) as well.

Then, consider an interaction of type 1.1 or 1.3 and not of type 1.2 involving an MSF of the first family, occurring at time t inside Δ. Then, the set of incoming waves consists of:

- ν_2 weak waves of the second family on the left of $x_{1,\ell}$, $\ell \in \{1, \ldots, n_1\}$;
- ν_1 weak waves of the first family on the left of $x_{1,\ell}$, $\ell \in \{1, \ldots, n_1\}$;
- the strong shock of the first family along the MSF in $x_{1,\ell}$, $\ell \in \{1, \ldots, n_1\}$.

Assume that p_i weak waves of the i-th family emerge from the interaction. Observe that, by the properties of the approximate Riemann solver, $p_1 \neq 0$ only if $\nu_2 \neq 0$. Call N^-, N^+, the weighted numbers of weak waves relative only to the set of incoming and outgoing waves.

Then, if $\ell = 1$, i.e. if the strong shock of the first family involved in the interaction belongs to the MSF of the first family coming farther from the left, we have
(7.18)
$$\begin{cases} N^- \geq (N_\varepsilon + 1)^{3(n_1+n_2)+4}\big[1 + W_1\big(v^{L_1^1}(t), v^{R_1^1}(t)\big) + \\ \qquad\qquad + W_2\big(v^{L_{n_2}^2}(t), v^{R_{n_2}^2}(t)\big)\big], \\ N^+ \leq N_\varepsilon(N_\varepsilon + 1)^{3n_1} + (N_\varepsilon + 1)^{3(n_1+n_2)+4}\big[W_1\big(v^{L_1^1}(t), v^{R_1^1}(t)\big) + \\ \qquad\qquad + W_2\big(v^{L_{n_2}^2}(t), v^{R_{n_2}^2}(t)\big)\big], \end{cases} \quad \text{if } \nu_2 = 0,$$

$$\begin{cases} N^- \geq (N_\varepsilon + 1)^{3(n_1+n_2)+4}\big[1 + W_1\big(v^{L_1^1}(t), v^{R_1^1}(t)\big) + \\ \qquad\qquad + W_2\big(v^{L_{n_2}^2}(t), v^{R_{n_2}^2}(t)\big)\big] \\ N^+ \leq N_\varepsilon(N_\varepsilon + 1)^{3(n_1+n_2)+4}\big[W_1\big(v^{L_1^1}(t), v^{R_1^1}(t)\big) + \\ \qquad\qquad + W_2\big(v^{L_{n_2}^2}(t), v^{R_{n_2}^2}(t)\big)\big] + N_\varepsilon(N_\varepsilon + 1)^{3n_1} \end{cases} \quad \text{if } \nu_2 \neq 0.$$

On the other hand, if $\ell > 1$, we have

(7.19)
$$\begin{cases} N^- \geq (N_\varepsilon + 1)^{3(n_1-\ell+1)+1} \\ N^+ \leq N_\varepsilon(N_\varepsilon + 1)^{3(n_1-\ell+1)} \end{cases} \quad \text{if } \nu_2 = 0,$$

$$\begin{cases} N^- \geq (N_\varepsilon + 1)^{3(n_1-\ell+2)} \\ N^= \leq N_\varepsilon(N_\varepsilon + 1)^{3(n_1-\ell+1)+1} + N_\varepsilon(N_\varepsilon + 1)^{3(n_1-\ell+1)} \end{cases} \quad \text{if } \nu_2 \neq 0.$$

In any case we get
$$N(t+) \leq N(t-) - 1$$
proving that N enjoys the properties III)-IV), which completes the proof of Lemma 7.1. $\square$

REMARK 7.1. Propositions 3.1-3.3 imply that if $\tilde{v} \in \mathcal{D}^{\varepsilon,\delta_2}$, $\varepsilon \leq \varepsilon'_0$, then the corresponding ε-approximate solution is well defined for all $t \geq 0$. This result can be extended to a larger set of initial data, not necessarily with bounded support. More precisely, if we define

(7.20) $\quad \widetilde{\mathcal{D}}^\varepsilon \doteq \Big\{ \tilde{v} : \mathbb{R} \to \mathbb{R}^2 : \tilde{v}$ is piecewise constant and

$$V^\varepsilon\left(\tilde{v} \cdot \chi_{[-K,K]}\right) + Q^\varepsilon\left(\tilde{v} \cdot \chi_{[-K,K]}\right) < \delta_2, \quad \forall\, K > 0 \Big\},$$

then, the previous analysis ensures that the ε-approximate solution $v(t,\cdot)$ with initial condition $\tilde{v} \in \widetilde{\mathcal{D}}^\varepsilon$ is well defined for all $t \geq 0$ as well.

7.3. Proof of Proposition 3.4.

Towards a proof of Proposition 3.4, we will first show that, for a certain class of ε-solutions, the set of interaction points not only has no limit point in the (t,x)-plane, but it is actually finite. The proof relies on the fact that, for t sufficiently large, these solutions approach the self-similar ε-solution of a Riemann problem.

LEMMA 7.2. *Let $\tilde{v} \in \widetilde{\mathcal{D}}^\varepsilon$. Assume that the ε-solution $\omega = \omega(t,x)$ of the Riemann problem with initial data*

(7.21) $\quad \omega(0,x) = \begin{cases} v^L \doteq \lim\limits_{x \to -\infty} \tilde{v}(x) & \text{if } x < 0, \\ v^R \doteq \lim\limits_{x \to +\infty} \tilde{v}(x) & \text{if } x > 0, \end{cases}$

does not contain two shocks of sizes σ_1, σ_2 with $|\sigma_i| > \sqrt{\varepsilon}/2$, $i = 1, 2$. Then, the ε-approximate solution $v(t,\cdot)$ with initial condition $\tilde{v}$ contains finitely many wavefronts and interaction points in the (t,x)-plane.

PROOF. By Proposition 3.2 the total number of MSFs appearing in v is finite. Hence, there exists a time $\tau > 0$ such that, beyond τ, the number of MSFs remains constant and no interaction occurs involving two or more of these shock fronts. Call

(7.22) $\quad x_{1,1}(t) < \cdots < x_{1,n_1}(t) < x_{2,1}(t) < \cdots < x_{2,n_2}(t) \qquad t \in [\tau, \infty[,$

the location of the MSF of of the first and second characteristic family. The proof is given in three steps:

Step 1. Assume that $n_1, n_2 \geq 1$, and consider the region

$$\widetilde{\Delta} \doteq \big\{(t,x) \;:\; t > \tau, \quad x_{1,n_1}(t) \leq x \leq x_{2,1}(t)\big\}.$$

CLAIM *The total number of interaction points of type* 1.1-1.4, 2.2-2.4 *outside $\widetilde{\Delta}$ is finite.*

PROOF OF THE CLAIM. First observe that, using the same arguments of the proof of Lemma 7.1, one can easily check that the functional

$$\widetilde{M}(t) \doteq \sum_{y_{1,\alpha}(t) < x_{1,n_1}(t)} W_1\big(w^{L^1_\alpha}, w^{R^1_\alpha}\big) +$$

$$+ \sum_{y_{2,\alpha}(t) > x_{2,1}(t)} W_2\big(w^{L^2_\alpha}, w^{R^2_\alpha}\big) + \sum_{i=1}^{2} \sum_{\ell=1}^{n_2} W_i\big(v^{L^i_\ell}, v^{R^i_\ell}\big) \qquad t \in [\tau, \infty[,$$

has the following properties:
   I) $\widetilde{M}(t) \in \mathbb{N}$, for all $t \in [\tau, \infty[$;
  II) $\widetilde{M}$ is non increasing at any time t;
 III) $\widetilde{M}$ is strictly decreasing at any time t where an interaction of type 1.2, 1.4, 2.2, 2.4, occurs outside $\widetilde{\Delta}$.

Hence, the total number of interaction points of type 1.2, 1.4, 2.2, 2.4 outside $\widetilde{\Delta}$ is finite. Next, to each weak wave of the i-th family $(w^{L^i_\alpha}, w^{R^i_\alpha})$, located outside $\widetilde{\Delta}$ at $x = y_{i,\alpha}(t)$, we associate a weight $\widetilde{P}_i(w^{L^i_\alpha}, w^{R^i_\alpha})$ defined as follows:

$(7.23)_1$

position $y_{1,\alpha}(t) \in$	weight $\widetilde{P}_1(w^{L^1_\alpha}, w^{R^1_\alpha})$
$]-\infty, x_{1,1}(t)[$	$W_1(w^{L^1_\alpha}, w^{R^1_\alpha})(N_\varepsilon + 1)^{3n_1+1}$
$]x_{1,\ell}(t), x_{1,\ell+1}(t)[,\ 1 \leq \ell < n_1$	$W_1(w^{L^1_\alpha}, w^{R^1_\alpha})(N_\varepsilon + 1)^{3(n_1-\ell)+1}$
$]x_{2,\ell}(t), x_{2,\ell+1}(t)[,\ 1 \leq \ell < n_2$	$W_1(w^{L^1_\alpha}, w^{R^1_\alpha})(N_\varepsilon + 1)^{3(\ell+1)}$
$]x_{2,n_2}(t), +\infty[$	$W_1(w^{L^1_\alpha}, w^{R^1_\alpha})(N_\varepsilon + 1)^{3(n_2+1)}$

$(7.23)_2$

position $y_{2,\alpha}(t) \in$	weight $\widetilde{P}_2(w^{L^2_\alpha}, w^{R^2_\alpha})$
$]-\infty, x_{1,1}(t)[$	$W_2(w^{L^2_\alpha}, w^{R^2_\alpha})(N_\varepsilon + 1)^{3(n_1+1)}$
$]x_{1,\ell}(t), x_{1,\ell+1}(t)[,\ 1 \leq \ell < n_1$	$W_2(w^{L^2_\alpha}, w^{R^2_\alpha})(N_\varepsilon + 1)^{3(n_1-\ell+1)}$
$]x_{2,\ell}(t), x_{2,\ell+1}(t)[,\ 1 \leq \ell < n_2$	$W_2(w^{L^2_\alpha}, w^{R^2_\alpha})(N_\varepsilon + 1)^{3\ell+1}$
$]x_{2,n_2}(t), +\infty[$	$W_2(w^{L^2_\alpha}, w^{R^2_\alpha})(N_\varepsilon + 1)^{3n_2+1}$.

For any $t \in [\tau, \infty[$ we define the *weighted number of weak waves present at time t inside $\widetilde{\Delta}$* as

(7.24)
$$\widetilde{N}(t) \doteq \sum_{i=1,2} \sum_\alpha \widetilde{P}_i(w^{L^i_\alpha}, w^{R^i_\alpha}) + \left[W_1(v^{L^1_{n_1}}, v^{R^1_{n_1}}) + W_2(v^{L^2_1}, v^{R^2_1})\right](N_\varepsilon + 1)^4$$

where the first sum ranges over the set of all weak waves located at time t inside $\widetilde{\Delta}$. Then, as in the proof of Lemma 7.1, one can check that the following properties hold

  IV) $\widetilde{N}(\tau) \in \mathbb{N}$, for all $t \in [\tau, \infty[$;
   V) $\widetilde{N}$ is non increasing at any time t where no interaction of type 1.2, 2.2, occurs outside $\widetilde{\Delta}$;
  VI) $\widetilde{N}$ is strictly decreasing at any time t where an interaction of type 1.1, 1.3, 2.1, 2.3, occurs outside $\widetilde{\Delta}$.

Thus, the Claim is established.

Step 2. By the above Claim, if $n_1, n_2 \geq 1$, there will be some time $\tau' \geq T - \delta$ sufficiently large so that, beyond τ', all possible interactions involving strong shocks take place inside the region $\widetilde{\Delta}$. We want to show now that, if $n_1, n_2 \geq 1$, indeed there is exactly one MSF present in the solution after τ', i.e. one has $n_1 = n_2 = 1$. To this end, observe that, by the properties of the wave-speed maps $\lambda_1^{\psi,\varepsilon}$, $\lambda_2^{\psi,\varepsilon}$ defined at (4.39), given any $1S$-wave $\sigma_1 > 0$ with left and right state v^{L^1}, v^{R^1}, and any $2S$-wave $\sigma_2 > 0$ with left and right state v^{L^2}, v^{R^2}, one has

(7.25)
$$\lambda_1^{1,\varepsilon+}(v^{L^1}) \geq \lambda_1^{\psi,\varepsilon}(v^{L^1}, \sigma_1) > \lambda_1^{1,\varepsilon+}(v^{R^1}) + C_\varepsilon,$$
$$\lambda_2^{2,\varepsilon+}(v^{R^2}) \leq \lambda_2^{\psi,\varepsilon}(v^{R^2}, \sigma_2) < \lambda_2^{2,\varepsilon+}(v^{L^2}) - C_\varepsilon,$$

for some constant $C_\varepsilon > 0$ depending on ε but not on v^{L^i}, v^{R^i} (see (3.34)-(3.35) in [**AM2**]). Similar estimates hold for iS-waves of negative size. Then, if $n_1 > 1$, because of (7.25) the strong shock along the MSF at x_{1,n_1-1} would eventually interact with some weak wave or with the strong shock along x_{1,n_1}, against the assumption. Therefore, it must be $n_1 = 1$ and, for similar reasons we find that $n_2 = 1$.

Step 3. The above analysis shows that, for $t > \tau$ sufficiently large, the solution v contains at most one MSF for each characteristic family. To conclude the proof of the lemma we are thus led to discuss the following three cases.

CASE 1: For $t \geq \tau$ sufficiently large the solution does not contain any strong shock, i.e. $n_1 = n_2 = 0$. Define the *weighted number of weak waves present at time t* as

$$\widetilde{N}(t) = \sum_{i=1,2} \sum_\alpha W_i(w^{L_\alpha^i}, w^{R_\alpha^i}) \qquad t \in [\tau, \infty[.$$

Here, the sum ranges over the set of all weak waves present in the solution at time t. First show, using $\widetilde{N}$, that the total number of interactions of type $1.4 - 2.4$ is finite. Then, since no strong shock is present in the solution after time τ, conclude as in the proof of Proposition 3.3 that the total number of wave-fronts and interaction points of any type in the t-x plane is finite as well.

CASE 2: For $t \geq \tau$ sufficiently large the solution contains only one strong shock. To fix the ideas, assume that this shock belongs to the first NGNL1 family and is located at $x = x_1(t)$ (the other case being entirely similar). Again, with similar arguments to those used in the proof of Lemma 7.1, one can first check that there is a finite number of interactions of type 1.2. Next, introducing the weights

$$\widetilde{P}_i(w^{L_\alpha^i}, w^{R_\alpha^i}) \doteq \begin{cases} W_1(w^{L_\alpha^1}, w^{R_\alpha^1})(N_\varepsilon + 1) & \text{if} \quad i = 1, \quad y_{1,\alpha}(t) < x_1(t) \\ W_1(w^{L_\alpha^1}, w^{R_\alpha^1}) & \text{if} \quad i = 1, \quad y_{1,\alpha}(t) > x_1(t) \\ W_2(w^{L_\alpha^2}, w^{R_\alpha^2})(N_\varepsilon + 1)^2 & \text{if} \quad i = 2, \quad y_{2,\alpha}(t) < x_1(t) \\ W_2(w^{L_\alpha^2}, w^{R_\alpha^2}) & \text{if} \quad i = 2, \quad y_{2,\alpha}(t) > x_1(t) \end{cases}$$

we define the functional

$$\widetilde{N}(t) \doteq \sum_{i=1,2} \sum_\alpha \widetilde{P}_i(w^{L_\alpha^i}, w^{R_\alpha^i})$$

where the sum ranges over the set of all weak waves present in the solution at time t. This functional can increase only when an interaction of type 1.2 occurs, and decreases at any time $t \geq \tau$ where an interaction of other type takes place. Therefore, we conclude as in Case 1 that the total number of interactions and wave fronts in the t-x plane is finite.

CASE 3: For $t \geq \tau$ sufficiently large the solution contains exactly two strong shocks (one for each family), located at $x_1(t) < x_2(t)$. By the same arguments of Case 2, we first check that the total number of interactions occurring outside the region

$$\widetilde{\Delta} \doteq \{(t,x) \; : \; t > \tau, \; x_1(t) < x < x_2(t)\}$$

is finite. Thus we deduce that, after some time τ' sufficiently large, no wave-front exists on the left of x_1 having speed larger than the shock at x_1 and no wave-front exists on the right of x_2 having speed smaller than the shock at x_2. Moreover, we may assume that, after τ', no interaction occurs outside the region $\widetilde{\Delta}$. Notice that, by construction, one has

$$\lambda_1^{\psi,\varepsilon}(w^L, w^R) > \lambda_1^{1,\varepsilon}(w^R)$$

whenever (w^L, w^R) is a 1W-wave with $\lfloor w^L \rfloor_1 \neq \lfloor w^R \rfloor_1$ (here we understand $\lambda_1^{1,\varepsilon}(w^R) = \lambda_1^{1,\varepsilon+}(w^R)$ if (w^L, w^R) has positive size, $\lambda_1^{1,\varepsilon}(w^R) = \lambda_1^{1,\varepsilon-}(w^R)$ otherwise). This and the above assumptions imply that, after time τ', each 1W-wave (w^L, w^R) at the left of $x_1(t)$ is a rarefaction wave-front. Of course, for similar reasons each 2W-wave (w^L, w^R) present at time $t > \tau'$ at the right of $x_2(t)$ is a rarefaction wave-front.

Therefore, for $t > \tau'$, the two states $v^L \doteq v(t, -\infty)$, $v(t, x_1(t)+)$ are connected by a 1-wave containing a strong shock of the first family, while the two states $v(t, x_2(t)-)$, $v^R \doteq v(t, +\infty)$ are connected by a 2-wave containing a strong shock of the second family. On the other hand, observe that by (7.25) every weak wave inside $\widetilde{\Delta}$ either hits one of the two strong shocks or else it cancels interacting with other weak waves of the same family. It follows that the interaction potential and the total strength of the waves inside the interval $]x_1(t), x_2(t)[$ both approach zero as $t \to +\infty$. Hence

$$(7.26) \qquad \lim_{t \to +\infty} \left| v(t, x_1(t)+) - v(t, x_2(t)-) \right| = 0.$$

Then, because of (7.26), and by the continuous dependence of the ε-solution of a Riemann problem on the initial data, we deduce that the solution ω of the Riemann problem with initial data v^L, v^R contains two strong shocks, against the assumptions. Hence Case 3 cannot occur, which completes the proof of Lemma 7.2. $\square$

PROOF OF PROPOSITION 3.4. Since $\bar{v}$ has bounded support, $\lim_{x \to +\infty} \bar{v}(x) = \lim_{x \to -\infty} \bar{v}(x) = 0$. The result then follows from Lemma 7.2. $\square$

REMARK 7.2. Observe that, relying on Lemma 7.2, one can easily show with the same arguments in [**BC1**] that Lemma 10 in [**BC1**] still holds. It follows that, if $v(t, x) = S_t^\varepsilon \tilde{v}(x)$, $\tilde{v} \in \tilde{\mathcal{D}}^\varepsilon$, and $\omega = \omega(t, x)$ is the ε-solution of the Riemann problem with initial data (7.21), then one has

$$(7.27) \qquad \lim_{\theta \to 0+} \int_{\mathbb{R}} \left| v\left(\frac{t}{\theta}, \frac{x}{\theta}\right) - \omega(t, x) \right| dx = 0 \qquad \forall\, t \geq 0.$$

8. Estimates on Shift Differentials

Aim of this section is to establish the $\mathbb{L}^1$ stability estimates on the ε-approximate solutions stated in Propositions 3.5-3.6. The same notations of Section 6.1 will be used throughout. Some more notations are introduced below. Let v be a piecewise constant function with jumps at $x_1, \ldots, x_N$. Assume that at each point x_α the Riemann problem is solved in terms of a single ε-admissible wave-front of size σ_α. Given an N-tuple of shift rates $\xi \doteq (\xi_1, \ldots, \xi_N)$ (representing displacements along the x-direction), for any pair of wave-fronts σ_α, σ_β in v, define the quantity
(8.1)

$$K_\xi(\sigma_\alpha, \sigma_\beta) \doteq \begin{cases} 0 & \text{if } \sigma_\alpha = \sigma_\beta, \\ K_3 |\sigma_\alpha \sigma_\beta| \left(|\xi_\alpha| + |\xi_\beta|\right) & \text{if } i_\alpha > i_\beta \text{ and } x_\alpha < x_\beta, \\ K_3 |\sigma_\alpha \sigma_\beta| \left(|\xi_\alpha| + |\xi_\beta|\right) & \text{if } i_\alpha = i_\beta \text{ and } \sigma_\alpha \sigma_\beta < 0, \\ K_2 |\sigma_\alpha \sigma_\beta| \left(|\xi_\alpha| + |\xi_\beta|\right) & \text{if } i_\alpha = i_\beta \text{ and } \sigma_\alpha \sigma_\beta > 0. \end{cases}$$

Then set

$$V_\xi(v) = V_\xi^\varepsilon(v) \doteq \sum_\alpha |\sigma_\alpha \xi_\alpha|,$$

$$(8.2) \qquad Q_\xi(v) = Q_\xi^\varepsilon(v) \doteq \sum_{(\sigma_\alpha, \sigma_\beta) \in \mathcal{A}} K_\xi(\sigma_\alpha, \sigma_\beta),$$

$$\Upsilon_\xi(v) = \Upsilon_\xi^\varepsilon(v) \doteq \left[V_\xi(v) + Q_\xi(v)\right] e^{K_4 Q^\varepsilon(v)}.$$

Here the summation defining Q_ξ ranges over all couples of *approaching* wave-fronts (cfr. Definition in Section 3) present in v, and K_2, K_3, K_4 are positive constants that will be appropriately chosen later, taking $K_3 > K_2$ in the case where at least one of the family is NGNL, and letting $K_2 = K_3$ in the classical case where both families are GNL or LD (cfr. Lemmas 8.9-8.10). With these conventions, the *weighted length* of the elementary path $\gamma : \theta \mapsto v^\theta$, $\theta \in\,]a, b[\,$, at (3.44)-(3.45), can be rewritten as

$$(8.3) \qquad \|\gamma\|_\varepsilon = (b - a) \cdot \Upsilon_\xi(v^\theta).$$

Now consider a family $\{v^\theta\}_{\theta \in\,]a,b[}$ of ε-approximate solutions (constructed as in Sections 3-4) satisfying the hypotheses of Proposition 3.5. In particular, let δ_2 be the constant given by Proposition 3.2. Assume that $v^\theta(t, \cdot) \in \mathcal{D}^{\varepsilon, 2\delta_2}$ for all θ, t, and that the wave-front configuration of each v^θ remains the same as the parameter θ ranges in $]a, b[$. Denote with $x_1^\theta(t) < \cdots < x_{N(t)}^\theta(t)$ the locations of the jumps in $v^\theta(t, \cdot)$. We want to study how the weighted length of the elementary path $\theta \mapsto v^\theta(t, \cdot)$ varies in time. To this purpose, for any given θ we consider the function

$$(8.4) \qquad t \mapsto \Upsilon_\xi(t) \doteq \Upsilon_{\xi(t)}\bigl(v^\theta(t, \cdot)\bigr)$$

where the $N(t)$-tuple of shift rates is given by

$$(8.5) \qquad \xi(t) = \bigl(\xi_1(t), \ldots, \xi_{N(t)}(t)\bigr), \qquad \xi_\alpha(t) = \frac{\partial x_\alpha^\theta(t)}{\partial \theta}.$$

In the same way we define the functions $t \mapsto V_\xi(t)$, $t \mapsto Q_\xi(t)$. Clearly, the map Υ_ξ in (8.4) is piecewise constant with discontinuities occurring only at the interaction times of v^θ. Our goal is to derive an a-priori bound on $\Upsilon_\xi(t)$, relying on the interaction estimates collected in Section 5. As in [**BC1**], one could try to show that, for a suitable choice of the constants K_2, K_3, K_4 in (8.2), assuming that the total variation of $v^\theta(t, \cdot)$ remains sufficiently small, the function at (8.4) be decreasing whenever two or more wave-fronts in v^θ interact. However, for approximate solutions v^θ of a NGNL system, the quantity Q_ξ and hence Υ_ξ may well increase after the same type of "bad" interactions pointed out at the beginning of Section 6 (see Example 6.1). Indeed, as it happens for Q, the summation defining Q_ξ contains all the products of wave-fronts belonging to the same family. Hence Q_ξ shall increase after any interaction that produces a composed wave consisting, say, of a rarefaction front σ^r and of a shock σ^s shifted at constant rates ξ^r, ξ^s, if the quantity $|\sigma^r \sigma^s|(|\xi^r|+|\xi^s|)$ is not comparable with the amount of "shift-interaction" $\sum_{i,j} |\sigma_i^- \sigma_j^-|(|\xi_i^-|+|\xi_j^-|)$ of the incoming fronts $\sigma_1^-, \ldots, \sigma_k^-$, shifted at constant rates $\xi_1^-, \ldots, \xi_k^-$.

Thus, in order to establish (3.47), we will use the same technique developed in Section 6. We shall first estimate the increase of $\Upsilon_\xi(t)$ at any time $t = t_0$ where an interaction of bad type takes place among the fronts of some $\mathcal{J}_0 \in \mathcal{B}(t_0)$. Then, we will show that such an increase of the functional Υ_ξ is counterbalanced by the decrease of Υ_ξ determined by (good) interactions occurring at later times $t > t_0$, and involving waves originated from $\mathcal{J}_0$. In this way we will prove that the total increase of the weighted length of the elementary path $\theta \mapsto v^\theta(t, \cdot)$, in any given time interval $[0, T]$, is uniformly bound by $\mathcal{O}(1)\, \Upsilon_\xi(T+) \cdot \varepsilon$.

REMARK 8.1. Throughout this section we will assume that:

1. at any time t, at most one interaction occurs;

2. each interaction involves at most two wave-fronts of v^θ.

This is not restrictive since, whenever three or more wave-fronts in v^θ interact at some point $P^\theta = (\bar{t}, \bar{x})$, for any fixed $\rho > 0$ one can construct a perturbed family of approximate solutions w_ρ^θ, defined on a sufficiently small strip $]\bar{t} - \delta, \bar{t} + \delta[$, that have the properties **1-2** and satisfy

$$(8.6) \qquad \begin{aligned} \Upsilon_{\xi(t)}\bigl(v^\theta(\bar{t}+\delta, \cdot)\bigr) - \rho &\leq \Upsilon_{\xi(t)}\bigl(w_\rho^\theta(\bar{t}+\delta, \cdot)\bigr), \\ \Upsilon_{\xi(t)}\bigl(w_\rho^\theta(\bar{t}-\delta, \cdot)\bigr) &\leq \Upsilon_{\xi(t)}\bigl(v^\theta(\bar{t}-\delta, \cdot)\bigr). \end{aligned}$$

Such perturbations can be obtained from v^θ by slightly changing the locations of the wave-fronts in v^θ, following the same procedure adopted in [**BC1**, Section 8].

Since, by Proposition 3.4, the set of interaction points of v^θ is finite, the above estimates, together with an a-priori bound on $\Upsilon_{\xi(t)}(w_\rho^\theta(t))$, yields an a-priori bound on $\Upsilon_{\xi(t)}(v^\theta(t))$ as desired.

8.1. Estimates on shifting interactions.

We consider here a point of interaction $P^0 = (t^0, x^0)$ between two wave-fronts of an approximate solution v of a system satisfying the assumption (**A**) (constructed as in Sections 3-4), and we study how the shifts (i.e. displacements in the x-direction) of the incoming fronts are related to the shifts of the outgoing ones. Let

$$x_1(t) = x^0 + \Lambda_1(t - t^0), \qquad x_2(t) = x^0 + \Lambda_2(t - t^0), \qquad \Lambda_1 < \Lambda_2,$$

be the locations of the two incoming fronts that interact together at P^0. Consider a family of approximate solutions $\theta \mapsto v^\theta$ obtained by shifting these two fronts at constant rates ξ_1, ξ_2. The corresponding wave-fronts in v^θ are thus located at

$$x_1^\theta(t) = x^0 + \Lambda_1(t - t^0) + \theta\xi_1, \qquad x_2^\theta(t) = x^0 + \Lambda_2(t - t^0) + \theta\xi_2,$$

and interact at the point $P^\theta = (t^\theta, x^\theta)$ given by

$$t^\theta = t^0 + \frac{\xi_1 - \xi_2}{\Lambda_2 - \Lambda_1}\theta, \qquad x^\theta = x^0 + \frac{\Lambda_2\xi_1 - \Lambda_1\xi_2}{\Lambda_2 - \Lambda_1}\theta.$$

It follows that a wave-front emerging from P^θ with speed Λ^+ will be shifted along the x-direction by the amount $\xi^+\theta$ with

(8.7) $$\xi^+ = \frac{(\Lambda^+ - \Lambda_1)\xi_2 - (\Lambda^+ - \Lambda_2)\xi_1}{\Lambda_2 - \Lambda_1}.$$

The next lemma provides the basic estimate on the change of the distance between two nearby approximate solutions (obtained one from the other by shifting the positions of the jumps) after an interaction time.

LEMMA 8.1. *Assume that the characteristic families are either GNL, or LD, or NGNL, and consider two wave-fronts with size σ', σ'' that interact together (σ' located on the left of σ''). Suppose that σ', σ'' are shifted by ξ', ξ'' along the x-direction. Let $\sigma_{i,\ell}^+$, $\ell = 1, \ldots, n_i$, $i = 1, 2$, be the sizes of the outgoing fronts, and call $\xi_{i,\ell}^+$ their displacements. Then, there exists some constant $C_{20} > 0$, independent of ε, σ', σ'', such that the following hold.*

1) *If the incoming fronts σ', σ'' belong to distinct characteristic families, one has*

(8.8)
$$\left[\sum_{\ell=1}^{n_1} |\sigma_{1,\ell}^+ \xi_{1,\ell}^+| - |\sigma''\xi''|\right]^+ + \left[\sum_{\ell=1}^{n_2} |\sigma_{2,\ell}^+ \xi_{2,\ell}^+| - |\sigma'\xi'|\right]^+ \leq C_{20}|\sigma'\sigma''|(|\xi'| + |\xi''|).$$

Moreover, if the outgoing wave of the first family consists of a single front σ_1^+, then one has

(8.9) $$\left|\sigma_1^+ \xi_1^+ - \sigma''\xi''\right| \leq C_{20}|\sigma'\sigma''|(|\xi'| + |\xi''|).$$

The same type of estimate holds in the case the outgoing wave of the second family consists of a single front σ_2^+.

II) *If both incoming fronts σ', σ'' belong to the i-th family, one has*

(8.10)
$$\left[\sum_{\ell=1}^{n_i}|\sigma_{i,\ell}^+\xi_{i,\ell}^+| - |\sigma'\xi'| - |\sigma''\xi''|\right]^+ + \sum_{\ell=1}^{n_j}|\sigma_{j,\ell}^+\xi_{j,\ell}^+| \leq C_{20}|\sigma'\sigma''|\left(|\xi'|+|\xi''|\right)$$

(here $[a]^+ = \max\{a,0\}$ denotes as usual the positive parts of a, and clearly $j \neq i$ in the second summation of (8.10)). Moreover, if the outgoing wave of the first family consists of a single front σ_1^+, then one has

(8.11) $$\left|\sigma_1^+\xi_1^+ - (\sigma'\xi' + \sigma''\xi'')\right| \leq C_{20}|\sigma'\sigma''|\left(|\xi'|+|\xi''|\right).$$

The same type of estimate holds in the case the outgoing wave of the second family consists of a single front σ_2^+.

PROOF.
1. Assume that the incoming fronts σ', σ'' belong to different characteristic families: σ' to the second family and σ'' to the first one. Call Λ', Λ'' the speeds of σ', σ'', and let $\Lambda_{i,\ell}^+$ be the speeds of the outgoing wave-fronts $\sigma_{i,\ell}^+$, $\ell=1,\ldots,n_i$, $i=1,2$. Observe that, if $\max\{|\sigma'|,|\sigma''|\} \leq \sqrt{\varepsilon}$, by Lemma 5.7 one has $\sum_{\ell=1}^{n_2}\sigma_{2,\ell}^+ = \sigma'$, $\sum_{\ell=1}^{n_1}\sigma_{1,j}^+ = \sigma''$, and each component of the approximate solution of the corresponding Riemann problem coincides with the exact (weak) solution of a scalar conservation law (see Remark 3.1 for the NGNL families, and [**BC1**, Remark 3] for the GNL or LD ones). The $\mathbb{L}^1$ contractive property of the semigroup generated by a scalar conservation law [**K**] then implies

$$\sum_{\ell=1}^{n_2}|\sigma_{2,\ell}^+\xi_{2,\ell}^+| \leq |\sigma'\xi'|, \qquad \sum_{\ell=1}^{n_1}|\sigma_{1,\ell}^+\xi_{1,\ell}^+| \leq |\sigma''\xi''|,$$

which proves (8.8). On the other hand, in the case the outgoing wave of the first family consists of a single front σ_1^+, one obtains (8.9) by a direct computation recalling that the wave speed of σ_1^+ is the averaged characteristic speed defined at (3.3) (see Remark 3.3). Next, assume that

(8.12) $$\max\{|\sigma'|,|\sigma''|\} > \sqrt{\varepsilon}$$

and consider the outgoing wave-fronts of the first family emerging from the interaction (the same analysis holds for the outgoing wave-fronts of the second family). Two cases can occur.

CASE 1: $n_1 = 1$, i.e. the outgoing wave of the first family consists of a single front. Set $\sigma_1^+ \doteq \sigma_{1,1}^+$, $\xi_1^+ \doteq \xi_{1,1}^+$ and $\Lambda_1^+ \doteq \Lambda_{1,1}^+$. By Lemmas 5.7, 5.10, it follows that

(8.13) $$|\sigma_1^+ - \sigma''| = \mathcal{O}(1)|\sigma'\sigma''|, \qquad |\Lambda_1^+ - \Lambda''| = \mathcal{O}(1)|\sigma'|.$$

On the other hand, the strict hyperbolicity assumption (2.1) implies

(8.14) $$\frac{1}{|\Lambda'-\Lambda''|} \leq \frac{1}{\lambda^{\min}}.$$

Thus, using the expression of ξ_1^+ given by (8.7), with $\Lambda_2 = \Lambda'$, $\Lambda_1 = \Lambda''$, and relying on (8.13)-(8.14), we get

$$
\begin{aligned}
|\sigma_1^+ \xi_1^+ - \sigma'' \xi''| &\leq |\sigma''||\xi_1^+ - \xi''| + |\sigma_1^+ - \sigma''||\xi_1^+| \\
&= |\sigma''|\left|\frac{(\Lambda_1^+ - \Lambda'')\xi' - (\Lambda_1^+ - \Lambda')\xi''}{\Lambda' - \Lambda''} - \xi''\right| + \\
&\quad + \mathcal{O}(1)|\sigma' \sigma''|\left|\frac{(\Lambda_1^+ - \Lambda'')\xi' - (\Lambda_1^+ - \Lambda')\xi''}{\Lambda' - \Lambda''}\right| \\
&\leq |\sigma''|\left|\frac{\Lambda_1^+ - \Lambda''}{\Lambda' - \Lambda''} \cdot (\xi' - \xi'')\right| + \mathcal{O}(1)|\sigma' \sigma''|(|\xi'| + |\xi''|) \\
&= \mathcal{O}(1)|\sigma' \sigma''|(|\xi'| + |\xi''|)
\end{aligned}
$$
(8.15)

proving (8.9). Clearly, the same type of estimate as (8.15) holds for $|\sigma_2^+ \xi_2^+ - \sigma'\xi'|$ in the case the outgoing wave of the second family consists of a single front. Hence, in the case the outgoing waves of both families consist of a single front, we recover (8.8) from (8.9).

CASE 2: The outgoing wave of the first family σ_1^+ consists of $n_1 > 1$ wavefronts. For definiteness, assume that the first family is NGNL and that σ_1^+ is a composed wave consisting of $n_1 - 1$ rarefaction fronts, say $\sigma_{1,\ell}^+$, $1 \leq \ell < n_1$, and of a shock σ_{1,n_1}^+ (the analysis of the case where σ_1^+ consists only of rarefaction fronts is entirely similar). Applying Lemmas 5.7, 5.10, and because of (8.14), we deduce

(8.16) $\quad \sigma_1^+ = \sigma'' + \mathcal{O}(1)|\sigma'\sigma''|, \qquad \left|\frac{\Lambda'' - \Lambda_{1,\ell}^+}{\Lambda' - \Lambda''}\right| = \mathcal{O}(1)|\sigma'| \qquad \forall\, 1 \leq \ell \leq n_1.$

Hence, using the expression of $\xi_{1,\ell}^+$ given by (8.7), with $\Lambda_2 = \Lambda'$, $\Lambda_1 = \Lambda''$, and relying on (8.16), we get

$$
\begin{aligned}
\sum_{\ell=1}^{n_1} |\sigma_{1,\ell}^+ \xi_{1,\ell}^+| &= \sum_{\ell=1}^{n_1} |\sigma_{1,\ell}^+|\left|\frac{(\Lambda_{1,\ell}^+ - \Lambda'')\xi' - (\Lambda_{1,\ell}^+ - \Lambda')\xi''}{\Lambda' - \Lambda''}\right| \\
&\leq |\sigma_1^+ \xi''| + \sum_{\ell=1}^{n_1} |\sigma_{1,\ell}^+|\left|\frac{\Lambda_{1,\ell}^+ - \Lambda''}{\Lambda' - \Lambda''} \cdot (\xi' - \xi'')\right| \\
&\leq |\sigma'' \xi''| + \mathcal{O}(1)|\sigma'\sigma''|(|\xi'| + |\xi''|).
\end{aligned}
$$
(8.17)

Clearly, the same type of estimate as (8.17) holds for $\sum_{\ell=1}^{n_2} |\sigma_{2,\ell}^+ \xi_{2,\ell}^+|$ in the case the outgoing wave of the second family σ_2^+ is a composed wave consisting of a shock and of a piecewise constant rarefaction fan, and hence we obtain (8.8).

2. Assume that both incoming fronts σ', σ'' belong to the first characteristic family (the case that σ', σ'' belong to the second characteristic family being entirely similar). As in **1**, call Λ', Λ'' the speeds of σ', σ'', and let $\Lambda_{i,\ell}^+$ be the speeds of the outgoing wave-fronts $\sigma_{i,\ell}^+$, $\ell = 1, \ldots, n_i$, $i = 1, 2$. Observe that, if $\max\{|\sigma'|, |\sigma''|\} \leq \sqrt{\varepsilon}$, by Lemma 5.1 one has $\sum_{\ell=1}^{n_1} \sigma_{1,\ell}^+ = \sigma' + \sigma''$, $\sigma_2^+ = 0$, and the first component of the approximate solution of the corresponding Riemann problem coincides with the exact (weak) solution of a scalar conservation law (see Remark 3.1 if the first family is NGNL, and [**BC1**, Remark 3] in the case the first family is GNL or LD). The $\mathbb{L}^1$ contractive property of the semigroup generated by a scalar conservation law [**K**] then implies

$$\sum_{\ell=1}^{n_1} |\sigma_{1,\ell}^+ \xi_{1,\ell}^+| \leq |\sigma' \xi'| + |\sigma'' \xi''|$$

which proves (8.10). On the other hand, in the case the outgoing wave of the first family consists of a single front σ_1^+, one obtains (8.11) by a direct computation recalling that the wave speed of σ_1^+ is the averaged characteristic speed defined at (3.3) (see Remark 3.3). Next, assume that (8.12) holds and observe that, by Lemma 5.1, we have

(8.18) $$\sum_{\ell=1}^{n_2} |\sigma_{2,\ell}^+| = \mathcal{O}(1)|\sigma'\sigma''||\Lambda' - \Lambda''|.$$

Hence, using the expression of $\xi_{2,\ell}^+$ given by (8.7), with $\Lambda_2 = \Lambda'$, $\Lambda_1 = \Lambda''$, and relying on (8.18), we get

(8.19)
$$\sum_{\ell=1}^{n_2} |\sigma_{2,\ell}^+ \xi_{2,\ell}^+| = \sum_{\ell=1}^{n_2} \left| \sigma_{2,\ell}^+ \frac{(\Lambda_{2,\ell}^+ - \Lambda'')\xi' - (\Lambda_{2,\ell}^+ - \Lambda')\xi''}{\Lambda' - \Lambda''} \right|$$
$$= \mathcal{O}(1) \sum_{\ell=1}^{n_2} |\sigma_{2,\ell}^+| \left| \frac{|\xi'| + |\xi''|}{\Lambda' - \Lambda''} \right|$$
$$= \mathcal{O}(1)|\sigma'\sigma''|\big(|\xi'| + |\xi''|\big).$$

Regarding the wave-fronts of the first family emerging from the interaction, two cases can occur:

CASE 1: $n_1 = 1$, i.e. the outgoing wave of the first family consists of a single front. Set $\sigma_1^+ \doteq \sigma_{1,1}^+$, $\xi_1^+ \doteq \xi_{1,1}^+$ and $\Lambda_1^+ \doteq \Lambda_{1,1}^+$. By Lemma 5.1 and Lemmas 5.4-5.6, it follows that

(8.20) $$\sigma_1^+ = \sigma' + \sigma'' + \mathcal{O}(1)|\sigma'\sigma''| \cdot |\Lambda' - \Lambda''|,$$

and

(8.21) $$\left|\frac{\Lambda_1^+ - \Lambda'}{\Lambda' - \Lambda''}\right| \leq \mathcal{O}(1), \qquad \left|\frac{\Lambda_1^+ - \Lambda''}{\Lambda' - \Lambda''}\right| \leq \mathcal{O}(1),$$
$$\left|\frac{(\sigma' + \sigma'') \cdot \Lambda_1^+ - \sigma' \cdot \Lambda' - \sigma'' \cdot \Lambda''}{\Lambda' - \Lambda''}\right| \leq \mathcal{O}(1)|\sigma'\sigma''|.$$

Hence, using the expression of ξ_1^+ given by (8.7), with $\Lambda_2 = \Lambda'$, $\Lambda_1 = \Lambda''$, and relying on (8.20)-(8.21), we deduce
(8.22)
$$|\sigma_1^+ \xi_1^+ - (\sigma' \xi' + \sigma'' \xi'')| \leq \left|\sigma'(\xi_1^+ - \xi') + \sigma''(\xi_1^+ - \xi'')\right| + \left|\sigma_1^+ - (\sigma' + \sigma'')\right| |\xi_1^+|$$
$$= \left|\frac{\sigma'(\Lambda_1^+ - \Lambda') + \sigma''(\Lambda_1^+ - \Lambda'')}{\Lambda' - \Lambda''}(\xi' - \xi'')\right| +$$
$$+ \mathcal{O}(1)|\sigma' \sigma''||\Lambda' - \Lambda''|\left|\frac{\Lambda_1^+ - \Lambda''}{\Lambda' - \Lambda''}\xi' - \frac{\Lambda_1^+ - \Lambda'}{\Lambda' - \Lambda''}\xi''\right|$$
$$= \mathcal{O}(1)|\sigma' \sigma''|(|\xi'| + |\xi''|)$$
proving (8.11). In turn (8.11), together with (8.19), yield (8.10).

CASE 2: The outgoing wave of the first family σ_1^+ consists of $n_1 > 1$ wavefronts, and hence the first family is NGNL. For definiteness, assume in particular that the first family is NGNL1 (the case of a NGNL2 family being entirely similar). In this case, by the properties of the approximate Riemann solver described in Section 3, the front incoming from the left σ' must be a shock, while σ'' is a rarefaction front discordant with σ'. Hence, we have $|\sigma''| \leq \varepsilon$ and $|\sigma'| > \sqrt{\varepsilon}$. Moreover, by Remark 5.2, one has $n_1 = 2$, i.e. the outgoing wave of the first family σ_1^+ consists of exactly two fronts: a rarefaction front $\widetilde{\sigma}_1^+ \doteq \sigma_{1,1}^+$, followed by a shock $\widehat{\sigma}_1^+ \doteq \sigma_{1,n_1}^+$. Set $\widetilde{\Lambda}_1^+ \doteq \Lambda_{1,1}^+$, $\widehat{\Lambda}_1^+ \doteq \Lambda_{1,2}^+$. Then, by Lemmas 5.1-5.2, it follows that

(8.23)
$$|\sigma'| = |\widetilde{\sigma}_1^+| + |\widehat{\sigma}_1^+| + |\sigma''| + \mathcal{O}(1)|\sigma' \sigma''| \cdot |\Lambda' - \Lambda''|,$$
$$|\widetilde{\sigma}_1^+| = \mathcal{O}(1)|\sigma'|^2, \qquad |\widehat{\sigma}_1^+| = \mathcal{O}(1)|\sigma'|,$$

while Lemma 5.5 implies
(8.24)
$$\left|\frac{\Lambda' - \widetilde{\Lambda}_1^+}{\Lambda' - \Lambda''}\right| = \mathcal{O}(1)\left|\frac{\sigma''}{\sigma'}\right|, \qquad \left|\frac{\Lambda'' - \widetilde{\Lambda}_1^+}{\Lambda' - \Lambda''}\right| \leq 1 + \mathcal{O}(1)\left|\frac{\sigma''}{\sigma'}\right|,$$
$$\left|\frac{\Lambda' - \widehat{\Lambda}_1^+}{\Lambda' - \Lambda''}\right| \leq \left|\frac{\sigma''}{\sigma'}\right| + \mathcal{O}(1)|\sigma''|, \qquad \left|\frac{\Lambda'' - \widehat{\Lambda}_1^+}{\Lambda' - \Lambda''}\right| \leq \frac{|\widehat{\sigma}_1^+| + |\sigma''|}{|\widehat{\sigma}_1^+|} + \mathcal{O}(1)|\sigma''|.$$

Hence, using the expression of ξ_1^+ given by (8.7), with $\Lambda_2 = \Lambda'$, $\Lambda_1 = \Lambda''$, and relying on (8.23)-(8.24), we obtain

$$|\sigma_{1,1}^+ \xi_{1,1}^+| + |\sigma_{1,2}^+ \xi_{1,2}^+| \leq |\widetilde{\sigma}_1^+|\left(\left|\frac{\Lambda'' - \widetilde{\Lambda}_1^+}{\Lambda' - \Lambda''}\right||\xi'| + \left|\frac{\Lambda' - \widetilde{\Lambda}_1^+}{\Lambda' - \Lambda''}\right||\xi''|\right) +$$
$$+ |\widehat{\sigma}_1^+|\left(\left|\frac{\Lambda'' - \widehat{\Lambda}_1^+}{\Lambda' - \Lambda''}\right||\xi'| + \left|\frac{\Lambda' - \widehat{\Lambda}_1^+}{\Lambda' - \Lambda''}\right||\xi''|\right)$$
$$\leq \left(|\widetilde{\sigma}_1^+| + |\widehat{\sigma}_1^+| + |\sigma''|\right)|\xi'| + \frac{|\widehat{\sigma}_1^+ \cdot \sigma''|}{|\sigma'|}|\xi''| +$$
$$+ \mathcal{O}(1)\left(\frac{|\widetilde{\sigma}_1^+ \cdot \sigma''|}{|\sigma'|} + |\widehat{\sigma}_1^+ \cdot \sigma''|\right)(|\xi'| + |\xi''|)$$

(8.25) $$\leq |\sigma' \xi'| + |\sigma'' \xi''| + \mathcal{O}(1)|\sigma' \sigma''|(|\xi'| + |\xi''|).$$

Together, (8.19) and (8.25) yield (8.10), which completes the proof of the lemma. □

8.2. Basic notations.

Consider an elementary path of approximate solutions $\theta \mapsto v^\theta(t,\cdot)$ satisfying the assumptions of Proposition 3.5, and let $\Upsilon_\xi(t)$ be the map in (8.4). We shall first derive the basic estimates on the change in the values of $\Upsilon_\xi(t)$ across times where interactions between two wave-fronts of v^θ take place, and then we will establish an a-priori bound on the increase of $\Upsilon_\xi(t)$ caused by the occurrence of some *interaction of bad type* (as defined in Section 6.1). Throughout the section, we will use the same notations introduced in Section 6.1, setting $\mathcal{J} \in \mathcal{B}^i(t)$, if the i-th family is NGNL and $\mathcal{J}$ is a *pair of fronts of bad type* (PFB) for the i-th family that interact at t (and letting $\mathcal{B}^i(t) \doteq \emptyset$ if either the i-th family is GNL, or LD, or the i-th family is NGNL but no interaction of bad type for the i-th family occurs at t), calling $\mathcal{N}^i(\mathcal{J},t)$ the set of i-fronts originated in the time interval $[t_0, t]$ by a given set of fronts $\mathcal{J}$ present in v^θ at t_0, denoting by $\mathcal{I}^i(\mathcal{J},t)$ the set of i-fronts interacting with $\mathcal{J}$ at t, and letting $\mathcal{C}^i(t)$ be the set of i-fronts canceled by an interaction occurring at t. For any pair of fronts $\mathcal{J}_0 \in \mathcal{B}(t_0)$, we will give two definitions of "*interaction time of good type*" for $\mathcal{J}_0$, that correspond to the smallest time intervals $[t_0, \widehat{t}_0]$ and $[t_0, \widehat{t}_0]$ along which the functionals $\Upsilon_\xi(t)$ and $V_\xi(t) + Q_\xi(t)$ decrease by a quantity comparable with the increases $\Upsilon_\xi(t_0+) - \Upsilon_\xi(t_0-)$ and $Q_\xi(t_0+) - Q_\xi(t_0-)$ caused by the bad interaction at t_0.

DEFINITION 8.1. Assume that the first characteristic family is NGNL1, and let $\mathcal{J}_0$ be a pair of fronts of bad type for the first family that interact together at t_0. We will call **interaction time of 1st good type** (ITG1) for $(\mathcal{J}_0, t_0)$ the first time $\widehat{t}_0 > t_0$ at which it occurs an interaction that *either* satisfies one of the five conditions of Definition 6.1, in which we replace the properties I.III) and II.III) of conditions I) and II) by:

I.III)' the wave $\sigma_{\widehat{t}_0}$ of the first family emerging from $(\widehat{t}_0, x_{\widehat{t}_0})$ is concordant with σ_0 and the total strength of waves of the first family that have interacted with fronts σ_h of $\mathcal{N}_S^1(\mathcal{J}_0, t)$, $t_0 \leq t \leq \widehat{t}_0$, is such that either

(8.26) $$\sum_{\substack{\sigma'_h \in \mathcal{I}^1(\sigma_h, t_h),\ 0<h\leq H \\ \sigma'_h \cdot \sigma_0 < 0}} |\sigma'_h| > \frac{e^{K_4 Q(t_0+)}}{16 C_{10}} \cdot \varepsilon,$$

or

(8.27) $$\sum_{\substack{\sigma'_h \in \mathcal{I}^1(\sigma_h, t_h),\ 0<h\leq H \\ \sigma'_h \cdot \sigma_0 > 0}} |\sigma'_h| > (20\, C_{10})^2\, e^{K_4 Q(t_0+)} \cdot \varepsilon,$$

(where C_{10} is the constant in (6.2), and K_4 is the constant in (8.2), cfr. Lemmas 8.9-8.10);

II.III)′ the total strength of waves of the second family that have interacted with fronts σ_h of $\mathcal{N}_S^1(\mathcal{J}_0, t)$, $t_0 \leq t \leq \widehat{t}_0$, is such that

$$(8.28) \qquad \sum_{\substack{\sigma_h' \in \mathcal{I}^2(\sigma_h, t_h), \\ 0 < h \leq H}} |\sigma_h'| > \frac{e^{K_4 Q(t_0+)}}{16 C_{10}} \cdot \varepsilon ;$$

or satisfies the following condition:

VI) the value of the functional Υ_ξ decreases in the interval $[t_0, \widehat{t}_0]$ by one fourth of the value assumed at t_0, i.e. there holds

$$(8.29) \qquad \Upsilon_\xi(\widehat{t}_0+) \leq \frac{\Upsilon_\xi(t_0-)}{4} .$$

As in Section 6.1, we will say that the interaction occurring at the ITG1 $\widehat{t}_0$ is an **interaction of 1st good type** for $(\mathcal{J}_0, t_0)$, and entirely similar definitions are given in the case we are considering a family of fronts of bad type for any k-th NGNL characteristic family that satisfies either one of (2.8)-(2.9) (i.e. that is either NGNL1 or NGNL2).

REMARK 8.2. Assume that the first family is NGNL1, let $\mathcal{J}_0 \in \mathcal{B}^1(t_0)$ be a pair of fronts that have an interaction of bad type for the first family at t_0, and let $\widehat{t}_0$ be the ITG1 for $(\mathcal{J}_0, t_0)$. Comparing the Definitions 6.1 and 8.1 of ITG and ITG1 for $(\mathcal{J}_0, t_0)$, it follows that the same conclusions derived in Remark 6.3 for the ITG apply to the ITG1 as well. Thus, in particular, the set $\mathcal{N}_S^1(\mathcal{J}_0, t)$ consists of a single shock σ_t^s for all $t < \widehat{t}_0$, and, in the case $\widehat{t}_0$ satisfies condition V) of Definition 8.1 (i.e. of Definition 6.1) but none of the other conditions I)-IV), the set $\mathcal{N}_R^1(\mathcal{J}_0, t)$, $t < \widehat{t}_0$ consists of a single rarefaction front of size $\sigma_t^r = \sigma_0^r$, concordant with σ_t^s. Moreover, in any case, the fronts of the first family cannot interact from the right with the fronts in $\mathcal{N}_R^1(\mathcal{J}_0, t)$ and from the left with the front in $\mathcal{N}_S^1(\mathcal{J}_0, t)$ in the time interval $]t_0, \widehat{t}_0[$. On the other hand, condition VI) of Definition 8.1 ensures that

$$(8.30) \qquad \Upsilon_\xi(t) > \frac{\Upsilon_\xi(t_0-)}{4} \qquad \forall\, t_0 < t < \widehat{t}_0 .$$

Clearly, similar properties hold in the case $\mathcal{J}_0 \in \mathcal{B}^k(t_0)$ for any k-th NGNL family that satisfies either one of (2.8)-(2.9).

DEFINITION 8.2. Assume that the first characteristic family is NGNL1, and let $\mathcal{J}_0$ be a pair of fronts of bad type for the first family that interact together at t_0. We will call **interaction time of 2nd good type (ITG2)** for $(\mathcal{J}_0, t_0)$ the first time $\widehat{t}_0' > t_0$ at which it occurs an interaction that satisfies one the following two conditions.

VII) The interaction point $(\widehat{t}_0', x_{\widehat{t}_0'})$ is connected with (t_0, x_0) by a polygonal line as in Definition 6.1-I), having properties I.I), I.II), and

    VII.III) Let ξ_1 be the shift rate of the outgoing shock σ_1 generated by the bad interaction at t_0, and let C_{10} be the constant at (6.2). Then, the total amount of (front)·(shift) products of waves of both families that interact with fronts $\sigma_h \in \mathcal{N}_S^1(\mathcal{J}_0, t_h)$, $t_0 < t_h \leq \widehat{t}_0$, satisfies one of the following three estimates:

(8.31)
$$\sum_{\substack{\sigma_h' \in \mathcal{I}^1(\sigma_h, t_h),\ 0<h\leq H \\ (\sigma_h \xi_h)\cdot(\sigma_h' \xi_h')>0}} |\sigma_h' \xi_h'| > (40 C_{10})^2 \cdot e^{K_4 Q(t_0+)} |\xi_1| \cdot \varepsilon,$$

(8.32)
$$\sum_{\substack{\sigma_h' \in \mathcal{I}^2(\sigma_h, t_h) \\ 0<h\leq H}} |\sigma_h' \xi_h'| > (40 C_{10})^2 \cdot e^{K_4 Q(t_0+)} |\xi_1| \cdot \varepsilon,$$

(8.33)
$$\sum_{\substack{\sigma_h' \in \mathcal{I}(\sigma_h, t_h) \\ 0<h\leq H}} |\sigma_h' \xi_h| > (40 C_{10})^2 \cdot e^{K_4 Q(t_0+)} |\xi_1| \cdot \varepsilon,$$

(where ξ_h, ξ_h' are the shift rates of σ_h, σ_h' and K_4 is the constant at (8.2), cfr. Lemmas 8.9-8.10). This property guarantees that the decrease of the shift-interaction functional Q_ξ at (8.2) caused by interactions occurring in the time interval $[t_0, \widehat{t}_0]$ is of order $2K_2|\sigma_1 \xi_1| \cdot \varepsilon \approx Q_\xi(t_0+) - Q_\xi(t_0-)$.

VIII) The interaction point $(\widehat{t}_0, x_{\widehat{t}_0})$ is connected with (t_0, x_0) by a polygonal line as in Definition 6.1-I), having properties i.i), i.ii), and

VIII.III) Let ξ_1 be as in VI.III). Then, the total amount of (front)·(shift) products of 1-waves that interact with fronts $\sigma_h \in \mathcal{N}_S^1(\mathcal{J}_0, t_h)$, $t_0 < t_h \leq \widehat{t}_0$, is such that

(8.34)
$$\sum_{\substack{\sigma_h' \in \mathcal{I}^1(\sigma_h, t_h),\ 0<h\leq H \\ (\sigma_h \xi_h)\cdot(\sigma_h' \xi_h')<0}} |\sigma_h' \xi_h'| > \frac{e^{K_4 Q(t_0+)}}{12} |\xi_1| \cdot \varepsilon.$$

This property guarantees that the decrease of the functional V_ξ at (8.2) caused by interactions occurring in the time interval $[t_0, \widehat{t}_0]$ between fronts of the first family is of order $2K_2|\sigma_1 \xi_1| \cdot \varepsilon \approx Q_\xi(t_0+) - Q_\xi(t_0-)$.

As for the ITG1, we will say that the interaction occurring at the ITG2 $\widehat{t}_0$ is an **interaction of 2nd good type** for $(\mathcal{J}_0, t_0)$, and entirely similar definitions are given in the case we are considering a family of fronts of bad type for any k-th NGNL characteristic family that is either NGNL1 or NGNL2.

We next introduce the analog of the Glimm-type interaction functionals defined in Section 6.1, which are modified for the presence of the shift rates and of the exponential term in Υ_ξ. Given a pair of fronts $\mathcal{J}_0 \in \mathcal{B}^i(t_0)$, $i \in \{1, 2\} \cap NGNL$, let ξ_0^r, ξ_0^s be the shift rates of the rarefaction front $\sigma_0^r \in \mathcal{N}_R^i(\mathcal{J}_0, t_0+)$, and of the shock $\sigma_0^s \in \mathcal{N}_S^i(\mathcal{J}_0, t_0+)$ of the i-th composed wave generated at t_0 (cfr. Remark 6.1). Moreover, in the case the ITG1 $\widehat{t}_0$ for $(\mathcal{J}_0, t_0)$ satisfies condition v) of Definition 8.1, let ξ_t^r, ξ_t^s denote the shift rates of the single rarefaction front $\sigma_t^r \in \mathcal{N}_R^i(\mathcal{J}_0, t)$ and of the shock $\sigma_t^s \in \mathcal{N}_S^i(\mathcal{J}_0, t)$ (cfr. Remark 8.2). Then, letting $\widehat{t}_0$ be the ITG2 for $(\mathcal{J}_0, t_0)$, and letting $E_i(\mathcal{J}_0; t)$ be the quantity in (6.11), we define

$$\widehat{\mathcal{E}}_{i,\xi}(\mathcal{J}_0; t) \doteq \begin{cases} E_i(\mathcal{J}_0; t) \cdot \min\{|\xi_t^r|, |\xi_t^s|\} \cdot e^{K_4 Q(t+)} & \text{if } \begin{cases} \widehat{t}_0 \text{ satisfies cond. v)} \\ \text{of Def. 8.1} \\ \text{and } \widehat{t}_0' \geq \widehat{t}_0, \end{cases} \\ E_i(\mathcal{J}_0; t_0) \cdot \min\{|\xi_0^r|, |\xi_0^s|\} \cdot e^{K_4 Q(t_0+)} & \text{otherwise}, \end{cases}$$

for any $t \in [t_0, \widehat{\tau}_0]$, $\widehat{\tau}_0 \doteq \min\{\widehat{t}_0, \widehat{t}_0'\}$, and we set $\widehat{\mathcal{E}}_{i,\xi}(\mathcal{J}_0; t) \doteq 0$ for all $t \in]\widehat{\tau}_0, \widehat{t}_0]$. Next, letting K_2, K_4 be the constants in (8.1)-(8.2), we define

$$(8.35) \quad \mathcal{E}_i(\mathcal{J}_0; t) \doteq \begin{cases} K_4 E_i(\mathcal{J}_0; t) \Upsilon_\xi(t-) + \\ + 2K_2 \widehat{\mathcal{E}}_{i,\xi}(\mathcal{J}_0; t) \end{cases} \quad \text{if } \begin{cases} \widehat{t}_0 \text{ satisfies cond. v)} \\ \text{of Def. 8.1} \end{cases} \\ \begin{cases} K_4 E_i(\mathcal{J}_0; t_0) \Upsilon_\xi(t_0-) + \\ + 2K_2 \widehat{\mathcal{E}}_{i,\xi}(\mathcal{J}_0; t) \end{cases} \quad \text{otherwise}, \end{cases}$$

for any $t \in [t_0, \widehat{t}_0]$, and we set $\mathcal{E}_{i,\xi}(\mathcal{J}_0; t) = \mathcal{E}_i(\mathcal{J}_0; t) \doteq 0$, for all $t > \widehat{t}_0$. On the other hand, if $\mathcal{J}_0$ is a pair of fronts that interact together at t_0 which is not PFB for the i-th family, i.e. if $\mathcal{J}_0 \notin \mathcal{B}^i(t_0)$, we set $\widehat{\mathcal{E}}_{i,\xi}(\mathcal{J}_0; t) = \mathcal{E}_i(\mathcal{J}_0; t) \doteq 0$, for all $t \geq t_0$. This means, in particular, that if the i-th family is GNL or LD, then the quantities $\widehat{\mathcal{E}}_{i,\xi}(\mathcal{J}_0; t), \mathcal{E}_i(\mathcal{J}_0; t)$ are always zero for any pair of interacting fronts $\mathcal{J}_0$, and for all t. The quantity $\mathcal{E}_i(\mathcal{J}_0; t)$ in (8.35) provides an upper bound for the increase of Υ_ξ occurring across t_0 due to the self-"shift interactions" among the new fronts of the i-th family generated at t_0. Instead, the functionals measuring the decrease of Υ_ξ caused by the (good) interactions occurring at later times $\tau > t_0$ and involving fronts of the i-th family originated from $\mathcal{J}_0$, are defined as follows.

For any pair of fronts σ', σ'' with shift rates ξ', ξ'', and for any time t, set

$$(8.36) \quad \begin{aligned} \mathcal{L}(\sigma', \sigma'', t) &\doteq K_4 |\sigma' \sigma''| \Upsilon_\xi(t-) \\ \widehat{\mathcal{L}}_\xi(\sigma', \sigma'', t) &\doteq K_\xi(\sigma', \sigma'') \cdot e^{K_4 Q(t+)}, \\ \widetilde{\mathcal{L}}_\xi(\sigma', \sigma'', t) &\doteq K_3 |\sigma' \sigma''| |\xi''| \cdot e^{K_4 Q(t+)}, \end{aligned}$$

where K_1, K_3 are the constants at (6.15) and (8.1), while K_ξ denotes the map in (8.1). Then, given a pair of fronts $\mathcal{J}_0$ of v^θ that interact together at t_0, for any time $\tau \geq t_0$, and for any i-th characteristic family, we define three type of functionals $\mathcal{F}_i, \widehat{\mathcal{F}}_{i,\xi}, \mathcal{G}_i, \widehat{\mathcal{G}}_{i,\xi}, \mathcal{H}_i, \widetilde{\mathcal{H}}_{i,\xi}$, analogous to the functionals F_i, G_i, H_i introduced in (6.12)-(6.14).

$$\mathcal{F}_i(\mathcal{J}_0; t_0, \tau) \doteq \frac{1}{4} \sum_{\substack{\sigma' \in \mathcal{N}_S^i(\mathcal{J}_0, t_k) \\ \sigma'' \in \mathcal{I}^i(\sigma', t_k) \\ \sigma' \cdot \sigma'' > 0 \\ t_0 \leq t_k \leq \tau}} \mathcal{L}(\sigma', \sigma'', t_k),$$

(8.37)
$$\widehat{\mathcal{F}}_{i,\xi}(\mathcal{J}_0; t_0, \tau) \doteq \frac{1}{4} \sum_{\substack{\sigma' \in \mathcal{N}_S^i(\mathcal{J}_0, t_k) \\ \sigma'' \in \mathcal{I}^i(\sigma', t_k) \\ \sigma' \cdot \sigma'' > 0 \\ t_0 \leq t_k \leq \tau}} \widehat{\mathcal{L}}_\xi(\sigma', \sigma'', t_k),$$

(8.38)
$$\mathcal{G}_i(\mathcal{J}_0; t_0, \tau) \doteq \frac{K_1}{4} \sum_{\substack{\sigma' \in \mathcal{N}_S^i(\mathcal{J}_0, t_k) \\ \sigma'' \in \mathcal{I}^i(\sigma', t_k) \\ \sigma' \cdot \sigma'' < 0 \\ t_0 \leq t_k \leq \tau}} \mathcal{L}(\sigma', \sigma'', t_k) + \frac{K_1}{8} \sum_{\substack{\sigma' \in \mathcal{N}_S^i(\mathcal{J}_0, t_k) \\ \sigma'' \in \mathcal{I}^j(\sigma', t_k),\, j \neq i \\ t_0 \leq t_k \leq \tau}} \mathcal{L}(\sigma', \sigma'', t_k),$$

$$\widehat{\mathcal{G}}_{i,\xi}(\mathcal{J}_0; t_0, \tau) \doteq \frac{1}{4} \sum_{\substack{\sigma' \in \mathcal{N}_S^i(\mathcal{J}_0, t_k) \\ \sigma'' \in \mathcal{I}^i(\sigma', t_k) \\ \sigma' \cdot \sigma'' < 0 \\ t_0 \leq t_k \leq \tau}} \widehat{\mathcal{L}}_\xi(\sigma', \sigma'', t_k) + \frac{1}{8} \sum_{\substack{\sigma' \in \mathcal{N}_S^i(\mathcal{J}_0, t_k) \\ \sigma'' \in \mathcal{I}^j(\sigma', t_k),\, j \neq i \\ t_0 \leq t_k \leq \tau}} \widehat{\mathcal{L}}_\xi(\sigma', \sigma'', t_k),$$

(8.39)
$$\mathcal{H}_i(\mathcal{J}_0; t_0, \tau) \doteq \frac{K_1}{4} \sum_{\substack{\sigma' \in \mathcal{C}^i(t_k) \cap N_R^i(\mathcal{J}_0, t_k) \\ \sigma'' \in \mathcal{N}_S^i(\mathcal{J}', t_k) \setminus \mathcal{I}(\sigma', t_k), \\ \mathcal{J}' \in \mathcal{B}^i(t),\, t < t_k \\ t_0 \leq t_k \leq \tau}} \mathcal{L}(\sigma', \sigma'', t_k) +$$

$$+ \frac{K_1}{4} \sum_{\substack{\sigma' \in C^j(t_k) \cap \mathcal{N}^j(\mathcal{N}_R^i(\mathcal{J}_0, t_h),\, t_k) \\ j \neq i,\; t_0 \leq t_h \leq t_k \leq \tau \\ \sigma'' \in \mathcal{N}_S^i(\mathcal{J}', t_k) \setminus \mathcal{I}(\sigma', t_k), \\ \mathcal{J}' \in \mathcal{B}^i(t),\, t < t_k \\ (\sigma', \sigma'') \in \mathcal{A}}} \mathcal{L}(\sigma', \sigma'', t_k),$$

$$\widetilde{\mathcal{H}}_{i,\xi}(\mathcal{J}_0; t_0, \tau) \doteq \frac{1}{4} \sum_{\substack{\sigma' \in \mathcal{C}^i(t_k) \cap N_R^i(\mathcal{J}_0, t_k) \\ \sigma'' \in \mathcal{N}_S^i(\mathcal{J}', t_k) \setminus \mathcal{I}(\sigma', t_k), \\ \mathcal{J}' \in \mathcal{B}^i(t),\, t < t_k \\ t_0 \leq t_k \leq \tau}} \widetilde{\mathcal{L}}_\xi(\sigma', \sigma'', t_k) +$$

$$+ \frac{1}{4} \sum_{\substack{\sigma' \in C^j(t_k) \cap \mathcal{N}^j(\mathcal{N}_R^i(\mathcal{J}_0, t_h), \, t_k) \\ j \neq i, \; t_0 \leq t_h \leq t_k \leq \tau \\ \sigma'' \in \mathcal{N}_S^i(\mathcal{J}', t_k) \setminus \mathcal{I}(\sigma', t_k), \\ \mathcal{J}' \in \mathcal{B}^i(t), \; t < t_k \\ (\sigma', \sigma'') \in \mathcal{A}}} \widetilde{\mathcal{L}}_\xi(\sigma', \sigma'', t_k).$$

In the above summations t_k are the interaction times for the fronts of v^θ, while $\mathcal{A}$ denotes the set of approaching waves (cfr. definition in Section 3) present in $v^\theta(t_k)$. Notice that, in the case where the i-th family is LD or GNL, by definition one has $\mathcal{B}^i(t) = \emptyset$, for all t. Hence, in this case, the functionals $\mathcal{H}_i(\mathcal{J}_0; t_0, \tau)$, $\widetilde{\mathcal{H}}_{i,\xi}(\mathcal{J}_0; t_0, \tau)$ in (8.39) are always zero for any pair of interacting fronts $\mathcal{J}_0$, and for any $\tau \geq t_0$.

We also need to define a further functional $\mathcal{M}_i$ which measures the decrease of V_ξ caused by interactions involving fronts of the i-th family originated from $\mathcal{J}_0$, in which the (front)·(shift) products of the incoming waves have opposite signs. Namely, letting $\mathcal{S}_i(t)$ denote the set of pairs of i-fronts that interact together at t generating a single i-front, we set

$$(8.40) \qquad \mathcal{M}_i(\mathcal{J}_0; t_0, \tau) \doteq \sum_{\substack{\sigma' \in \mathcal{N}_S^i(\mathcal{J}_0, t_k) \\ \sigma'' \in \mathcal{I}^i(\sigma', t_k) \\ (\sigma', \sigma'') \in \mathcal{S}_i(t_k) \\ (\sigma' \xi') \cdot (\sigma'' \xi'') < 0 \\ t_0 \leq t_k \leq \tau}} \frac{\min\{|\sigma' \xi'|, |\sigma'' \xi''|\}}{8} \cdot e^{K_4 Q(t_k+)}.$$

REMARK 8.3. Consider two wave-fronts $\sigma_{i,1}, \sigma_{j,2}$ that interact together at t_0, with $\sigma_{i,1}$ belonging to the i-th family and $\sigma_{j,2}$ belonging to the j-th family. Let $\xi_{i,1}, \xi_{j,2}$ be the corresponding shift rates. Then, if we compute the above functional for $\mathcal{J}_0 \doteq \{\sigma_{i,1}, \sigma_{j,2}\}$ precisely at t_0, we obtain
(8.41)
$$\mathcal{F}_k(\mathcal{J}_0; t_0, t_0) = \begin{cases} \dfrac{\mathcal{L}(\sigma_{i,1}, \sigma_{j,2}, t_0)}{4} & \text{if} \quad i = j = k, \quad \sigma_{i,1} \cdot \sigma_{j,2} > 0, \\ 0 & \text{otherwise,} \end{cases}$$

$$\widehat{\mathcal{F}}_{k,\xi}(\mathcal{J}_0; t_0, t_0) = \begin{cases} \dfrac{\widehat{\mathcal{L}}_\xi(\sigma_{i,1}, \sigma_{j,2}, t_0)}{4} & \text{if} \quad i = j = k, \quad \sigma_{i,1} \cdot \sigma_{j,2} > 0, \\ 0 & \text{otherwise,} \end{cases}$$

(8.42)
$$\mathcal{G}_k(\mathcal{J}_0;t_0,t_0) = \begin{cases} \dfrac{K_1 \mathcal{L}(\sigma_{i,1}, \sigma_{j,2}, t_0)}{4} & \text{if} \quad i=j=k, \quad \sigma_{i,1}\cdot\sigma_{j,2}<0, \\ \dfrac{K_1 \mathcal{L}(\sigma_{i,1}, \sigma_{j,2}, t_0)}{8} & \text{if} \quad i\neq j, \\ 0 & \text{otherwise,} \end{cases}$$

$$\widehat{\mathcal{G}}_{k,\xi}(\mathcal{J}_0;t_0,t_0) = \begin{cases} \dfrac{\widehat{\mathcal{L}}_\xi(\sigma_{i,1}, \sigma_{j,2}, t_0)}{4} & \text{if} \quad i=j=k, \quad \sigma_{i,1}\cdot\sigma_{j,2}<0, \\ \dfrac{\widehat{\mathcal{L}}_\xi(\sigma_{i,1}, \sigma_{j,2}, t_0)}{8} & \text{if} \quad i\neq j, \\ 0 & \text{otherwise,} \end{cases}$$

(8.43)
$$\mathcal{H}_k(\mathcal{J}_0;t_0,t_0) = \sum_{\substack{\sigma_{k,\ell}\in\mathcal{C}^k(t_0) \\ \sigma'\in\mathcal{N}_S^k(\mathcal{J}',t_0)\setminus\mathcal{I}(\mathcal{J},t_0) \\ \mathcal{J}'\in\mathcal{B}^k(t), \, t<t_0}} \frac{K_1 \mathcal{L}(\sigma_{k,\ell}, \sigma', t_0)}{4} + \sum_{\substack{\sigma_{h,p}\in\mathcal{C}^h(t_0),\, h\neq k \\ \sigma'\in\mathcal{N}_S^k(\mathcal{J}',t_0)\setminus\mathcal{I}(\mathcal{J},t_0) \\ (\sigma_{p,m},\sigma')\in\mathcal{A} \\ \mathcal{J}'\in\mathcal{B}^k(t), \, t<t_0}} \frac{K_1 \mathcal{L}(\sigma_{h,p}, \sigma', t_0)}{4},$$

$$\widetilde{\mathcal{H}}_{k,\xi}(\mathcal{J}_0;t_0,t_0) = \sum_{\substack{\sigma_{k,\ell}\in\mathcal{C}^k(t_0) \\ \sigma'\in\mathcal{N}_S^k(\mathcal{J}',t_0)\setminus\mathcal{I}(\mathcal{J},t_0) \\ \mathcal{J}'\in\mathcal{B}^k(t), \, t<t_0}} \frac{\widetilde{\mathcal{L}}_\xi(\sigma_{k,\ell}, \sigma', t_0)}{4} + \sum_{\substack{\sigma_{h,p}\in\mathcal{C}^h(t_0),\, h\neq k \\ \sigma'\in\mathcal{N}_S^k(\mathcal{J}',t_0)\setminus\mathcal{I}(\mathcal{J},t_0) \\ (\sigma_{p,m},\sigma')\in\mathcal{A} \\ \mathcal{J}'\in\mathcal{B}^k(t), \, t<t_0}} \frac{\widetilde{\mathcal{L}}_\xi(\sigma_{h,p}, \sigma', t_0)}{4},$$

(8.44)
$$\mathcal{M}_k(\mathcal{J}_0;t_0,t_0) = \begin{cases} \dfrac{\min\{|\sigma_{i,1}\xi_{i,1}|,|\sigma_{j,2}\xi_{j,2}|\}e^{K_4 Q(t_0+)}}{8} & \text{if} \quad \begin{cases} i=j=k, \\ (\sigma_{i,1},\sigma_{j,2})\in\mathcal{S}_k(t_0), \\ (\sigma_{i,1}\xi_{i,1})\cdot(\sigma_{j,2}\xi_{j,2})<0, \end{cases} \\ 0 & \text{otherwise.} \end{cases}$$

Notice that, as observed above, in the case where the i-th family is GNL or LD, we have $\mathcal{H}_i(\mathcal{J}_0;t_0,t_0) = \widetilde{\mathcal{H}}_{i,\xi}(\mathcal{J}_0;t_0,t_0) = 0$.

8.3. Main estimates.

LEMMA 8.2. *For a 2×2 system satisfying the assumption* (**A**), *there exists constants $K_4 > K_3 > K_2$ and $0 < \bar{\delta} < \delta_2$, independent of ε, such that the following holds. Let $\{v^\theta\}_{\theta\in\,]a,b[}$ be a family of ε-solutions satisfying the assumptions of Proposition 3.5, and such that $v^\theta(0) \in \mathcal{D}^{\varepsilon,\bar{\delta}}$ for all $\theta \in\,]a,b[$, where $\mathcal{D}^{\varepsilon,\bar{\delta}}$ denotes the domain at (3.33) with $\delta = \bar{\delta}$. Let $\Upsilon_\xi = \Upsilon_\xi(t)$ be the functional defined at*

(8.2)-(8.4). Assume that two wave-fronts σ_1^-, σ_2^- of v^θ interact together at time $t_0 \in \,]0, T]$, and set $\mathcal{J}_0 \doteq \{\sigma_1^-, \sigma_2^-\}$. Then, one has

(8.45)
$$\Delta \Upsilon_\xi(t_0) \leq -2 \sum_{i=1}^{2} \left[\mathcal{F}_i(\mathcal{J}_0; t_0, t_0) + \mathcal{G}_i(\mathcal{J}_0; t_0, t_0) + \mathcal{H}_i(\mathcal{J}_0; t_0, t_0) + \mathcal{M}_i(\mathcal{J}_0; t_0, t_0) \right] +$$
$$- 2 \sum_{i=1}^{2} \left[\widehat{\mathcal{F}}_{i,\xi}(\mathcal{J}_0; t_0, t_0) + \widehat{\mathcal{G}}_{i,\xi}(\mathcal{J}_0; t_0, t_0) + \widetilde{\mathcal{H}}_{i,\xi}(\mathcal{J}_0; t_0, t_0) \right] +$$
$$+ \sum_{i \in NGNL} \mathcal{E}_i(\mathcal{J}_0; t_0).$$

PROOF.
1. Recalling the notation

(8.46)
$$\hat{K}(\sigma_\alpha, \sigma_\beta) \doteq \begin{cases} |\sigma_\alpha \cdot \sigma_\beta| & \text{if } \sigma_\alpha, \sigma_\beta \text{ belong to the same family} \\ & \text{and are concordant,} \\ K_1 |\sigma_\alpha \cdot \sigma_\beta| & \text{otherwise,} \end{cases}$$

introduced in Section 6 in connection with any pair of waves σ_α, σ_β, it is convenient to rewrite (8.45) in the equivalent form
(8.47)
$$\Delta V_\xi(t_0) + \Delta Q_\xi(t_0) + \left(V_\xi(t_0-) + Q_\xi(t_0-) \right) \left(1 - e^{-K_4 \Delta Q(t_0)} \right) \leq$$
$$\leq -\frac{1}{2} \left(K_4 \hat{K}(\sigma_1^-, \sigma_2^-) \cdot \Upsilon_\xi(t_0-) \cdot e^{-K_4 Q(t_0+)} + K_\xi(\sigma_1^-, \sigma_2^-) \right) +$$
$$- 2e^{-K_4 Q(t_0+)} \cdot \left(\sum_{i=1}^{2} \left[\mathcal{H}_i(\mathcal{J}_0; t_0, t_0) + \widetilde{\mathcal{H}}_{i,\xi}(\mathcal{J}_0; t_0, t_0) + \mathcal{M}_i(\mathcal{J}_0; t_0, t_0) \right] \right) +$$
$$+ e^{-K_4 Q(t_0+)} \cdot \left(\sum_{i \in NGNL} \mathcal{E}_i(\mathcal{J}_0; t_0) \right).$$

Observe that, taking $K_3 > K_2 > 4C_{20}$, by Lemma 8.2 and (8.44) we have

(8.48) $$\Delta V_\xi(t_0) \leq -2e^{-K_4 Q(t_0+)} \cdot \left(\sum_{i=1}^{2} \mathcal{M}_i(\mathcal{J}_0; t_0, t_0) \right) + \frac{K_\xi(\sigma_1^-, \sigma_2^-)}{4}.$$

On the other hand, applying Lemma 6.1, we obtain

(8.49)
$$\left(V_\xi(t_0-) + Q_\xi(t_0-) \right) \left(1 - e^{-K_4 \Delta Q(t_0)} \right) \leq$$
$$\leq K_4 \left(V_\xi(t_0-) + Q_\xi(t_0-) \right) e^{-K_4 \Delta Q(t_0)} \cdot \Delta Q(t_0) \leq$$
$$\leq -\left(\frac{3\hat{K}(\sigma_1^-, \sigma_2^-)}{4} + 2 \sum_{i=1}^{2} H_i(\mathcal{J}_0; t_0, t_0) \right) \cdot K_4 \Upsilon_\xi(t_0-) \cdot e^{-K_4 Q(t_0+)} +$$

$$+ \bigg(\sum_{i \in NGNL} \Xi_i(\mathcal{J}_0; t_0) \, E_i(\mathcal{J}_0; t_0) \bigg) \cdot K_4 \, \Upsilon_\xi(t_0-) \cdot e^{-K_4 \, Q(t_0+)}$$

(Ξ_i being the map defined in Section 6.1). Therefore, recalling that, by (8.35) and by the definition of the map Ξ_i in Section 6.1, one has

$$\sum_{i \in NGNL} \Big(\mathcal{E}_i(\mathcal{J}_0; t_0) - 2K_2 \, \widehat{\mathcal{E}}_{i,\xi}(\mathcal{J}_0; t_0) \Big) = \bigg(\sum_{i \in NGNL} \Xi_i(\mathcal{J}_0; t_0) \, E_i(\mathcal{J}_0; t_0) \bigg) \cdot K_4 \, \Upsilon_\xi(t_0-),$$

using (8.48)-(8.49) we derive
(8.50)
$$\Delta V_\xi(t_0) + \Big(V_\xi(t_0-) + Q_\xi(t_0-) \Big) \Big(1 - e^{-K_4 \Delta Q(t_0))} \Big) \leq$$

$$\leq - \bigg(\frac{3\hat{K}(\sigma_1^-, \sigma_2^-)}{4} + 2 \sum_{i=1}^{2} \mathcal{H}_i(\mathcal{J}_0; t_0, t_0) \bigg) \cdot K_4 \, \Upsilon_\xi(t_0-) \cdot e^{-K_4 \, Q(t_0+)} +$$

$$- 2 \bigg(\sum_{i=1}^{2} \mathcal{M}_i(\mathcal{J}_0; t_0, t_0) \bigg) \cdot e^{-K_4 \, Q(t_0+)} +$$

$$+ \sum_{i \in NGNL} \Big(\mathcal{E}_i(\mathcal{J}_0; t_0) - 2K_2 \, \widehat{\mathcal{E}}_{i,\xi}(\mathcal{J}_0; t_0) \Big) \cdot e^{-K_4 \, Q(t_0+)} + \frac{K_\xi(\sigma_1^-, \sigma_2^-)}{4}.$$

Thanks to (8.50) we deduce that, in order to establish (8.47), it suffices to show
(8.51)
$$\Delta Q_\xi(t_0) \leq -\frac{3K_\xi(\sigma_1^-, \sigma_2^-)}{4} - 2e^{-K_4 \, Q(t_0+)} \cdot \bigg(\sum_{i=1}^{2} \Big[\mathcal{H}_i(\mathcal{J}_0; t_0, t_0) + \widetilde{\mathcal{H}}_{i,\xi}(\mathcal{J}_0; t_0, t_0) \Big] \bigg) +$$

$$+ \bigg(\frac{\hat{K}(\sigma_1^-, \sigma_2^-)}{4} + 2 \sum_{i=1}^{2} \mathcal{H}_i(\mathcal{J}_0; t_0, t_0) \bigg) \cdot K_4 \, \Upsilon_\xi(t_0-) \cdot e^{-K_4 \, Q(t_0+)} +$$

$$+ \bigg(\sum_{i \in NGNL} \widehat{\mathcal{E}}_{i,\xi}(\mathcal{J}_0; t_0) \bigg) \cdot 2K_2 \, e^{-K_4 \, Q(t_0+)}.$$

2. We shall prove (8.51) assuming that the incoming fronts σ_1^-, σ_2^- either belong to different characteristic families or belong both to the first family (the case where σ_1^-, σ_2^- belong both to the second family being entirely similar). Thus, we are led to consider the following three cases.

CASE 1: σ_1^- and σ_2^- belong to the first characteristic family and are concordant;

CASE 2: σ_1^- and σ_2^- belong to the first characteristic family and are discordant;

CASE 3: σ_1^- belongs to the first characteristic family and σ_2^- belongs to second one.

Call ξ_i^-, $i = 1, 2$ the shift rates of the incoming fronts σ_i^-, $i = 1, 2$, and let Λ_i^- be their speeds.

CASE 1. Since the two incoming fronts σ_1^-, σ_2^- are concordant waves of the first family, by the properties of the approximate Riemann solver described in Section 3 it follows that the outgoing wave of the first family is a single front σ_1^+ concordant with σ_1^-, σ_2^- (no matter if the first family is GNL or NGNL). Let ξ_1^+ be its displacement, call $\sigma_{2,\ell}^+$, $1 \leq \ell \leq n_2$ the outgoing fronts of the second family, and let $\xi_{2,\ell}^+$ be the corresponding shift rates. Take

$$(8.52) \qquad \bar\delta < \min\left\{ \sqrt{\frac{\ln 2}{4K_1 K_4}},\; \frac{1}{32(C_1+C_3)C_{20}K_3} \right\}$$

(C_1, C_3, C_{20} being the constants of Lemmas 5.1, 5.7, 8.1), and observe that, since $v^\theta(0) \in \mathcal{D}^{\varepsilon,\bar\delta}$, by Proposition 3.4 one has

$$(8.53) \qquad V(t) \leq 2\bar\delta < 1, \qquad e^{K_4 Q(t)} \leq e^{4K_1 K_4 \bar\delta^2} < 2 \qquad \forall\, t,$$

which, in particular, implies $|\Lambda_1^- - \Lambda_2^-| < 1$. Thus, applying Lemma 5.1 and Lemma 8.1, we deduce

$$(8.54) \qquad \left||\sigma_1^+| - |\sigma_1^-| - |\sigma_2^-|\right| + \sum_{\ell=1}^{n_2} |\sigma_{2,\ell}^+| \leq C_1 |\sigma_1^- \sigma_2^-|,$$

$$(8.55)$$
$$\left\lfloor |\sigma_1^+ \xi_1^+| - |\sigma_1^- \xi_1^-| - |\sigma_2^- \xi_2^-| \right\rfloor^+ + \sum_{\ell=1}^{n_2} |\sigma_{2,\ell}^+ \xi_{2,\ell}^+| \leq C_{20} |\sigma_1^- \sigma_2^-| \left(|\xi_1^-| + |\xi_2^-|\right).$$

Moreover, since σ_1^+ is concordant with both incoming fronts σ_1^-, σ_2^-, no cancellation occurs after this interaction, i.e. $(\mathcal{C}^1(t_0) \cup \mathcal{C}^2(t_0)) \cap \mathcal{J}_0 = \emptyset$ (cfr. definition of $\mathcal{C}^i(t)$ at Section 6.1), and hence, by (8.43), one has

$$(8.56) \quad \mathcal{H}_1(\mathcal{J}_0; t_0, t_0) = \widetilde{\mathcal{H}}_{1,\xi}(\mathcal{J}_0; t_0, t_0) = \mathcal{H}_2(\mathcal{J}_0; t_0, t_0) = \widetilde{\mathcal{H}}_{2,\xi}(\mathcal{J}_0; t_0, t_0) = 0.$$

Then, assuming $K_4 > 16 C_1 K_3 > 16 C_1 K_2$, and relying on (8.52)-(8.55), we derive the following estimates on the terms of $\Delta Q_\xi(t_0)$:
$$(8.57)$$
$$\sum_{\ell=1}^{n_2} \sum_{(\sigma_{2,\ell}^+, \sigma_\alpha) \in \mathcal{A}} K_\xi(\sigma_{2,\ell}^+, \sigma_\alpha) \leq$$

$$\leq K_3 \left[\left(\sum_{\ell=1}^{n_2} |\sigma_{2,\ell}^+| \right) \cdot \left(\sum_{\sigma_\alpha \notin \mathcal{J}_0} |\sigma_\alpha \xi_\alpha| \right) + \left(\sum_{\sigma_\alpha \notin \mathcal{J}_0} |\sigma_\alpha| \right) \cdot \left(\sum_{\ell=1}^{n_2} |\sigma_{2,\ell}^+ \xi_{2,\ell}^+| \right) \right]$$

$$\leq C_1 K_3\, \hat{K}(\sigma_1^-, \sigma_2^-) \cdot V_\xi(t_0-) + \frac{C_{20} K_3}{K_2} V(t_0-) \cdot K_\xi(\sigma_1^-, \sigma_2^-)$$

$$\leq \frac{K_4\, \hat{K}(\sigma_1^-, \sigma_2^-)}{8} \cdot V_\xi(t_0-)\, e^{-K_4 \Delta Q(t_0)} + \frac{K_\xi(\sigma_1^-, \sigma_2^-)}{16},$$

(8.58)
$$\sum_{\substack{\ell,m \\ \ell \neq m}} K_\xi(\sigma_{2,\ell}^+, \sigma_{2,m}^+) \leq K_2 \left(\sum_{\ell=1}^{n_2} |\sigma_{2,\ell}^+| \right) \cdot \left(\sum_{\ell=1}^{n_2} |\sigma_{2,\ell}^+ \xi_{2,\ell}^+| \right)$$
$$\leq C_1 C_{20} |\sigma_1^- \sigma_2^-| \cdot K_\xi(\sigma_1^-, \sigma_2^-)$$
$$\leq \frac{K_\xi(\sigma_1^-, \sigma_2^-)}{16},$$

(8.59)
$$\sum_{\sigma_\alpha \notin \mathcal{J}_0} \left[K_\xi(\sigma_1^+, \sigma_\alpha) - K_\xi(\sigma_1^-, \sigma_\alpha) - K_\xi(\sigma_2^-, \sigma_\alpha) \right] \leq$$
$$\leq K_3 \Big\lfloor |\sigma_1^+| - |\sigma_1^-| - |\sigma_2^-| \Big\rfloor^+ \cdot \left(\sum_{\sigma_\alpha \notin \mathcal{J}_0} |\sigma_\alpha \xi_\alpha| \right) +$$
$$+ K_3 \Big\lfloor |\sigma_1^+ \xi_1^+| - |\sigma_1^- \xi_1^-| - |\sigma_2^- \xi_2^-| \Big\rfloor^+ \cdot \left(\sum_{\sigma_\alpha \notin \mathcal{J}_0} |\sigma_\alpha| \right)$$
$$\leq C_1 K_3 \hat{K}(\sigma_1^-, \sigma_2^-) \cdot V_\xi(t_0-) + \frac{C_{20} K_3}{K_2} V(t_0-) \cdot K_\xi(\sigma_1^-, \sigma_2^-)$$
$$\leq \frac{K_4 \hat{K}(\sigma_1^-, \sigma_2^-)}{8} \cdot V_\xi(t_0-) e^{-K_4 \Delta Q(t_0)} + \frac{K_\xi(\sigma_1^-, \sigma_2^-)}{16}.$$

Thus, using (8.57)-(8.59), we obtain

(8.60)
$$\Delta Q_\xi(t_0) = \sum_{\sigma_\alpha \notin \mathcal{J}_0} \left[K_\xi(\sigma_1^+, \sigma_\alpha) - K_\xi(\sigma_1^-, \sigma_\alpha) - K_\xi(\sigma_2^-, \sigma_\alpha) \right] - K_\xi(\sigma_1^-, \sigma_2^-) +$$
$$+ \sum_{\ell=1}^{n_2} \sum_{(\sigma_{2,\ell}^+, \sigma_\alpha) \in \mathcal{A}} K_\xi(\sigma_{2,\ell}^+, \sigma_\alpha) + \sum_{\substack{\ell,m \\ \ell \neq m}} K_\xi(\sigma_{2,\ell}^+, \sigma_{2,m}^+) \leq$$
$$\leq -\frac{3 K_\xi(\sigma_1^-, \sigma_2^-)}{4} + \frac{K_4 \hat{K}(\sigma_1^-, \sigma_2^-)}{4} \cdot V_\xi(t_0-) e^{-K_4 \Delta Q(t_0)}$$

which, because of (8.56), yields (8.51), completing the proof of Case 1.

3. Case 2. To fix the ideas, assume that σ_1^- is located on the left of σ_2^- and that $|\sigma_1^-| > |\sigma_2^-|$, the other cases being entirely similar or easier. Then, by the properties of the approximate Riemann solver described in Section 3, it follows that the front incoming from the left σ_1^- must be a shock, while σ_2^- is a rarefaction front discordant with σ_1^- (no matter if the first family is GNL or NGNL). Let $\sigma_{i,\ell}^+$, $\ell = 1, \ldots, n_i$, $i = 1, 2$, be the sizes of the outgoing fronts, denote by $\Lambda_{i,\ell}^+$ their speeds, and call $\xi_{i,\ell}^+$ their shift rates. Observe that, since the outgoing wave $\sigma_1^+ \doteq \sum_{\ell=1}^{n_1} \sigma_{1,\ell}^+$

is discordant with the incoming front σ_2^-, in this case one has $\mathcal{C}^1(t_0) = \{\sigma_2^-\}$, $\mathcal{C}^2(t_0) = \emptyset$, and hence, by (8.36), (8.43), there holds
(8.61)
$$e^{-K_4 Q(t_0+)} \cdot \left(\sum_{i=1}^{2} \left[\mathcal{H}_i(\mathcal{J}_0; t_0, t_0) + \widetilde{\mathcal{H}}_{i,\xi}(\mathcal{J}_0; t_0, t_0) \right] \right) \leq$$
$$\leq \frac{K_1 K_4}{4} \cdot \left(\sum_{\sigma_\alpha \notin \mathcal{J}_0} |\sigma_2^- \sigma_\alpha| \right) \cdot \Upsilon_\xi(t_0-) \cdot e^{-K_4 Q(t_0+)} + \frac{K_3}{4} |\sigma_2^-| \cdot \left(\sum_{\sigma_\alpha \notin \mathcal{J}_0} |\sigma_\alpha \xi_\alpha| \right)$$
$$\leq K_4 \left(\sum_{i=1}^{2} H_i(\mathcal{J}_0; t_0 t_0) \right) \cdot \Upsilon_\xi(t_0-) \cdot e^{-K_4 Q(t_0+)} + \frac{K_3}{4} |\sigma_2^-| \cdot \left(\sum_{\sigma_\alpha \notin \mathcal{J}_0} |\sigma_\alpha \xi_\alpha| \right).$$

Moreover, applying Lemma 5.1 and Lemma 8.1, we deduce

(8.62)
$$\left| |\sigma_1^+| - |\sigma_1^-| + |\sigma_2^-| \right| + \sum_{\ell=1}^{n_2} |\sigma_{2,\ell}^+| \leq C_1 |\sigma_1^- \sigma_2^-|,$$

(8.63)
$$\left\lfloor \sum_{\ell=1}^{n_1} |\sigma_{1,\ell}^+ \xi_{1,\ell}^+| - |\sigma_1^- \xi_1^-| - |\sigma_2^- \xi_2^-| \right\rfloor^+ + \sum_{\ell=1}^{n_2} |\sigma_{2,\ell}^+ \xi_{2,\ell}^+| \leq C_{20} |\sigma_1^- \sigma_2^-| \left(|\xi_1^-| + |\xi_2^-| \right).$$

Then, assuming $K_4 > 16 C_1 K_3 > 16 C_1 K_2$, and relying on (8.52)-(8.53), (8.62), (8.63), by the same computations performed in Case 1 we derive (8.57)-(8.58), and

(8.64)
$$\sum_{\sigma_\alpha \notin \mathcal{J}_0} \left[\sum_{\ell=1}^{n_1} K_\xi(\sigma_{1,\ell}^+, \sigma_\alpha) - K_\xi(\sigma_1^-, \sigma_\alpha) - K_\xi(\sigma_2^-, \sigma_\alpha) \right] \leq$$
$$\leq K_3 \left| |\sigma_1^+| - |\sigma_1^-| + |\sigma_2^-| \right|^+ \cdot \left(\sum_{\sigma_\alpha \notin \mathcal{J}_0} |\sigma_\alpha \xi_\alpha| \right) - K_3 |\sigma_2^-| \cdot \left(\sum_{\sigma_\alpha \notin \mathcal{J}_0} |\sigma_\alpha \xi_\alpha| \right) +$$
$$+ K_3 \left\lfloor \sum_{\ell=1}^{n_1} |\sigma_{1,\ell}^+ \xi_{1,\ell}^+| - |\sigma_1^- \xi_1^-| - |\sigma_2^- \xi_2^-| \right\rfloor^+ \cdot \left(\sum_{\sigma_\alpha \notin \mathcal{J}_0} |\sigma_\alpha| \right) \leq$$
$$\leq C_1 K_3 V_\xi(t_0-) \cdot \hat{K}(\sigma_1^-, \sigma_2^-) - K_3 |\sigma_2^-| \cdot \left(\sum_{\sigma_\alpha \notin \mathcal{J}_0} |\sigma_\alpha \xi_\alpha| \right) +$$
$$+ \frac{C_{20} K_3}{K_2} V(t_0-) \cdot K_\xi(\sigma_1^-, \sigma_2^-) \leq$$
$$\leq \frac{K_4 \hat{K}(\sigma_1^-, \sigma_2^-)}{8} \cdot V_\xi(t_0-) e^{-K_4 \Delta Q(t_0)} + \frac{K_\xi(\sigma_1^-, \sigma_2^-)}{16} +$$
$$- K_3 |\sigma_2^-| \cdot \left(\sum_{\sigma_\alpha \notin \mathcal{J}_0} |\sigma_\alpha \xi_\alpha| \right).$$

Therefore, if $n_1 = 1$, i.e. if the outgoing wave of the first family consists of a single front, using (8.57)-(8.58), (8.64), we obtain

(8.65)
$$\Delta Q_\xi(t_0) = \sum_{\sigma_\alpha \notin \mathcal{J}_0} \left[K_\xi(\sigma_1^+, \sigma_\alpha) - K_\xi(\sigma_1^-, \sigma_\alpha) - K_\xi(\sigma_2^-, \sigma_\alpha) \right] - K_\xi(\sigma_1^-, \sigma_2^-) +$$
$$+ \sum_{\ell=1}^{n_2} \sum_{(\sigma_{2,\ell}^+, \sigma_\alpha) \in \mathcal{A}} K_\xi(\sigma_{2,\ell}^+, \sigma_\alpha) + \sum_{\substack{\ell, m \\ \ell \neq m}} K_\xi(\sigma_{2,\ell}^+, \sigma_{2,m}^+)$$
$$\leq -\frac{3 K_\xi(\sigma_1^-, \sigma_2^-)}{4} + \frac{K_4 \hat{K}(\sigma_1^-, \sigma_2^-)}{4} \cdot V_\xi(t_0-) e^{-K_4 \Delta Q(t_0)} +$$
$$- K_3 |\sigma_2^-| \cdot \left(\sum_{\sigma_\alpha \notin \mathcal{J}_0} |\sigma_\alpha \xi_\alpha| \right)$$

which, together with (8.61), yields (8.51). On the other hand, if $n_1 > 1$, i.e. if the outgoing wave of the first family σ_1^+ is a composed wave, by the properties of the approximate Riemann solver described in Section 3 it follows that the first family must be NGNL, and that σ_1^+ consists of $n_1 - 1$ rarefaction fronts, say $\sigma_{1,\ell}^+$, $1 \leq \ell < n_1$, and of a shock σ_{1,n_1}^+. In this case, in order to establish (8.51), we need an a-priori bound on the additional terms $K_\xi(\sigma_{1,\ell}^+, \sigma_{1,m}^+)$ appearing in $Q_\xi(t_0+)$.

4. We shall distinguish two subcases.

I) Assume that $\mathcal{J}_0 \notin \mathcal{B}^1(t_0)$. Then, by Remark 6.2, one has

(8.66)
$$\sum_{\ell=1}^{n_1} |\sigma_{1,\ell}^+| [|\sigma_1^+| - |\sigma_{1,\ell}^+|] \leq C_{13} |\sigma_1^- \sigma_2^-|.$$

On the other hand, applying Lemmas 5.2, 5.5, we deduce that there exists some constant $C_{21} > 0$, independent of ε, σ_1^-, σ_2^-, such that

$$\left| \frac{\Lambda_{1,\ell}^+ - \Lambda_i^-}{\Lambda_1^- - \Lambda_2^-} \right| \leq \left| \frac{\Lambda_{1,n_1}^+ - \Lambda_i^-}{\Lambda_1^- - \Lambda_2^-} \right| \leq C_{21} \qquad \forall\, \ell, i.$$

Thus, recalling the expression (8.7) of $\xi_{1,\ell}^+$, we obtain

(8.67)
$$|\xi_{1,\ell}^+| \leq \max \left\{ \left| \frac{\Lambda_{1,\ell}^+ - \Lambda_i^-}{\Lambda_1^- - \Lambda_2^-} \right| ; \; i = 1, 2 \right\} \cdot (|\xi_1^-| + |\xi_2^-|)$$
$$\leq C_{21} \cdot (|\xi_1^-| + |\xi_2^-|) \qquad \forall\, \ell.$$

Then, using (8.66)-(8.67), assuming $K_3 > 16 K_2 C_{13} C_{21}$, and recalling (8.1), since σ_1^-, σ_2^- are discordant fronts we derive

(8.68)
$$\sum_{\substack{\ell,m \\ \ell \neq m}} K_\xi(\sigma_{1,\ell}^+, \sigma_{1,m}^+) \leq 2K_2 \left(\sum_{\ell=1}^{n_1} |\sigma_{1,\ell}^+|[|\sigma_1^+|-|\sigma_{1,\ell}^+|] \right) \cdot \max \left\{ |\xi_{1,\ell}^+| \; ; \; 1 \leq \ell \leq n_1 \right\}$$
$$\leq 2K_2 C_{13} C_{21} |\sigma_1^- \sigma_2^-| \cdot (|\xi_1^-| + |\xi_2^-|)$$
$$\leq \frac{K_\xi(\sigma_1^-, \sigma_2^-)}{16}.$$

Hence, relying on (8.57)-(8.58), (8.64), (8.68), we obtain

(8.69)
$$\Delta Q_\xi(t_0) = \sum_{\sigma_\alpha \notin \mathcal{J}_0} \left[K_\xi(\sigma_1^+, \sigma_\alpha) - K_\xi(\sigma_1^-, \sigma_\alpha) - K_\xi(\sigma_2^-, \sigma_\alpha) \right] - K_\xi(\sigma_1^-, \sigma_2^-) +$$
$$+ \sum_{\ell=1}^{n_2} \sum_{(\sigma_{2,\ell}^+, \sigma_\alpha) \in \mathcal{A}} K_\xi(\sigma_{2,\ell}^+, \sigma_\alpha) + \sum_{i=1}^{2} \sum_{\substack{\ell,m \\ \ell \neq m}} K_\xi(\sigma_{i,\ell}^+, \sigma_{i,m}^+)$$
$$\leq -\frac{3 K_\xi(\sigma_1^-, \sigma_2^-)}{4} + \frac{K_4 \hat{K}(\sigma_1^-, \sigma_2^-)}{4} \cdot V_\xi(t_0-) e^{-K_4 \Delta Q(t_0)} +$$
$$- K_3 |\sigma_2^-| \cdot \left(\sum_{\sigma_\alpha \notin \mathcal{J}_0} |\sigma_\alpha \xi_\alpha| \right)$$

which, together with (8.61), yields (8.51).

II) Assume that $\mathcal{J}_0 \in \mathcal{B}^1(t_0)$ and, for definiteness, suppose that the first family is NGNL1 (the case of a NGNL2 family being entirely similar). In this case, by Remark 6.1, the outgoing wave of the first family is a composed wave consisting of a single rarefaction front $\sigma_1^{+,r}$ and of a shock $\sigma_1^{+,s}$ of strength $\geq \varepsilon$. Let $\Lambda_1^{+,r}, \Lambda_1^{+,s}$, be their speeds and call $\xi_1^{+,r}, \xi_1^{+,s}$ their shift rates. Notice that, by the properties of the approximate Riemann solver described in Section 3, the front incoming from the left σ_1^- must be a shock, while σ_2^- is a rarefaction front discordant with σ'. Then, using the expansions of the wave speeds $\lambda_1^{\psi,\varepsilon}$ provided by Lemma 4.2, and relying on the expansions of the exact shock speed λ_1^s collected in [**AM2**, Remark 2.2], we deduce that there exists some constant $C_{22} > 0$, independent of $\varepsilon, \sigma_1^-, \sigma_2^-$, such that

$$\left| \frac{\Lambda_1^{+,r} - \Lambda_1^{+,s}}{\Lambda_1^- - \Lambda_2^-} \right| \leq C_{22} \frac{|\sigma_2^-|}{\varepsilon}.$$

Thus, recalling the expression (8.7) of $\xi_1^{+,r}, \xi_1^{+,s}$, we get

(8.70)
$$|\xi_1^{+,r} - \xi_1^{+,s}| \leq \left| \frac{\Lambda_1^{+,r} - \Lambda_1^{+,s}}{\Lambda_1^- - \Lambda_2^-} (\xi_2^- - \xi_1^-) \right|$$
$$\leq C_{22} \frac{|\sigma_2^-|}{\varepsilon} \cdot (|\xi_1^-| + |\xi_2^-|).$$

Then, using (8.69), assuming $K_3 > 8K_2 C_{22}$, and recalling (8.1), since σ_1^-, σ_2^- are discordant fronts we obtain
(8.71)
$$K_\xi(\sigma_1^{+,r}, \sigma_1^{+,s}) \leq 2K_2 \widehat{\mathcal{E}}_{1,\xi}(\mathcal{J}_0; t_0) \cdot e^{-K_4 Q(t_0+)} + K_2 |\sigma_i^{+,r} \sigma_i^{+,s}| \cdot |\xi_1^{+,r} - \xi_1^{+,s}|$$
$$\leq 2K_2 \widehat{\mathcal{E}}_{1,\xi}(\mathcal{J}_0; t_0) \cdot e^{-K_4 Q(t_0+)} + K_2 C_{22} |\sigma_1^- \sigma_2^-| \cdot (|\xi_1^-| + |\xi_2^-|)$$
$$\leq 2K_2 \widehat{\mathcal{E}}_{1,\xi}(\mathcal{J}_0; t_0) \cdot e^{-K_4 Q(t_0+)} + \frac{K_\xi(\sigma_1^-, \sigma_2^-)}{16}.$$

Hence, relying on (8.57)-(8.58), (8.64), (8.71), by the same computation in (8.69) we derive

(8.72)
$$\Delta Q_\xi(t_0) \leq -\frac{3K_\xi(\sigma_1^-, \sigma_2^-)}{4} + \frac{K_4 \hat{K}(\sigma_1^-, \sigma_2^-)}{4} \cdot V_\xi(t_0-) e^{-K_4 \Delta Q(t_0)} +$$
$$- K_3 |\sigma_2^-| \cdot \left(\sum_{\sigma_\alpha \notin \mathcal{J}_0} |\sigma_\alpha \xi_\alpha| \right) + 2K_2 \widehat{\mathcal{E}}_{i,\xi}(\mathcal{J}_0; t_0) \cdot e^{-K_4 Q(t_0+)}$$

which, together with (8.61), yields (8.51), completing the proof of Case 2.

5. CASE 3. Let $\Lambda_{i,\ell}^+$ be the speeds of the outgoing fronts $\sigma_{i,\ell}^+$, $\ell = 1, \ldots, n_i$, $i = 1, 2$, call $\xi_{i,\ell}^+$ their shift rates, and denote by $\sigma_i^+ \doteq \sum_{\ell=1}^{n_i} \sigma_{i,\ell}^+$, $i = 1, 2$ the total size of the outgoing waves. Then, applying Lemma 5.7 and Lemma 8.1, we deduce

(8.73)
$$\sum_{i=1}^2 \left| |\sigma_1^+| - |\sigma_i^-| \right| \leq C_3 |\sigma_1^- \sigma_2^-|,$$

(8.74)
$$\sum_{i=1}^2 \left\lfloor \sum_{\ell=1}^{n_i} |\sigma_{i,\ell}^+ \xi_{i,\ell}^+| - |\sigma_i^- \xi_i^-| \right\rfloor^+ \leq C_{20} |\sigma_1^- \sigma_2^-| (|\xi_1^-| + |\xi_2^-|).$$

Observe that in this case the outgoing wave σ_i^+ of each i-th family is concordant with the incoming front σ_i^-, and hence no cancellation occurs after this interaction, i.e. $(\mathcal{C}^1(t_0) \cup \mathcal{C}^2(t_0)) \cap \mathcal{J}_0 = \emptyset$ (cfr. definition of $\mathcal{C}^i(t)$ at Section 6.1). Thus, by (8.43), there holds (8.56). Then, assuming $K_4 > 16C_3 K_3 > 16C_3 K_2$, and relying on (8.52)-(8.53), (8.73)-(8.74), we derive

(8.75)
$$\sum_{i=1}^2 \sum_{\sigma_\alpha \notin \mathcal{J}_0} \left[\sum_{\ell=1}^{n_i} K_\xi(\sigma_{i,\ell}^+, \sigma_\alpha) - K_\xi(\sigma_i^-, \sigma_\alpha) \right] \leq$$
$$\leq K_3 \left(\sum_{i=1}^2 \lfloor |\sigma_i^+| - |\sigma_i^-| \rfloor^+ \right) \cdot \left(\sum_{\sigma_\alpha \notin \mathcal{J}_0} |\sigma_\alpha \xi_\alpha| \right) +$$
$$+ K_3 \left(\sum_{i=1}^2 \left\lfloor \sum_{\ell=1}^{n_i} |\sigma_{i,\ell}^+ \xi_{i,\ell}^+| - |\sigma_i^- \xi_i^-| \right\rfloor^+ \right) \cdot \left(\sum_{\sigma_\alpha \notin \mathcal{J}_0} |\sigma_\alpha| \right)$$
$$\leq C_3 K_3 V_\xi(t_0-) \cdot \hat{K}(\sigma_1^-, \sigma_2^-) + \frac{C_{20} K_3}{K_2} V(t_0-) \cdot K_\xi(\sigma_1^-, \sigma_2^-) \leq$$

$$\leq \frac{K_4 \widehat{K}(\sigma_1^-, \sigma_2^-)}{8} \cdot V_\xi(t_0-) \, e^{-K_4 \Delta Q(t_0)} + \frac{K_\xi(\sigma_1^-, \sigma_2^-)}{16}.$$

On the other hand, applying (6.5) (if $\mathcal{J}_0 \notin \mathcal{B}(t_0)$) and Lemma 5.8 (in the case where one of the families is NGNL and $\mathcal{J}_0 \in \mathcal{B}(t_0)$), together with Lemma 5.10, and using the expansions of the wave speeds λ_i^s, $\lambda_i^{\psi,\varepsilon}$ collected in [**AM2**, Remark 2.2] and in Lemma 4.2, with the same computations performed in Case 2 we obtain
(8.76)
$$\sum_{i \in NGNL} \sum_{\substack{\ell, m \\ \ell \neq m}} K_\xi(\sigma_{i,\ell}^+, \sigma_{i,m}^+) \leq \left(\sum_{i=1}^{2} \widehat{\mathcal{E}}_{i,\xi}(\mathcal{J}_0; t_0) \right) \cdot 2K_2 \, e^{-K_4 Q(t_0+)} + \frac{K_\xi(\sigma_1^-, \sigma_2^-)}{16}.$$

Thus, relying on (8.75)-(8.76), we derive

(8.77)
$$\Delta Q_\xi(t_0) \leq \sum_{i=1}^{2} \sum_{\sigma_\alpha \notin \mathcal{J}_0} \left[\sum_{\ell=1}^{n_i} K_\xi(\sigma_{i,\ell}^+, \sigma_\alpha) - K_\xi(\sigma_i^-, \sigma_\alpha) \right] - K_\xi(\sigma_1^-, \sigma_2^-) +$$
$$+ \sum_{i=1}^{2} \sum_{\substack{\ell, m \\ \ell \neq m}} K_\xi(\sigma_{i,\ell}^+, \sigma_{i,m}^+)$$
$$\leq -\frac{3 K_\xi(\sigma_1^-, \sigma_2^-)}{4} + \frac{K_4 \widehat{K}(\sigma_1^-, \sigma_2^-)}{8} \cdot V_\xi(t_0-) \, e^{-K_4 \Delta Q(t_0)} +$$
$$+ \left(\sum_{i \in NGNL} \widehat{\mathcal{E}}_{i,\xi}(\mathcal{J}_0; t_0) \right) \cdot e^{-K_4 Q(t_0+)}$$

which, together with (8.56), yields (8.51), concluding the proof of Case 3, and hence completing the proof of the lemma as well. □

REMARK 8.4. Given a PFB $\mathcal{J}_0 \in \mathcal{B}^k(t_0)$, $k \in \{1,2\} \cap \text{NGNL}$, consider a pair of fronts $\mathcal{J}$ that interact together at $t \in \,]t_0, \widehat{t}_0\,[$, and satisfy the same assumptions of Lemma 8.2. Then, recalling the definition (8.35) of $\mathcal{E}_k$, from the proof of Lemma 8.2 we deduce the further estimate

(8.78)
$$\Delta \Upsilon_\xi(t) \leq -2 \sum_{i=1}^{2} \Big[\mathcal{F}_i(\mathcal{J}; t, t) + \mathcal{G}_i(\mathcal{J}; t, t) + \mathcal{H}_i(\mathcal{J}; t, t) + \mathcal{M}_i(\mathcal{J}; t, t) \Big] +$$
$$- 2 \sum_{i=1}^{2} \Big[\widehat{\mathcal{F}}_{i,\xi}(\mathcal{J}; t, t) + \widehat{\mathcal{G}}_{i,\xi}(\mathcal{J}; t, t) + \widetilde{\mathcal{H}}_{j,\xi}(\mathcal{J}; t, t) \Big] +$$
$$+ \sum_{i \in NGNL} \mathcal{E}_i(\mathcal{J}; t) + \Big[\mathcal{E}_k(\mathcal{J}_0; t+) - \mathcal{E}_k(\mathcal{J}_0; t-) \Big].$$

REMARK 8.5. Consider two concordant fronts of a k-th NGNL1 characteristic family that interact together at t_0: a rarefaction front σ' and a shock σ'' (σ' located on the left of σ''). Let v^L be the left state of σ', and call ξ', ξ'' the shift rates of σ', σ''. Since this interaction produces a single front of the k-th family, it follows that $\mathcal{J}_0 \doteq \{\sigma', \sigma''\} \notin \mathcal{B}(t_0)$. Moreover, no cancellation occurs, i.e. $(\mathcal{C}^1(t_0) \cup \mathcal{C}^2(t_0)) \cap$

$\mathcal{J}_0 = \emptyset$ (cfr. definition of $\mathcal{C}^i(t)$ at Section 6.1). Hence, by (8.35), (8.42), (8.43), one has

$$\mathcal{E}_i(\mathcal{J}_0; t_0) = \mathcal{G}_i(\mathcal{J}_0; t_0, t_0) = \widehat{\mathcal{G}}_{i,\xi}(\mathcal{J}_0; t_0, t_0) = 0,$$
$$\mathcal{H}_i(\mathcal{J}_0; t_0, t_0) = \widetilde{\mathcal{H}}_{i,\xi}(\mathcal{J}_0; t_0, t_0) = 0, \qquad i = 1, 2.$$

Then, in the same setting of Lemma 8.2 (possibly assuming that $v^\theta(0) \in \mathcal{D}^{\varepsilon,\overline{\delta}'}$ for some $\overline{\delta}' < \delta$), and relying on (5.2)-(5.3), (5.10)-(5.11), one can easily obtain the sharper estimates

$$\Delta V_\xi(t_0) \leq -e^{-K_4 Q(t_0-)} \cdot \mathcal{M}_k(\mathcal{J}_0; t_0, t_0) +$$
$$+ \frac{K_2 |\sigma''| \cdot \min\{|\sigma' + \sigma'' - \nu_k^\varepsilon(v^L)|, |\sigma'|\}}{8(80 \, C_{10})^2} \cdot (|\xi'| + |\xi''|),$$

$$\Delta Q_\xi(t_0) \leq -K_\xi(\sigma', \sigma'') + \frac{K_2 |\sigma''| \cdot \min\{|\sigma' + \sigma'' - \nu_k^\varepsilon(v^L)|, |\sigma'|\}}{8(80 \, C_{10})^2} \cdot (|\xi'| + |\xi''|) +$$
$$+ \frac{K_4 |\sigma'||\sigma' + \sigma'' - \nu_k^\varepsilon(v^L)|}{8(80 \, C_{10})^2} \cdot \Upsilon_\xi(t_0-) \cdot e^{-K_4 Q(t_0-)},$$

(8.79)
$$\Delta \Upsilon_\xi(t_0) \leq -K_4 \Upsilon_\xi(t_0-) \cdot |\sigma'| \left(|\sigma''| - \frac{|\sigma' + \sigma'' - \nu_k^\varepsilon(v^L)|}{4(80 \, C_{10})^2} \right) +$$
$$- K_2 e^{K_4 Q(t_0-)} \cdot |\sigma''| \left(|\sigma'| - \frac{\min\{|\sigma' + \sigma'' - \nu_k^\varepsilon(v^L)|, |\sigma'|\}}{4(80 \, C_{10})^2} \right) \cdot (|\xi'| + |\xi''|) +$$
$$- \mathcal{M}_k(\mathcal{J}_0; t_0, t_0)$$

(C_{10} denoting the constant at (6.2)). The above estimates will be useful in dealing with the *interactions of good type*. Clearly, the same estimates hold in the case we consider two concordant fronts of a k-th NGNL2 family interacting together: a shock σ' and a rarefaction σ''.

8.4. Estimates on the increase of the shift-interaction functional.

We establish here several lemmas providing the basic estimates on the increase $\mathcal{E}_i$ of the shift-interaction functional Υ_ξ, due to the occurrence of interactions of bad type, which are the analog of Lemmas 6.6-6.11 derived in Section 6.3.

LEMMA 8.3. *There exist $K_3 > K_2$ and $0 < \overline{\delta}' \leq \overline{\delta} < \delta_2$ such that, in the same setting of Lemma 8.2, assuming that $v^\theta(0) \in \mathcal{D}^{\varepsilon,\overline{\delta}'}$ for all $\theta \in \,]a, b[$, the following holds. Given a PFB of v^θ $\mathcal{J}_0 \in \mathcal{B}^k(t_0)$, $k \in \{1,2\} \cap NGNL$, suppose that the interaction time of **1st good type** $\widehat{t}_0$ for $(\mathcal{J}_0, t_0)$ satisfies either one of the conditions* I)-II) *of Definition 8.1, and does not satisfy condition* VI). *Let $\widehat{t}'_0$ be the interaction time of **2nd good type** for $(\mathcal{J}_0, t_0)$ (according with Definition 8.2)* and set $\widehat\tau_0 \doteq \min\{\widehat{t}_0, \widehat{t}_0'\}$. Then, one has

$$\mathcal{E}_k(\mathcal{J}_0; t_0) - 2K_2 \widehat{\mathcal{E}}_{k,\xi}(\mathcal{J}_0; t_0) \leq \Big[\mathcal{F}_k(\mathcal{J}_0; t_0, \widehat{t}_0) - \mathcal{F}_k(\mathcal{J}_0; t_0, t_0)\Big] +$$
$$+ \Big[\mathcal{G}_k(\mathcal{J}_0; t_0, \widehat{t}_0) - \mathcal{G}_k(\mathcal{J}_0; t_0, t_0)\Big],$$

(8.80)
$$2K_2 \widehat{\mathcal{E}}_{k,\xi}(\mathcal{J}_0; t_0) \leq \Big[\widehat{\mathcal{F}}_{k,\xi}(\mathcal{J}_0; t_0, \widehat\tau_0) - \widehat{\mathcal{F}}_{k,\xi}(\mathcal{J}_0; t_0, t_0)\Big] +$$
$$+ \Big[\widehat{\mathcal{G}}_{k,\xi}(\mathcal{J}_0; t_0, \widehat\tau_0) - \widehat{\mathcal{G}}_{k,\xi}(\mathcal{J}_0; t_0, t_0)\Big] +$$
$$+ \Big[\mathcal{M}_k(\mathcal{J}_0; t_0, \widehat\tau_0) - \mathcal{M}_k(\mathcal{J}_0; t_0, t_0)\Big],$$

($\mathcal{E}_k$, $\mathcal{F}_k$, $\widehat{\mathcal{F}}_{k,\xi}$, $\mathcal{G}_k$, $\widehat{\mathcal{G}}_{k,\xi}$, $\mathcal{M}_k$ being the functionals defined in (8.35), (8.37)-(8.38), and (8.40)).

PROOF.
1. To fix the ideas, we shall consider a PFB for the first family $\mathcal{J}_0 \in \mathcal{B}^1(t_0)$, and we shall assume that the first characteristic family is NGNL1, the other cases being entirely similar. Throughout the proof we adopt the same notations of Lemma 6.6 and thus call σ_h, $1 \leq h \leq H$, the wave-fronts of a polygonal line $x(t)$, $t_0 \leq t \leq \widehat{t}_0$, connecting the interaction points (t_0, x_0), $(\widehat{t}_0, x_{\widehat{t}_0})$, starting with the shock σ_1 generated by the bad interaction at t_0, and satisfying either one of the conditions I)-II) of Definition 8.1. Towards a proof of (8.80), we will first derive an upper bound on $\mathcal{E}_1(\mathcal{J}_0; t_0) - 2K_2 \widehat{\mathcal{E}}_{1,\xi}(\mathcal{J}_0; t_0)$. Namely, we will show that, setting

(8.81)
$$\sum\nolimits^{\mathrm{i}} \doteq \frac{K_1 K_4}{4}\left[\sum_{\substack{\sigma_h'' \in \mathcal{I}^1(\sigma_h, t_h) \\ 0 < h \leq H \\ \sigma_h'' \cdot \sigma_0 < 0}} |\sigma_h \, \sigma_h''| \, \Upsilon_\xi(t_h-)\right] + \frac{K_1 K_4}{8}\left[\sum_{\substack{\sigma_h'' \in \mathcal{I}^2(\sigma_h, t_h), \\ 0 < h \leq H}} |\sigma_h \, \sigma_h''| \, \Upsilon_\xi(t_h-)\right]$$
$$+ \frac{K_4}{4}\left[\sum_{\substack{\sigma_h'' \in \mathcal{I}^1(\sigma_h, t_h) \\ 0 < h \leq H \\ \sigma_h'' \cdot \sigma_0 > 0}} |\sigma_h \, \sigma_h''| \, \Upsilon_\xi(t_h-)\right],$$

there holds

(8.82)
$$\sum\nolimits^{\mathrm{i}} \geq \mathcal{E}_1(\mathcal{J}_0; t_0) - 2K_2 \widehat{\mathcal{E}}_{1,\xi}(\mathcal{J}_0; t_0).$$

Observe that, by Definition 8.1, at every interaction time t_h, $h < H$, the estimates (8.26), (8.28) are not verified, and hence, thanks to (8.52)-(8.53), we deduce

(8.83)
$$\sum_{\substack{\sigma_h'' \in \mathcal{I}^1(\sigma_h, t_h),\ 0 < h < H \\ \sigma_h'' \cdot \sigma_0 < 0}} |\sigma_h''| \leq \frac{\varepsilon}{8 C_{10}}, \qquad \sum_{\substack{\sigma_h'' \in \mathcal{I}^2(\sigma_h, t_h), \\ 0 < h < H}} |\sigma_h''| \leq \frac{\varepsilon}{8 C_{10}}.$$

Then, relying on (6.67), (8.83), by the same arguments of Lemma 6.6 we deduce

$$|\sigma_h| > \frac{|\sigma_1|}{(5C_{10})^2} \qquad \forall \, 1 < h \leq H. \tag{8.84}$$

Observe now that, since $\widehat{t}_0$ satisfies either one of the conditions I)-II) of Definition 8.1, it follows that one among the estimates (8.26)-(8.28) is verified at $t = t_H$. Thus, recalling that by (6.15) one has $K_1 > 2(80)^2 \, C_{10}^3$, and since, by (6.11) there holds

$$E_1(\mathcal{J}_0; t_0) = |\sigma_1 \cdot \sigma'| \leq |\sigma_1| \cdot \varepsilon, \qquad \sigma' \in \mathcal{N}_R^1(\mathcal{J}_0, t_0+), \tag{8.85}$$

relying on one of the estimates (8.26)-(8.28), and on (8.30), (8.84), we obtain

$$\tag{8.86}
\begin{aligned}
\Sigma^{\mathrm{i}} &\geq \frac{K_1 K_4}{16} \cdot \Upsilon_\xi(t_0-) \Bigg[\sum_{\substack{\sigma_h'' \in \mathcal{I}^1(\sigma_h, t_h) \\ 0 < h \leq H \\ \sigma_h'' \cdot \sigma_0 < 0}} |\sigma_h \, \sigma_h''| + \sum_{\substack{\sigma_h'' \in \mathcal{I}^2(\sigma_h, t_h), \\ 0 < h \leq H}} |\sigma_h \, \sigma_h''| \Bigg] + \\
&\quad + \frac{K_4}{8} \cdot \Upsilon_\xi(t_0-) \Bigg[\sum_{\substack{\sigma_h'' \in \mathcal{I}^1(\sigma_h, t_h) \\ 0 < h \leq H \\ \sigma_h'' \cdot \sigma_0 > 0}} |\sigma_h \, \sigma_h''| \Bigg] \\
&\geq \frac{K_4}{4} \cdot \Upsilon_\xi(t_0-) \cdot \frac{|\sigma_1|}{(5C_{10})^2} \cdot (40 \, C_{10})^2 \cdot \varepsilon \\
&\geq K_4 \, \Upsilon_\xi(t_0-) \, |\sigma_1| \cdot \varepsilon \\
&\geq K_4 \, \Upsilon_\xi(t_0-) \, E_i(\mathcal{J}_0; t_0)
\end{aligned}$$

which, because of (8.35), proves (8.82). Thus, observing that

$$\big[\mathcal{F}_1(\mathcal{J}_0; t_0, \widehat{t}_0) - \mathcal{F}_1(\mathcal{J}_0; t_0, t_0)\big] + \big[\mathcal{G}_1(\mathcal{J}_0; t_0, \widehat{t}_0) - \mathcal{G}_1(\mathcal{J}_0; t_0, t_0)\big] \geq \Sigma^{\mathrm{i}}, \tag{8.87}$$

from (8.82) we deduce the first inequality in (8.80).

2. We will next establish an upper bound on $\widehat{\mathcal{E}}_{1,\xi}(\mathcal{J}_0; t_0)$. Let $\widehat{t}_0$ be the interaction time of **2nd good type** for $(\mathcal{J}_0, t_0)$ (according with Definition 8.2). Observe that, since the set $\mathcal{N}_S^1(\mathcal{J}_0, t)$ consists of a single shock for all $t < \widehat{t}_0$ (cfr. Remark 8.2), if $\widehat{t}_0 < \widehat{t}_0$ the polygonal line of Definition 8.2 connecting the interaction points $(\widehat{t}_0, x_{\widehat{t}_0})$ and (t_0, x_0) is contained in $x(\cdot)$. Thus, we are led to consider the following three cases.

CASE 1: $\widehat{t}_0 \geq \widehat{t}_0$.

CASE 2: $\widehat{t}_0 = t_{\widehat{H}}$ for some $\widehat{H} < H$, and condition VII) of Definition 8.2 is verified.

CASE 3: $\widehat{t}_0 = t_{\widehat{H}}$ for some $\widehat{H} < H$, and condition VIII) of Definition 8.2 is verified.

CASE 1. Since $\widehat{t}'_0 \geq \widehat{t}_0$, by Definition 8.2 at every interaction time t_h, $h < H$, the estimates (8.31)-(8.34) are not verified, and hence, by (8.53), there holds

$$
\sum_{\substack{\sigma''_h \in \mathcal{I}^1(\sigma_h, t_h),\ 0<h<H \\ (\sigma_h\,\xi_h)\cdot(\sigma''_h\,\xi''_h)>0}} |\sigma''_h\,\xi''_h| \leq (40 C_{10})^2 |\xi_1| \cdot \varepsilon,
$$

$$
\sum_{\substack{\sigma''_h \in \mathcal{I}^2(\sigma_h, t_h) \\ 0<h<H}} |\sigma''_h\,\xi''_h| \leq (40 C_{10})^2 |\xi_1| \cdot \varepsilon,
$$

(8.88)

$$
\sum_{\substack{\sigma''_h \in \mathcal{I}(\sigma_h, t_h) \\ 0<h<H}} |\sigma''_h\,\xi_h| \leq (40 C_{10})^2 |\xi_1| \cdot \varepsilon,
$$

$$
\sum_{\substack{\sigma''_h \in \mathcal{I}^1(\sigma_h, t_h),\ 0<h<H \\ (\sigma_h\,\xi_h)\cdot(\sigma''_h\,\xi''_h)<0}} |\sigma''_h\,\xi''_h| \leq \frac{1}{6} |\xi_1| \cdot \varepsilon.
$$

Recall that, by Remark 6.1, the outgoing shock generated by the bad interaction at t_0 has strength $|\sigma_1| \geq \varepsilon$, and observe that the 1-wave emerging after every interaction at $t_h, 0 \leq h < H$ consists of a single front (see Remark 8.2). To fix the ideas, assume that $\sigma_1 \xi_1 > 0$. Then, taking

(8.89)
$$\overline{\delta}' < \min\left\{\overline{\delta},\ \frac{1}{40^3\, C_{10}^2\, C_{20}}\right\},$$

relying on (8.53), (8.88), and applying Lemma 8.1, with the same type of computations performed in Lemma 6.6 we find
(8.90)
$$
\sigma_h\,\xi_h \geq \sigma_1\,\xi_1 - \frac{6}{5}\left[\sum_{\substack{\sigma''_k \in \mathcal{I}^1(\sigma_k, t_k) \\ 0<k<h \\ (\sigma_k\,\xi_k)\cdot(\sigma''_k\,\xi''_k)<0}} |\sigma''_k\,\xi''_k|\right] - \frac{1}{40^3\, C_{10}^2}\left[\sum_{\substack{\sigma''_k \in \mathcal{I}(\sigma_k, t_k) \\ 0<k<h}} |\sigma''_k\,\xi_k|\right] +
$$

$$
- \frac{1}{40^3\, C_{10}^2}\left[\sum_{\substack{\sigma''_k \in \mathcal{I}^1(\sigma_k, t_k) \\ 0<k<h \\ (\sigma_k\,\xi_k)\cdot(\sigma''_k\,\xi''_k)>0}} |\sigma''_k\,\xi''_k| + \sum_{\substack{\sigma''_k \in \mathcal{I}^2(\sigma_k, t_k) \\ 0<k<h}} |\sigma''_k\,\xi''_k|\right]
$$

$$
\geq \sigma_1\,\xi_1 - \frac{2}{5}|\xi_1| \cdot \varepsilon
$$

$$
\geq \frac{3}{5}\sigma_1\,\xi_1 \qquad \forall\, 1 < h \leq H.
$$

On the other hand, since $\widehat{t}_0$ satisfies either one of the conditions I)-II) of Definition 8.1, it follows that one among the estimates (8.26)-(8.28) is verified at $t = t_H$.

Then, relying on (8.90), we will show that an upper bound on $2K_2 \widehat{\mathcal{E}}_{1,\xi}(\mathcal{J}_0; t_0)$ is provided by

(8.91)
$$\sum^{\mathrm{ii}} \doteq \frac{K_3}{4}\left[\sum_{\substack{\sigma_h'' \in \mathcal{I}^1(\sigma_h, t_h) \\ 0 < h \leq H \\ \sigma_h'' \cdot \sigma_0 < 0}} |\sigma_h \, \sigma_h''| \, |\xi_h| \cdot e^{K_4 Q(t_h+)}\right] +$$
$$+ \frac{K_3}{4}\left[\sum_{\substack{\sigma_h'' \in \mathcal{I}^2(\sigma_h, t_h), \\ 0 < h \leq H}} |\sigma_h \, \sigma_h''| \, |\xi_h| \cdot e^{K_4 Q(t_h+)}\right] +$$
$$+ \frac{K_2}{4}\left[\sum_{\substack{\sigma_h'' \in \mathcal{I}^1(\sigma_h, t_h) \\ 0 < h \leq H \\ \sigma_h'' \cdot \sigma_0 > 0}} |\sigma_h \, \sigma_h''| \, |\xi_h| \cdot e^{K_4 Q(t_h+)}\right]$$

(where ξ_h are the shift rates of σ_h). Indeed, assuming $K_3 > 640\, C_{10} K_2$, and using (8.53), (8.85), (8.90), and one of the estimates (8.26)-(8.28), we derive

(8.92)
$$\sum^{\mathrm{ii}} \geq \frac{K_3}{20} |\sigma_1 \xi_1| \left[\sum_{\substack{\sigma_h'' \in \mathcal{I}^1(\sigma_h, t_h) \\ 0 < h \leq H \\ \sigma_h'' \cdot \sigma_0 < 0}} |\sigma_h''| + \sum_{\substack{\sigma_h'' \in \mathcal{I}^2(\sigma_h, t_h), \\ 0 < h \leq H}} |\sigma_h''|\right] + \frac{K_2}{20} |\sigma_1 \xi_1| \left[\sum_{\substack{\sigma_h'' \in \mathcal{I}^1(\sigma_h, t_h) \\ 0 < h \leq H \\ \sigma_h'' \cdot \sigma_0 > 0}} |\sigma_h''|\right]$$

$$\geq 2 K_2 \, |\sigma_1 \xi_1| \, e^{K_4 Q(t_0+)} \cdot \varepsilon$$

$$\geq 2 K_2 \, \widehat{\mathcal{E}}_{1,\xi}(\mathcal{J}_0; t_0).$$

Thus, since in this case one has $\widehat{\tau}_0 = \widehat{t}_0$, observing that
(8.93)
$$\left[\widehat{\mathcal{F}}_{1,\xi}(\mathcal{J}_0; t_0, \widehat{t}_0) - \widehat{\mathcal{F}}_{1,\xi}(\mathcal{J}_0; t_0, t_0)\right] + \left[\widehat{\mathcal{G}}_{1,\xi}(\mathcal{J}_0; t_0, \widehat{t}_0) - \widehat{\mathcal{G}}_{1,\xi}(\mathcal{J}_0; t_0, t_0)\right] \geq \sum^{\mathrm{ii}},$$

from (8.92), we obtain the second inequality in (8.80).

3. CASE 2. Since $\widehat{t}_0 = t_{\widehat{H}} < \widehat{t}_0$, we cannot derive the estimate (8.90). Thus, instead of proving (8.92), we will show that an upper bound on $2K_2 \widehat{\mathcal{E}}_{1,\xi}(\mathcal{J}_0; t_0)$ is provided by

(8.94)
$$\sum^{\mathrm{iii}} \doteq \frac{K_2}{4}\left[\sum_{\substack{\sigma_h'' \in \mathcal{I}(\sigma_h, t_h) \\ 0 < h \leq \widehat{H}}} |\sigma_h \, \sigma_h''| \cdot (|\xi_h| + |\xi_h''|) \cdot e^{K_4 Q(t_h+)}\right]$$

(where ξ_h, ξ_h'' are the shift rates of σ_h, σ_h''). In fact, since $\widehat{t}_0 = t_{\widehat{H}}$ satisfies condition VII) of Definition 8.2, relying on one of the estimates (8.31)-(8.33), and on

(8.53), (8.84), we find
(8.95)
$$\sum\nolimits^{\text{iii}} \geq \frac{K_2}{4} \cdot \frac{|\sigma_1|}{(5C_{10})^2} \cdot \left[\sum_{\substack{\sigma_h'' \in \mathcal{I}^1(\sigma_h, t_h) \\ 0 < h \leq \widehat{H} \\ (\sigma_h \xi_h) \cdot (\sigma_h'' \xi_h'') > 0}} |\sigma_h'' \xi_h''| + \sum_{\substack{\sigma_h'' \in \mathcal{I}^2(\sigma_h, t_h), \\ 0 < h \leq \widehat{H}}} |\sigma_h'' \xi_h''| + \sum_{\substack{\sigma_h'' \in \mathcal{I}(\sigma_h, t_h), \\ 0 < h \leq \widehat{H}}} |\sigma_h'' \xi_h| \right]$$

$$\geq \frac{K_2}{4} \cdot \frac{|\sigma_1|}{(5C_{10})^2} \cdot 2 \, (10 C_{10})^2 \cdot e^{K_4 Q(t_0+)} \, |\xi_1| \cdot \varepsilon$$

$$\geq 2 K_2 \cdot e^{K_4 Q(t_0+)} \, |\sigma_1 \xi_1| \cdot \varepsilon$$

$$\geq 2 K_2 \, \widehat{\mathcal{E}}_{1,\xi}(\mathcal{J}_0; t_0) \, .$$

Then, observing that in this case one has $\widehat{\tau}_0 = \widehat{t}_0'$, and that, assuming $K_3 > 2 K_2$, there holds
(8.96)
$$\left[\widehat{\mathcal{F}}_{1,\xi}(\mathcal{J}_0; t_0, \widehat{t}_0') - \widehat{\mathcal{F}}_{1,\xi}(\mathcal{J}_0; t_0, t_0) \right] + \left[\widehat{\mathcal{G}}_{1,\xi}(\mathcal{J}_0; t_0, \widehat{t}_0') - \widehat{\mathcal{G}}_{1,\xi}(\mathcal{J}_0; t_0, t_0) \right] \geq \sum\nolimits^{\text{iii}},$$

that, together with (8.95), yields the second inequality in (8.80).

4. CASE 3. We will show that, taking

(8.97)
$$\overline{\delta}' < \min \left\{ \overline{\delta}, \frac{1}{384 \, K_2} \right\},$$

the quantity

(8.98)
$$\sum\nolimits^{\text{iv}} \doteq \sum_{\substack{\sigma_h'' \in \mathcal{I}^1(\sigma_h, t_h), \, 0 < h \leq \widehat{H} \\ (\sigma_h \xi_h) \cdot (\sigma_h'' \xi_h'') < 0}} \frac{\min \left\{ |\sigma_h \xi_h|, \, |\sigma_h'' \xi_h''| \right\}}{8}$$

provides an upper bound on $2 K_2 \, \widehat{\mathcal{E}}_{1,\xi}(\mathcal{J}_0; t_0)$. To fix the ideas assume $\sigma_1 \xi_1 > 0$. Observe that, with the same computation in (8.90), we deduce the estimates

$$\sigma_h \xi_h \geq \frac{3}{5} \sigma_1 \xi_1 > 0 \qquad \forall \, 1 < h \leq \widehat{H} \, ,$$

$$\sigma_{\widehat{H}+1} \xi_{\widehat{H}+1} \geq \frac{3}{5} \sigma_1 \xi_1 - \frac{6}{5} |\sigma_{\widehat{H}+1} \xi_{\widehat{H}+1}|,$$

which, in turn, by Lemma 8.1, yield the implications

(8.99)
$$\left. \begin{array}{l} \sigma_h'' \in \mathcal{I}^1(\sigma_h, t_h) \\ 0 < h < \widehat{H} \\ (\sigma_h \xi_h) \cdot (\sigma_h'' \xi_h'') < 0 \end{array} \right\} \quad \Longrightarrow \quad \min \left\{ |\sigma_h \xi_h|, |\sigma_h'' \xi_h''| \right\} = |\sigma_h'' \xi_h''|,$$

$$\sigma_{\widehat{H}+1} \xi_{\widehat{H}+1} > 0 \quad \Longrightarrow \quad \min \left\{ |\sigma_{\widehat{H}} \xi_{\widehat{H}}|, |\sigma_{\widehat{H}}'' \xi_{\widehat{H}}''| \right\} = |\sigma_{\widehat{H}}'' \xi_{\widehat{H}}''|,$$

$$\sigma_{\widehat{H}+1} \xi_{\widehat{H}+1} < 0 \quad \Longrightarrow \quad |\sigma_{\widehat{H}} \xi_{\widehat{H}}| > \frac{\sigma_1 \xi_1}{2} \, .$$

Moreover, observe that, since $v^\theta(0) \in \mathcal{D}^{\varepsilon,\bar{\delta}}$, by Proposition 3.4 one has $|\sigma_1| < 2\bar{\delta}'$. Then, relying on (8.34), (8.53), (8.85), (8.97), (8.99), and since $|\sigma_1| \geq \varepsilon$ (cfr. Remark 6.1), we derive

(8.100)
$$\sum^{\mathrm{iv}} \geq \min\left\{ \frac{|\sigma_1 \xi_1|}{16}, \sum_{\substack{\sigma_h'' \in \mathcal{I}^1(\sigma_h, t_h),\ 0 < h \leq \widehat{H} \\ (\sigma_h \xi_h) \cdot (\sigma_h'' \xi_h'') < 0}} \frac{|\sigma_h'' \xi_h''|}{8} \right\}$$
$$\geq \frac{|\xi_1| \cdot \varepsilon}{96}$$
$$\geq 4 K_2 \bar{\delta}' |\xi_1| \cdot \varepsilon$$
$$\geq 2 K_2 \widehat{\mathcal{E}}_{1,\xi}(\mathcal{J}_0; t_0).$$

Then, observing that in this case one has $\widehat{\tau}_0 = \widehat{t}_0'$, and that

(8.101)
$$\left[\mathcal{M}_1(\mathcal{J}_0; t_0, \widehat{t}_0') - \mathcal{M}_1(\mathcal{J}_0; t_0, t_0) \right] \geq \sum^{\mathrm{iv}},$$

from (8.100), we obtain the second inequality in (8.80). This completes the proof of the lemma. □

LEMMA 8.4. *There exist $K_3 > K_2$ and $0 < \bar{\delta}' \leq \bar{\delta} < \delta_2$ such that, in the same setting of Lemma 8.2, assuming that $v^\theta(0) \in \mathcal{D}^{\varepsilon,\bar{\delta}'}$ for all $\theta \in]a, b[$, the following holds. Given a PFB of v^θ $\mathcal{J}_0 \in \mathcal{B}^k(t_0)$, $k \in \{1,2\} \cap NGNL$, suppose that the interaction time of **1st good type** $\widehat{t}_0$ for $(\mathcal{J}_0, t_0)$ satisfies either one of the conditions III)-IV) of Definition 8.1, and does not satisfy condition VI). Let $\widehat{t}_0'$ be the interaction time of **2nd good type** for $(\mathcal{J}_0, t_0)$ (according with Definition 8.2) and set $\widehat{\tau}_0 \doteq \min\{\widehat{t}_0, \widehat{t}_0'\}$. Then, one has*

$$\mathcal{E}_k(\mathcal{J}_0; t_0) - 2K_2 \widehat{\mathcal{E}}_{k,\xi}(\mathcal{J}_0; t_0) \leq \left[\mathcal{F}_k(\mathcal{J}_0; t_0, \widehat{t}_0) - \mathcal{F}_k(\mathcal{J}_0; t_0, t_0) \right] +$$
$$+ \left[\mathcal{G}_k(\mathcal{J}_0; t_0, \widehat{t}_0) - \mathcal{G}_k(\mathcal{J}_0; t_0, t_0) \right] +$$
$$+ \left[\mathcal{H}_k(\mathcal{J}_0; t_0, \widehat{t}_0) - \mathcal{H}_k(\mathcal{J}_0; t_0, t_0) \right],$$

(8.102)
$$2K_2 \widehat{\mathcal{E}}_{k,\xi}(\mathcal{J}_0; t_0) \leq \left[\widehat{\mathcal{F}}_{k,\xi}(\mathcal{J}_0; t_0, \widehat{\tau}_0) - \widehat{\mathcal{F}}_{k,\xi}(\mathcal{J}_0; t_0, t_0) \right] +$$
$$+ \left[\widehat{\mathcal{G}}_{k,\xi}(\mathcal{J}_0; t_0, \widehat{\tau}_0) - \widehat{\mathcal{G}}_{k,\xi}(\mathcal{J}_0; t_0, t_0) \right] +$$
$$+ \left[\widetilde{\mathcal{H}}_{k,\xi}(\mathcal{J}_0; t_0, \widehat{\tau}_0) - \widetilde{\mathcal{H}}_{k,\xi}(\mathcal{J}_0; t_0, t_0) \right] +$$
$$+ \left[\mathcal{M}_k(\mathcal{J}_0; t_0, \widehat{\tau}_0) - \mathcal{M}_k(\mathcal{J}_0; t_0, t_0) \right],$$

($\mathcal{E}_k$, $\widehat{\mathcal{E}}_{k,\xi}$, $\mathcal{F}_k$, $\widehat{\mathcal{F}}_{k,\xi}$, $\mathcal{G}_k$, $\widehat{\mathcal{G}}_{k,\xi}$, $\mathcal{H}_k$, $\widetilde{\mathcal{H}}_{k,\xi}$ $\mathcal{M}_k$ being the functionals defined in (8.35)-(8.40)).

PROOF.

1. As in the previous lemma, we shall consider only a PFB for the first family $\mathcal{J}_0 \in \mathcal{B}^1(t_0)$, and we shall assume that the first family is NGNL1, the other cases being entirely similar. Let σ'_h, $1 \leq h \leq H'$, be the wave-fronts of a polygonal line $x'(t)$, $t_0 \leq t \leq \bar{t}$, $\bar{t} \leq \widehat{t}_0$, connecting the interaction points (t_0, x_0), $(\bar{t}, x_{\bar{t}})$, starting with the rarefaction front σ'_1 of the composed wave generated by the bad interaction at t_0, and satisfying either one of the conditions III)-IV) of Definition 8.1 (i.e. of Definition 6.1). As in the previous lemma, we will first derive an upper bound on $\mathcal{E}_1(\mathcal{J}_0; t_0) - 2K_2 \widehat{\mathcal{E}}_{1,\xi}(\mathcal{J}_0; t_0)$. Namely, we will show that, letting (t'_h, x'_h), $0 \leq h \leq H'$, be the nodes of $x'(t)$, $t \in [t_0, \bar{t}]$, denoting $\mathcal{J}_h$, $0 \leq h \leq H'$ the pair of fronts interacting at (t'_h, x'_h), calling σ_h the shock in the set $\mathcal{N}^1_S(\mathcal{J}_0, t'_h)$ (consisting of a single front: see Remark 8.2), and setting

$$(8.103) \quad \sum{}^{\mathrm{v}} \doteq \begin{cases} \dfrac{K_1 K_4}{4} \left[\displaystyle\sum_{\substack{\sigma''_h \in \mathcal{I}^1(\sigma'_h, t'_h) \cap \mathcal{C}^1(t'_h) \\ t_0 < t'_h < t'_{H'}}} |\sigma''_h \sigma_h| \cdot \Upsilon_\xi(t'_h-) \right] + \\ \quad + \dfrac{K_1 K_4}{4} |\sigma'_{H'} \sigma_{H'}| \cdot \Upsilon_\xi(t'_{H'}-) \\ \hfill \text{if cond. III) is verified,} \\[1em] \dfrac{K_1 K_4}{4} \left[\displaystyle\sum_{\substack{\sigma''_k \in \mathcal{C}^2(t''_k) \cap \mathcal{N}^2(\mathcal{J}_h, t''_k) \\ t_0 \leq t'_h < t''_k \leq \widehat{t}_0}} |\sigma''_k \sigma_k| \cdot \Upsilon_\xi(t''_k-) \right] \\ \hfill \text{if cond. IV) is verified,} \end{cases}$$

there holds

$$(8.104) \qquad \sum{}^{\mathrm{v}} \geq \mathcal{E}_1(\mathcal{J}_0; t_0) - 2K_2 \widehat{\mathcal{E}}_{1,\xi}(\mathcal{J}_0; t_0).$$

Indeed, by the analysis developed in the proof of Lemma 6.7 it follows that one of the following two cases occurs.

a) $\widehat{t}_0$ satisfies condition III) of Definition 6.1, $\bar{t} = \widehat{t}_0$, $\sigma'_{H'} \in \mathcal{C}^1(\widehat{t}_0)$, and there holds

$$(8.105) \qquad |\sigma'_{H'}| + \sum_{\substack{\sigma''_h \in \mathcal{I}^1(\sigma'_h, t'_h) \cap \mathcal{C}^1(t'_h) \\ t_0 < t'_h < t'_{H'}}} |\sigma''_h| \geq \frac{|\sigma'_1|}{2}.$$

b) $\widehat{t}_0$ satisfies condition IV) of Definition 6.1, and there holds

$$(8.106) \qquad \sum_{\substack{\sigma''_k \in \mathcal{C}^2(t''_k) \cap \mathcal{N}^2(\mathcal{J}_h, t''_k) \\ t_0 \leq t'_h < t''_k \leq \widehat{t}_0}} |\sigma''_k| > |\sigma'_1|$$

where $t_k'' > t_h'$ are the times at which take place some cancellation of a front in $\mathcal{N}^2(\mathcal{J}_h)$ that was approaching a front in $\mathcal{N}_S^1(\mathcal{J}_0, t)$, $t \in \,]t_0, \widehat{t}_0]$.

On the other hand, since the shock σ_h in the set $\mathcal{N}_S^1(\mathcal{J}_0, t_h')$ belongs to a polygonal line $x(t)$, $t_0 \leq t \leq t_h' < \widehat{t}_0$, satisfying conditions I.I)-I.II) of Definition 6.1, and not verifying the estimates (8.26), (8.28), by the same arguments of the proof of Lemma 8.3 we deduce the bound (8.84). Clearly, the same bound holds also for the shock σ_k in the set $\mathcal{N}_S^1(\mathcal{J}_0, t_k'')$, $t_k'' > t_h'$. Thus, observing that by (6.15) one has $K_1 > (40\,C_{10})^2$, and relying on (8.30), (8.84), and on one of the estimates (8.105)-(8.106), we obtain

$$(8.107) \quad \sum^{\mathrm{v}} \geq \frac{K_1 K_4}{8} \cdot \Upsilon_\xi(t_0-) \cdot \frac{|\sigma_1|}{(5 C_{10})^2} \cdot \frac{|\sigma_1'|}{2}$$

$$\geq K_4 \Upsilon_\xi(t_0-) \cdot |\sigma_1 \sigma_1'|$$

which, because of (8.35), (8.85), proves (8.104). Thus, observing that

$$(8.108) \quad \left[\mathcal{H}_1(\mathcal{J}_0; t_0, \widehat{t}_0) - \mathcal{H}_1(\mathcal{J}_0; t_0, t_0)\right] \geq \sum^{\mathrm{v}},$$

from (8.104) we obtain the first inequality in (8.102).

2. We next establish an upper bound on $\widehat{\mathcal{E}}_{1,\xi}(\mathcal{J}_0; t_0)$. Let $\widehat{t}_0'$ be the interaction time of **2nd good type** for $(\mathcal{J}_0, t_0)$ (according with Definition 8.2), and consider the following three cases.

CASE 1: $\widehat{t}_0' \geq \widehat{t}_0$.
CASE 2: $\widehat{t}_0' = t_{\widehat{H}}'$ for some $\widehat{H} < H'$, and condition VII) of Definition 8.2 is verified.
CASE 3: $\widehat{t}_0' = t_{\widehat{H}}'$ for some $\widehat{H} < H'$, and condition VIII) of Definition 8.2 is verified.

In Case 1, since the estimate (8.90) of the previous lemma is still valid, we will show that an upper bound on $2K_2 \widehat{\mathcal{E}}_{1,\xi}(\mathcal{J}_0; t_0)$ is provided by

$$(8.109) \quad \sum^{\mathrm{vi}} \doteq \begin{cases} \dfrac{K_3}{4}\left[\displaystyle\sum_{\substack{\sigma_h'' \in \mathcal{I}^1(\sigma_h', t_h') \cap \mathcal{C}^1(t_h') \\ t_0 < t_h' < t_{H'}'}} |\sigma_h'' \sigma_h|\,|\xi_h| \cdot e^{K_4 Q(t_h'+)}\right] + \\ \qquad + \dfrac{K_3}{4} |\sigma_{H'}' \sigma_{H'}|\,|\xi_{H'}| \cdot e^{K_4 Q(t_{H'}'+)} \qquad \text{if} \quad \text{cond. iii) is verified,} \\[1em] \dfrac{K_3}{4}\left[\displaystyle\sum_{\substack{\sigma_k'' \in \mathcal{C}^2(t_k'') \cap \mathcal{N}^2(\mathcal{J}_h, t_k'') \\ t_0 \leq t_h' < t_k'' \leq \widehat{t}_0}} |\sigma_k'' \sigma_k|\,|\xi_h| \cdot e^{K_4 \Delta Q(t_k'')}\right] \\ \qquad \qquad \qquad \qquad \qquad \text{if} \quad \text{cond. iv) is verified.} \end{cases}$$

Indeed, assuming $K_3 > 160 K_2$, and using (8.53), (8.85), (8.90), and one of the estimates (8.105)-(8.106), we derive

(8.110)
$$\sum\nolimits^{\text{vi}} \geq \frac{K_3}{4} \cdot \frac{3}{5} |\sigma_1 \xi_1| \cdot \frac{|\sigma_1'|}{2} \cdot \frac{e^{K_4 Q(t_0+)}}{2}$$
$$\geq 2K_2 |\sigma_1 \sigma_1'| |\xi_1| e^{K_4 Q(t_0+)}$$
$$\geq 2K_2 \widehat{\mathcal{E}}_{1,\xi}(\mathcal{J}_0; t_0).$$

Thus, since in this case one has $\widehat{\tau}_0 = \widehat{t}_0$, observing that

(8.111)
$$\left[\widetilde{\mathcal{H}}_{1,\xi}(\mathcal{J}_0; t_0, \widehat{t}_0) - \widetilde{\mathcal{H}}_{1,\xi}(\mathcal{J}_0; t_0, t_0)\right] \geq \sum\nolimits^{\text{vi}},$$

from (8.110) we obtain the second inequality in (8.102).

Concerning the Cases 2-3, observe that, if $\widehat{t}_0' < \widehat{t}_0$, the estimates (8.95) and (8.100) of Lemma 8.3 are still valid, and one has $\widehat{\tau}_0 = \widehat{t}_0'$. Then, in Case 2 we derive the second inequality in (8.102) from (8.95), relying on (8.96), while in Case 3 we recover the second inequality in (8.102) from (8.100)-(8.101). This completes the proof of the lemma. $\square$

LEMMA 8.5. *There exist $K_3 > K_2$ and $0 < \overline{\delta}' \leq \overline{\delta} < \delta_2$ such that, in the same setting of Lemma 8.2, assuming that $v^\theta(0) \in \mathcal{D}^{\varepsilon, \overline{\delta}'}$ for all $\theta \in]a, b[$, the following holds. Given a PFB of v^θ $\mathcal{J}_0 \in \mathcal{B}^k(t_0)$, $k \in \{1, 2\} \cap NGNL1$, suppose that the interaction time of **1st good type** $\widehat{t}_0$ for $(\mathcal{J}_0, t_0)$ satisfies condition* V) *of Definition 8.1, and none of the other conditions* I)-IV), VI). *Let σ^s, σ^r be, respectively, the shock in $\mathcal{N}_S^k(\mathcal{J}_0, \widehat{t}_0)$ and the wave-front in $\mathcal{N}_R^k(\mathcal{J}_0, \widehat{t}_0)$ that interact together at $\widehat{t}_0$ (σ^s located on the right of σ^r) according with condition* V), *and let ξ^r, ξ^s be their shift rates. Call v^L the left state of σ^r. Let $\widehat{t}_0'$ be the interaction time of **2nd good type** for $(\mathcal{J}_0, t_0)$ (according with Definition 8.2). Then, one has*

(8.112)
$$\frac{K_4 |\sigma^r| |\sigma^r + \sigma^s - \nu_k^\varepsilon(v^L)|}{4(80 C_{10})^2} \cdot \Upsilon_\xi(\widehat{t}_0-) \leq \left[\mathcal{F}_k(\mathcal{J}_0; t_0, \widehat{t}_0-) - \mathcal{F}_k(\mathcal{J}_0; t_0, t_0)\right] +$$
$$+ \left[\mathcal{G}_k(\mathcal{J}_0; t_0, \widehat{t}_0) - \mathcal{G}_k(\mathcal{J}_0; t_0, t_0)\right],$$

where C_{10} denotes the constant at (6.2), and $\mathcal{F}_k, \mathcal{G}_k$ are the functionals defined in (8.37)-(8.38). Moreover, if $\widehat{t}_0' \geq \widehat{t}_0$, then there holds

$$\frac{K_2 |\sigma^s| \cdot \min\{|\sigma^r + \sigma^s - \nu_k^\varepsilon(v^L)|, |\sigma^r|\} e^{K_4 Q(\widehat{t}_0-)}}{4(80 C_{10})^2} \cdot (|\xi^r| + |\xi^s|) \leq$$

(8.113)
$$\leq K_2 |\sigma^r \sigma^s| \cdot (|\xi^r| + |\xi^s|) e^{K_4 Q(\widehat{t}_0-)} - 2K_2 \widehat{\mathcal{E}}_{k,\xi}(\mathcal{J}_0; \widehat{t}_0-) +$$
$$+ \left[\widehat{\mathcal{F}}_{k,\xi}(\mathcal{J}_0; t_0, \widehat{t}_0-) - \widehat{\mathcal{F}}_{k,\xi}(\mathcal{J}_0; t_0, t_0)\right] +$$
$$+ \left[\widehat{\mathcal{G}}_{k,\xi}(\mathcal{J}_0; t_0, \widehat{t}_0) - \widehat{\mathcal{G}}_{k,\xi}(\mathcal{J}_0; t_0, t_0)\right] +$$
$$+ \left[\mathcal{M}_k(\mathcal{J}_0; t_0, \widehat{t}_0-) - \mathcal{M}_k(\mathcal{J}_0; t_0, t_0)\right],$$

otherwise, one has

$$
\begin{aligned}
(8.114) \quad 2K_2 \widehat{\mathcal{E}}_{k,\xi}(\mathcal{J}_0; t_0) \leq & \left[\widehat{\mathcal{F}}_{k,\xi}(\mathcal{J}_0; t_0, \widehat{t}_0') - \widehat{\mathcal{F}}_{k,\xi}(\mathcal{J}_0; t_0, t_0)\right] + \\
& + \left[\widehat{\mathcal{G}}_{k,\xi}(\mathcal{J}_0; t_0, \widehat{t}_0') - \widehat{\mathcal{G}}_{k,\xi}(\mathcal{J}_0; t_0, t_0)\right] + \\
& + \left[\mathcal{M}_k(\mathcal{J}_0; t_0, \widehat{t}_0') - \mathcal{M}_k(\mathcal{J}_0; t_0, t_0)\right],
\end{aligned}
$$

*($\widehat{\mathcal{E}}_{k,\xi}$, $\widehat{\mathcal{F}}_{k,\xi}$, $\widehat{\mathcal{G}}_{k,\xi}$, $\mathcal{M}_k$ being the functionals defined in (8.35), (8.37)-(8.38), (8.40)). The same type of estimates hold if we are considering a PFB for a k-th NGNL2 family $\mathcal{J}_0 \in \mathcal{B}^k(t_0)$, and $\widehat{t}_0$ is an interaction time of **2nd good type** for $(\mathcal{J}_0, t_0)$, satisfying condition* v) *of Definition 8.1.*

PROOF.
1. As in the previous lemmas, we consider only the case of a PFB for the first family $\mathcal{J}_0 \in \mathcal{B}^1(t_0)$, assuming that the first family is NGNL1, the other cases being entirely similar. Let σ_1, σ_1' be, respectively, the shock and the rarefaction front generated by the bad interaction for the first family at t_0. According with condition v) of Definition 8.1 (i.e. of Definition 6.1), there is a polygonal line $x(t)$, $t_0 \leq t \leq \widehat{t}_0$ having the properties I.I)-I.II), which starts with σ_1 and terminates with the shock $\sigma^s \in \mathcal{N}_S^1(\mathcal{J}_0, \widehat{t}_0)$ that interacts at $\widehat{t}_0$ with the front $\sigma^r \in \mathcal{N}_R^1(\mathcal{J}_0, \widehat{t}_0)$. Let σ_h, $0 < h \leq H$, be the fronts of $x(\cdot)$, where $\sigma_H = \sigma^s$. We will first derive the estimate in (8.112). Namely, we will show that, setting
(8.115)
$$
\sum^{\text{vii}} \doteq \frac{K_1 K_4}{4}\left[\sum_{\substack{\sigma_h'' \in \mathcal{I}^1(\sigma_h, t_h) \\ 0 < h < H \\ \sigma_h'' \cdot \sigma_0 < 0}} |\sigma_h \sigma_h''| \Upsilon_\xi(t_h-)\right] + \frac{K_1 K_4}{8}\left[\sum_{\substack{\sigma_h'' \in \mathcal{I}^2(\sigma_h, t_h), \\ 0 < h < H}} |\sigma_h \sigma_h''| \Upsilon_\xi(t_h-)\right] +
$$
$$
+ \frac{K_4}{4}\left[\sum_{\substack{\sigma_h'' \in \mathcal{I}^1(\sigma_h, t_h) \\ 0 < h < H \\ \sigma_h'' \cdot \sigma_0 > 0}} |\sigma_h \sigma_h''| \Upsilon_\xi(t_h-)\right],
$$

there holds

$$
(8.116) \quad \sum^{\text{vii}} \geq \frac{K_4 |\sigma^r||\sigma^r + \sigma^s - \nu_1^\varepsilon(v^L)|}{4(80\, C_{10})^2} \cdot \Upsilon_\xi(\widehat{t}_0-).
$$

Indeed, with the same computation (6.89) performed in the proof of Lemma 6.8, using (6.2), we deduce

$$
(8.117) \quad |\sigma^r + \sigma^s - \nu_1^\varepsilon(v^L)| \leq C_{10} \sum_{\substack{\sigma_h'' \in \mathcal{I}^2(\sigma_h, t_h) \\ 0 < h < H}} |\sigma_h''| + \sum_{\substack{\sigma_h''' \in \mathcal{I}^1(\sigma_h, t_h),\, \sigma_h''' > 0 \\ 0 < h < H}} |\sigma_h'''|.
$$

On the other hand, as observed in the previous lemma, since the shock σ_h belongs to a polygonal line satisfying conditions I.I)-I.II) of Definition 6.1, and not verifying the estimates (8.26), (8.28), the bound (8.84) is still verified. Hence, recalling that $\sigma^r \leq \varepsilon \leq \sigma_1$ (cfr. Remark 6.1), we have

$$\sigma_h > \frac{\sigma^r}{(5C_{10})^2} \qquad \forall\, 1 \leq h \leq H. \tag{8.118}$$

Then observing that by (6.15) one has $K_1 > C_{10}$, and relying on (8.30), (8.117)-(8.118), we obtain (8.116). Thus, since we have
$$\left[\mathcal{F}_1(\mathcal{J}_0; t_0, \widehat{t}_0-) - \mathcal{F}_1(\mathcal{J}_0; t_0, t_0)\right] + \left[\mathcal{G}_1(\mathcal{J}_0; t_0, \widehat{t}_0) - \mathcal{G}_1(\mathcal{J}_0; t_0, t_0)\right] \geq \sum\nolimits^{\text{vii}}, \tag{8.119}$$

from (8.116) we derive (8.112).

2. We next establish the inequalities (8.113)-(8.114). Let $\widehat{t}_0'$ be the interaction time of **2nd good type** for $(\mathcal{J}_0, t_0)$ (according with Definition 8.2), and consider the following three cases.

CASE 1: $\widehat{t}_0' \geq \widehat{t}_0$.

CASE 2: $\widehat{t}_0' = t_{\widehat{H}}$ for some $\widehat{H} < H$, and condition VII) of Definition 8.2 is verified.

CASE 3: $\widehat{t}_0' = t_{\widehat{H}}$ for some $\widehat{H} < H$, and condition VIII) of Definition 8.2 is verified.

In Case 1 observe that, relying on (8.53), (8.88)-(8.89), recalling that $\sigma_1 \geq \varepsilon$ (cfr. Remark 6.1), and applying Lemma 8.1, with the same type of computations in (8.90) we obtain

$$\begin{aligned}
|\sigma^s \xi^s| &\leq |\sigma_h \xi_h| + \frac{7}{5}(20\, C_{10})^2 |\xi_1| \cdot \varepsilon \\
&\leq |\sigma_h \xi_h| + \frac{7}{3}(20\, C_{10})^2 |\sigma_h \xi_h| \\
&\leq (40\, C_{10})^2 |\sigma_h \xi_h| \qquad \forall\, 1 \leq h \leq H.
\end{aligned} \tag{8.120}$$

Then, setting

$$\sum\nolimits^{\text{viii}} \doteq \frac{K_2}{4}\left[\sum_{\substack{\sigma_h'' \in \mathcal{I}(\sigma_h, t_h) \\ 0 < h < H}} |\sigma_h\, \sigma_h''| \cdot (|\xi_h| + |\xi_h''|) \cdot e^{K_4 Q(t_h+)}\right] \tag{8.121}$$

(where ξ_h, ξ_h'' are the shift rates of σ_h, σ_h''), and using (8.53), (8.117), (8.120), we derive

$$\begin{aligned}
\frac{K_2 |\sigma^s| \cdot \min\{|\sigma^r + \sigma^s - \nu_1^\varepsilon(v^L)|, |\sigma^r|\}\, e^{K_4 Q(\widehat{t}_0-)}}{4(80\, C_{10})^2} \cdot (|\xi^r| + |\xi^s|) &\leq \\
\leq \frac{K_2 |\sigma^s|\, |\sigma^r + \sigma^s - \nu_1^\varepsilon(v^L)|\, e^{K_4 Q(\widehat{t}_0-)}}{2(80\, C_{10})^2} \cdot |\xi^s| &+ \\
+ K_2 |\sigma^s \sigma^r|\, e^{K_4 Q(\widehat{t}_0-)} \cdot \big(\max\{|\xi^r|, |\xi^s|\} - \min\{|\xi^r|, |\xi^s|\}\big) &\leq \\
\leq \sum\nolimits^{\text{viii}} + K_2 |\sigma^s \sigma^r|\, e^{K_4 Q(\widehat{t}_0-)} \cdot \big(\max\{|\xi^r|, |\xi^s|\} - \min\{|\xi^r|, |\xi^s|\}\big). &
\end{aligned} \tag{8.122}$$

On the other hand, since $\widehat{t}_0$ satisfies condition V) of Definition 8.1, and because $\widehat{t}_0' \geq \widehat{t}_0$, recalling (8.35) we write

(8.123)
$$K_2 |\sigma^s \sigma^r| e^{K_4 Q(\widehat{t}_0-)} \cdot \left(\max\{|\xi^r|, |\xi^s|\} - \min\{|\xi^r|, |\xi^s|\} \right) =$$
$$= K_2 |\sigma^r \sigma^s| e^{K_4 Q(\widehat{t}_0-)} \cdot \left(|\xi^r| + |\xi^s| \right) - 2K_2 \, \widehat{\mathcal{E}}_{1,\xi}(\mathcal{J}_0; \widehat{t}_0-).$$

Thus, observing that
(8.124)
$$\left[\widehat{\mathcal{F}}_{1,\xi}(\mathcal{J}_0; t_0, \widehat{t}_0-) - \widehat{\mathcal{F}}_{1,\xi}(\mathcal{J}_0; t_0, t_0) \right] + \left[\widehat{\mathcal{G}}_{1,\xi}(\mathcal{J}_0; t_0, \widehat{t}_0) - \widehat{\mathcal{G}}_{1,\xi}(\mathcal{J}_0; t_0, t_0) \right] \geq \sum^{\text{viii}},$$

and combining together (8.122)-(8.123), we obtain (8.113).

Concerning the Cases 2-3, observe that, if $\widehat{t}_0' < \widehat{t}_0$, the estimates (8.95) and (8.100) of Lemma 8.3 are still valid, and there hold (8.96), (8.101). Then, in Case 2, we derive (8.114) from (8.95), and relying on (8.96), while, in Case 3, we recover (8.114) from (8.100), (8.101). This completes the proof of the lemma. $\square$

LEMMA 8.6. *In the same setting of Lemma 8.2, let $K_3 > K_2$ and $0 < \overline{\delta}' \leq \overline{\delta}$ be the constants given by Lemmas 8.3-8.5. Consider a PFB of v^θ $\mathcal{J}_0 \in \mathcal{B}^k(t_0)$, $k \in \{1,2\} \cap NGNL$, and assume that the interaction time of **1st good type** $\widehat{t}_0$ for $(\mathcal{J}_0, t_0)$ either satisfies one of the conditions I)-II) of Definition 8.1 and none of the other conditions III)-VI), or satisfies condition V) of Definition 8.1 while does not satisfy condition VI). Let $\widehat{t}_0'$ be the interaction time of **2nd good type** for $(\mathcal{J}_0, t_0)$ (according with Definition 8.2), and let $\widehat{\mathcal{J}}_0, \widehat{\mathcal{J}}_0'$ denote the pairs of fronts interacting at $\widehat{t}_0$, and at $\widehat{t}_0'$. Then, if $\widehat{t}_0' \geq \widehat{t}_0$, one has*
(8.125)
$$\Delta \Upsilon_\xi(\widehat{t}_0) \leq \left[\mathcal{F}_k(\mathcal{J}_0; t_0, \widehat{t}_0-) - \mathcal{F}_k(\mathcal{J}_0; t_0, t_0) \right] + \left[\widehat{\mathcal{F}}_{k,\xi}(\mathcal{J}_0; t_0, \widehat{t}_0-) - \widehat{\mathcal{F}}_{k,\xi}(\mathcal{J}_0; t_0, t_0) \right] +$$
$$+ \left[\mathcal{G}_k(\mathcal{J}_0; t_0, \widehat{t}_0-) - \mathcal{G}_k(\mathcal{J}_0; t_0, t_0) \right] + \left[\widehat{\mathcal{G}}_{k,\xi}(\mathcal{J}_0; t_0, \widehat{t}_0-) - \widehat{\mathcal{G}}_{k,\xi}(\mathcal{J}_0; t_0, t_0) \right] +$$
$$+ \left[\mathcal{M}_k(\mathcal{J}_0; t_0, \widehat{t}_0-) - \mathcal{M}_k(\mathcal{J}_0; t_0, t_0) \right] +$$
$$- \mathcal{E}_k(\mathcal{J}_0; \widehat{t}_0-) + \sum_{i \in NGNL} \mathcal{E}_i(\widehat{\mathcal{J}}_0; \widehat{t}_0),$$

otherwise, there hold:
(8.126)
$$\Delta \Upsilon_\xi(\widehat{t}_0) \leq \left[\mathcal{F}_k(\mathcal{J}_0; t_0, \widehat{t}_0-) - \mathcal{F}_k(\mathcal{J}_0; t_0, t_0) \right] + \left[\mathcal{G}_k(\mathcal{J}_0; t_0, \widehat{t}_0-) - \mathcal{G}_k(\mathcal{J}_0; t_0, t_0) \right] +$$
$$- \mathcal{E}_k(\mathcal{J}_0; \widehat{t}_0-) + \sum_{i \in NGNL} \mathcal{E}_i(\widehat{\mathcal{J}}_0; \widehat{t}_0),$$

$$\Delta \Upsilon_\xi(\widehat{t}_0') \leq \left[\widehat{\mathcal{F}}_{k,\xi}(\mathcal{J}_0; t_0, \widehat{t}_0'-) - \widehat{\mathcal{F}}_{k,\xi}(\mathcal{J}_0; t_0, t_0) \right] + \left[\widehat{\mathcal{G}}_{k,\xi}(\mathcal{J}_0; t_0, \widehat{t}_0'-) - \widehat{\mathcal{G}}_{k,\xi}(\mathcal{J}_0; t_0, t_0) \right] +$$
$$+ \left[\mathcal{M}_k(\mathcal{J}_0; t_0, \widehat{t}_0'-) - \mathcal{M}_k(\mathcal{J}_0; t_0, t_0) \right] +$$
$$- 2K_2 \widehat{\mathcal{E}}_{k,\xi}(\mathcal{J}_0; \widehat{t}_0'-) + \mathcal{E}_i(\widehat{\mathcal{J}}_0'; \widehat{t}_0'), \qquad i \neq k, \; i \in NGNL.$$

PROOF. As in the previous lemmas, we consider only the case of a PFB for the first family $\mathcal{J}_0 \in \mathcal{B}^1(t_0)$, assuming that the first family is NGNL1, the other cases being entirely similar. Suppose first that the ITG1 $\widehat{t}_0'$ for $(\mathcal{J}_0, t_0)$ satisfies either one of conditions I)-II) of Definition 8.1 and does not satisfy conditions III)-VI). Notice that, in this case, by (8.35) there holds

(8.127) $$\left.\begin{array}{l}\mathcal{E}_1(\mathcal{J}_0; \widehat{t}_0'-) = \mathcal{E}_1(\mathcal{J}_0; t_0) - 2K_2 \widehat{\mathcal{E}}_{1,\xi}(\mathcal{J}_0; t_0) \\ \widehat{\mathcal{E}}_{1,\xi}(\mathcal{J}_0; \widehat{t}_0'-) = \widehat{\mathcal{E}}_{1,\xi}(\mathcal{J}_0; t_0)\end{array}\right\} \quad \text{if} \quad \widehat{t}_0' < \widehat{t}_0,$$

$$\mathcal{E}_1(\mathcal{J}_0; \widehat{t}_0'-) = \mathcal{E}_1(\mathcal{J}_0; t_0) \quad \text{if} \quad \widehat{t}_0' \geq \widehat{t}_0.$$

Moreover, observe that, if $\widehat{t}_0' < \widehat{t}_0$, the interaction between the fronts of $\widehat{\mathcal{J}}_0'$ at $\widehat{t}_0'$ is not of bad type for the first family since the set $\mathcal{N}_S^1(\mathcal{J}_0, t)$ consists of a single shock for all $t < \widehat{t}_0$ (cfr. Remark 8.2). Hence, one has

(8.128) $$\mathcal{E}_1(\widehat{\mathcal{J}}_0'; \widehat{t}_0') = 0 \quad \text{if} \quad \widehat{t}_0' < \widehat{t}_0.$$

Therefore, if the ITG1 $\widehat{t}_0'$ for $(\mathcal{J}_0, t_0)$ satisfies either one of conditions I)-II) of Definition 8.1 and does not satisfy conditions III)-VI), the estimates (8.125)-(8.126) are obtained applying Lemmas 8.2-8.3 and relying on (8.127)-(8.128). Next, consider the case that $\widehat{t}_0'$ satisfies condition V) of Definition 8.1 (i.e. of Definition 6.1), but not condition VI). Let σ^s, σ^r be, respectively, the shock in $\mathcal{N}_S^1(\mathcal{J}_0, \widehat{t}_0')$ and the front in $\mathcal{N}_R^1(\mathcal{J}_0, \widehat{t}_0')$ that interact together at $\widehat{t}_0'$, let ξ^r, ξ^s be their shift rates, and call v^L the left state of σ^s. Observe that, if $\widehat{t}_0' \geq \widehat{t}_0$, by (6.11), (8.35), one has

(8.129) $$\mathcal{E}_1(\mathcal{J}_0; \widehat{t}_0'-) = K_4 |\sigma^r \sigma^s| \Upsilon_\xi(\widehat{t}_0'-) + 2K_2 \widehat{\mathcal{E}}_{1,\xi}(\mathcal{J}_0; \widehat{t}_0'-).$$

Then, using (8.79), (8.112)-(8.113), (8.129), we find

$$\Delta \Upsilon_\xi(\widehat{t}_0') \leq -K_4 \Upsilon_\xi(\widehat{t}_0'-) \cdot |\sigma^r| \left(|\sigma^s| - \frac{|\sigma^r + \sigma^s - \nu_1^\varepsilon(v^L)|}{4(80\, C_{10})^2} \right) +$$

$$- K_2 e^{K_4 Q(\widehat{t}_0'-)} \cdot |\sigma^s| \left(|\sigma^r| - \frac{\min\{|\sigma^r + \sigma^s - \nu_1^\varepsilon(v^L)|, |\sigma^r|\}}{4(80\, C_{10})^2} \right) \cdot (|\xi^r| + |\xi^s|)$$

$$\leq -\mathcal{E}_1(\mathcal{J}_0; \widehat{t}_0'-) +$$
$$+ \left[\mathcal{F}_1(\mathcal{J}_0; t_0, \widehat{t}_0'-) - \mathcal{F}_1(\mathcal{J}_0; t_0, t_0) \right] + \left[\widehat{\mathcal{F}}_{1,\xi}(\mathcal{J}_0; t_0, \widehat{t}_0'-) - \widehat{\mathcal{F}}_{1,\xi}(\mathcal{J}_0; t_0, t_0) \right] +$$
$$+ \left[\mathcal{G}_1(\mathcal{J}_0; t_0, \widehat{t}_0'-) - \mathcal{G}_1(\mathcal{J}_0; t_0, t_0) \right] + \left[\widehat{\mathcal{G}}_{1,\xi}(\mathcal{J}_0; t_0, \widehat{t}_0'-) - \widehat{\mathcal{G}}_{1,\xi}(\mathcal{J}_0; t_0, t_0) \right] +$$
$$+ \left[\mathcal{M}_1(\mathcal{J}_0; t_0, \widehat{t}_0'-) - \mathcal{M}_1(\mathcal{J}_0; t_0, t_0) \right],$$

which proves (8.125). On the other hand, in the case $\widehat{t}_0' < \widehat{t}_0$, by (6.11), (8.35), we have

(8.130) $$\begin{aligned}\mathcal{E}_1(\mathcal{J}_0; \widehat{t}_0'-) &= K_4 |\sigma^r \sigma^s| \Upsilon_\xi(\widehat{t}_0'-), \\ \widehat{\mathcal{E}}_{1,\xi}(\mathcal{J}_0; \widehat{t}_0'-) &= \widehat{\mathcal{E}}_{1,\xi}(\mathcal{J}_0; t_0).\end{aligned}$$

Hence, using (8.79), (8.112), (8.130), we get
(8.131)
$$\Delta \Upsilon_\xi(\widehat{t}_0) \leq -K_4 \, \Upsilon_\xi(\widehat{t}_0-) \cdot |\sigma^r| \left(|\sigma^s| - \frac{|\sigma^r + \sigma^s - \nu_1^\varepsilon(v^L)|}{4(80\,C_{10})^2} \right)$$

$$\leq -\mathcal{E}_1(\mathcal{J}_0; \widehat{t}_0-) +$$
$$+ \Big[\mathcal{F}_1(\mathcal{J}_0; t_0, \widehat{t}_0-) - \mathcal{F}_1(\mathcal{J}_0; t_0, t_0) \Big] + \Big[\mathcal{G}_1(\mathcal{J}_0; t_0, \widehat{t}_0-) - \mathcal{G}_1(\mathcal{J}_0; t_0, t_0) \Big],$$

while, applying Lemma 8.2, and relying on (8.114), (8.128), (8.130), we derive

(8.132)
$$\Delta \Upsilon_\xi(\widehat{t}_0') \leq -\Big[\widehat{\mathcal{F}}_{1,\xi}(\mathcal{J}_0; t_0, \widehat{t}_0') + \widehat{\mathcal{G}}_{1,\xi}(\mathcal{J}_0; t_0, \widehat{t}_0') \Big] + \mathcal{E}_2(\widehat{\mathcal{J}}_0'; \widehat{t}_0') \leq$$

$$\leq \Big[\widehat{\mathcal{F}}_{1,\xi}(\mathcal{J}_0; t_0, \widehat{t}_0-) - \widehat{\mathcal{F}}_{1,\xi}(\mathcal{J}_0; t_0, t_0) \Big] + \Big[\widehat{\mathcal{G}}_{1,\xi}(\mathcal{J}_0; t_0, \widehat{t}_0-) - \widehat{\mathcal{G}}_{1,\xi}(\mathcal{J}_0; t_0, t_0) \Big] +$$

$$+ \Big[\mathcal{M}_1(\mathcal{J}_0; t_0, \widehat{t}_0-) - \mathcal{M}_1(\mathcal{J}_0; t_0, t_0) \Big] +$$

$$- 2 K_2 \widehat{\mathcal{E}}_{1,\xi}(\mathcal{J}_0; \widehat{t}_0-) + \mathcal{E}_2(\widehat{\mathcal{J}}_0'; \widehat{t}_0'),$$

where $\mathcal{E}_2(\widehat{\mathcal{J}}_0'; \widehat{t}_0') = 0$ if the second family is GNL or LD (cfr. definitions in Section 8.2). Together (8.131)-(8.132) yield (8.126), thus completing the proof of the lemma. $\square$

LEMMA 8.7. *In the same setting of Lemma 8.2, let $K_3 > K_2$ and $0 < \overline{\delta}' \leq \overline{\delta}$ be the constants given by Lemmas 8.3-8.5. Consider a PFB of v^θ $\mathcal{J}_0 \in \mathcal{B}^k(t_0)$, $k \in \{1,2\} \cap NGNL$, and assume that the interaction time of* **1st good type** $\widehat{t}_0$ *for $(\mathcal{J}_0, t_0)$ satisfies either one of the conditions* III)-IV) *of Definition 8.1, and does not satisfy conditions* V)-VI). *Let $\widehat{t}_0'$ be the interaction time of* **2nd good type** *for $(\mathcal{J}_0, t_0)$ (according with Definition 8.2), and let $\widehat{\mathcal{J}}_0$, $\widehat{\mathcal{J}}_0'$ denote the pairs of fronts interacting at $\widehat{t}_0$, and at $\widehat{t}_0'$. Then, if $\widehat{t}_0' \geq \widehat{t}_0$, one has*
(8.133)
$$\Delta \Upsilon_\xi(\widehat{t}_0) \leq \Big[\mathcal{F}_k(\mathcal{J}_0; t_0, \widehat{t}_0-) - \mathcal{F}_k(\mathcal{J}_0; t_0, t_0) \Big] + \Big[\widehat{\mathcal{F}}_{k,\xi}(\mathcal{J}_0; t_0, \widehat{t}_0-) - \widehat{\mathcal{F}}_{k,\xi}(\mathcal{J}_0; t_0, t_0) \Big] +$$

$$+ \Big[\mathcal{G}_k(\mathcal{J}_0; t_0, \widehat{t}_0-) - \mathcal{G}_k(\mathcal{J}_0; t_0, t_0) \Big] + \Big[\widehat{\mathcal{G}}_{k,\xi}(\mathcal{J}_0; t_0, \widehat{t}_0-) - \widehat{\mathcal{G}}_{k,\xi}(\mathcal{J}_0; t_0, t_0) \Big] +$$

$$+ \Big[\mathcal{H}_k(\mathcal{J}_0; t_0, \widehat{t}_0-) - \mathcal{H}_k(\mathcal{J}_0; t_0, t_0) \Big] + \Big[\widetilde{\mathcal{H}}_{k,\xi}(\mathcal{J}_0; t_0, \widehat{t}_0-) - \widetilde{\mathcal{H}}_{k,\xi}(\mathcal{J}_0; t_0, t_0) \Big] +$$

$$+ \Big[\mathcal{M}_k(\mathcal{J}_0; t_0, \widehat{t}_0-) - \mathcal{M}_k(\mathcal{J}_0; t_0, t_0) \Big] +$$

$$- \mathcal{E}_k(\mathcal{J}_0; \widehat{t}_0-) + \sum_{i \in NGNL} \mathcal{E}_i(\widehat{\mathcal{J}}_0; \widehat{t}_0),$$

otherwise, there hold
(8.134)
$$\Delta \Upsilon_\xi(\widehat{t}_0') \leq \left[\mathcal{F}_k(\mathcal{J}_0; t_0, \widehat{t}_0-) - \mathcal{F}_k(\mathcal{J}_0; t_0, t_0)\right] + \left[\mathcal{G}_k(\mathcal{J}_0; t_0, \widehat{t}_0-) - \mathcal{G}_k(\mathcal{J}_0; t_0, t_0)\right] +$$
$$+ \left[\mathcal{H}_k(\mathcal{J}_0; t_0, \widehat{t}_0-) - \mathcal{H}_k(\mathcal{J}_0; t_0, t_0)\right] +$$
$$- \mathcal{E}_k(\mathcal{J}_0; \widehat{t}_0-) + \sum_{i \in NGNL} \mathcal{E}_i(\widehat{\mathcal{J}}_0; \widehat{t}_0),$$

$$\Delta \Upsilon_\xi(\widehat{t}_0') \leq \left[\widehat{\mathcal{F}}_{k,\xi}(\mathcal{J}_0; t_0, \widehat{t}_0-) - \widehat{\mathcal{F}}_{k,\xi}(\mathcal{J}_0; t_0, t_0)\right] + \left[\widehat{\mathcal{G}}_{k,\xi}(\mathcal{J}_0; t_0, \widehat{t}_0-) - \widehat{\mathcal{G}}_{k,\xi}(\mathcal{J}_0; t_0, t_0)\right] +$$
$$+ \left[\widetilde{\mathcal{H}}_{k,\xi}(\mathcal{J}_0; t_0, \widehat{t}_0-) - \widetilde{\mathcal{H}}_{k,\xi}(\mathcal{J}_0; t_0, t_0)\right] + \left[\mathcal{M}_k(\mathcal{J}_0; t_0, \widehat{t}_0-) - \mathcal{M}_k(\mathcal{J}_0; t_0, t_0)\right] +$$
$$- 2K_2 \widehat{\mathcal{E}}_{k,\xi}(\mathcal{J}_0; \widehat{t}_0-) + \mathcal{E}_i(\widehat{\mathcal{J}}_0'; \widehat{t}_0), \qquad i \neq k, \ i \in NGNL.$$

PROOF. Since the ITG1 $\widehat{t}_0$ for $(\mathcal{J}_0, t_0)$ does not satisfy condition v) of Definition 8.1, then, by (8.35), one has

(8.135)
$$\mathcal{E}_1(\mathcal{J}_0; \widehat{t}_0-) = \begin{cases} \mathcal{E}_1(\mathcal{J}_0; t_0) & \text{if } \widehat{t}_0' \geq \widehat{t}_0, \\ \mathcal{E}_1(\mathcal{J}_0; t_0) - 2K_2 \widehat{\mathcal{E}}_{1,\xi}(\mathcal{J}_0; t_0) & \text{if } \widehat{t}_0' < \widehat{t}_0, \end{cases}$$
$$\widehat{\mathcal{E}}_{1,\xi}(\mathcal{J}_0; \widehat{t}_0-) = \widehat{\mathcal{E}}_{1,\xi}(\mathcal{J}_0; t_0).$$

Hence, the estimates (8.133)-(8.134) are obtained applying Lemmas 8.2, 8.4, and using (8.135). □

LEMMA 8.8. *For a 2×2 system satisfying the assumption* (**A**), *there exists some constant C_{23} independent on ε so that, in the same setting of Lemma 8.2, and letting $\mathcal{E}_k$ be the functional defined in (8.35), there holds*

(8.136)
$$\sum_{k \in NGNL} \sum_{\substack{\mathcal{J} \in \mathcal{B}^k(s) \\ 0 < s \leq t}} \mathcal{E}_k(\mathcal{J}; t+) \leq C_{23} \left(\Upsilon_\xi(t-) + \Upsilon_\xi(t+)\right) \cdot \varepsilon \qquad \forall \, t > 0.$$

PROOF. Observe that, given any $\mathcal{J}_0 \in \mathcal{B}^k(t_0)$, $k \in \{1, 2\} \cap NGNL$, and letting $\widehat{t}_0$ be the interaction time of **1st good type** for $(\mathcal{J}_0, t_0)$, by definition (8.35) we have $\mathcal{E}_k(\mathcal{J}_0, t) = 0$ for all $t > \widehat{t}_0$. On the other hand, by Remark 8.2, we know that, as long as $t < \widehat{t}_0$, the set $\mathcal{N}_S^k(\mathcal{J}_0; t)$ consists of a single wave-front, say σ_t, that cannot be involved in another interaction of bad type for the k-th family. Thus, one has

$$\mathcal{J}', \mathcal{J}'' \in \bigcup_{s \leq t} \mathcal{B}^k(s), \quad \mathcal{J}' \neq \mathcal{J}''$$

$$\mathcal{E}_k(\mathcal{J}', t) \neq 0, \quad \mathcal{E}_k(\mathcal{J}'', t) \neq 0,$$

(8.137)
$$\mathcal{N}_S^k(\mathcal{J}', t) = \{\sigma_t'\}, \quad \mathcal{N}_S^k(\mathcal{J}'', t) = \{\sigma_t''\}$$

$$\Downarrow$$

$$\sigma_t' \neq \sigma_t''.$$

Notice that, since $\sigma_t \in \mathcal{N}_S^k(\mathcal{J}_0; t)$, $t < \widehat{t}_0$, belongs to a polygonal line with the properties I)-II) of Definition 8.1, by the arguments in the proof of Lemma 8.3 we deduce a bound as in (8.84), i.e., if $\sigma_{t_0}^+$ denotes the shock generated by the bad interaction at t_0, there holds

(8.138)
$$|\sigma_{t_0}^+| \leq (5C_{10})^2 |\sigma_t| \quad \forall \, t \in [t_0, \widehat{t}_0[.$$

Moreover, if we let $\widehat{t}'_0$ be the interaction time of **2nd good type** for $(\mathcal{J}_0, t_0)$, and set $\widehat{\tau}_0 \doteq \min\{\widehat{t}_0, \widehat{t}'_0\}$, relying again on the arguments of the proof of Lemma 8.3 we derive a bound as in (8.80), i.e. there holds

(8.139)
$$|\sigma_{t_0}^+ \xi_{t_0}^+| \leq \frac{5}{3} |\sigma_t \xi_t| \quad \forall \, t \in [t_0, \widehat{\tau}_0[,$$

where $\xi_{t_0}^+$, ξ_t are the shift rates of the fronts $\sigma_{t_0}^+$, σ_t. Hence, using (8.30), (8.53), (8.138)-(8.139), and recalling (8.35), we find

(8.140)
$$\widehat{\mathcal{E}}_{k,\varepsilon}(\mathcal{J}_0; t) \leq 2\varepsilon \cdot \max\{|\sigma_{t_0}^+ \xi_{t_0}^+|, |\sigma_t \xi_t|\}$$
$$\leq \varepsilon \cdot \frac{10}{3} |\sigma_t \xi_t|,$$
$$\mathcal{E}_k(\mathcal{J}_0; t) - 2K_2 \widehat{\mathcal{E}}_{i,\varepsilon}(\mathcal{J}_0; t) \leq K_4 \Upsilon_\varepsilon(t-) \varepsilon \cdot \max\{|\sigma_{t_0}^+|, |\sigma_t|\}$$
$$\leq \varepsilon \cdot K_4 (5C_{10})^2 \Upsilon_\varepsilon(t-) |\sigma_t|,$$
$$\forall \, t \geq t_0.$$

From (8.53), (8.137), (8.140), recalling (8.2), we derive

$$\sum_{k \in NGNL} \sum_{\substack{\mathcal{J} \in \mathcal{B}^k(s) \\ 0 < s \leq t}} \mathcal{E}_k(\mathcal{J}; t+) \leq \left(\frac{20}{3} K_2 V_\varepsilon(t+) + 2K_4 (5C_{10})^2 \, \overline{\delta} \, \Upsilon_\varepsilon(t-)\right) \cdot \varepsilon \quad \forall \, t > 0,$$

which clearly yields (8.136). $\square$

8.5. Proof of Propositions 3.5-3.6.

We are now ready to prove the main lemmas which allow to establish Proposition 3.5. In the following, in connection with any pair of fronts $\mathcal{J}_0 \in \mathcal{B}^i(t_0)$, $i \in \{1, 2\} \cap NGNL$, that have an interaction of bad type at t_0, we define the map $\Xi_i(\mathcal{J}_0; \cdot)$ as in (6.9), i.e. we set $\Xi_i(\mathcal{J}_0; t) = 1$ if $t \in [t_0, \widehat{t}_0[$, and $\Xi_i(\mathcal{J}_0; t) = 0$ if $t \geq \widehat{t}_0$, where $\widehat{t}_0$ denotes the interaction time of **1st good type** for $(\mathcal{J}_0, t_0)$

(according with Definition 8.1). Moreover, given $\tau_1 < \tau_2$, and a pair of fronts $\mathcal{J} = \{\sigma_1, \sigma_2\}$ that interact together at τ_2, we will write

$$(8.141) \qquad \mathcal{J} \in \Gamma(\tau_1, \tau_2) \quad \Longleftrightarrow \quad \sigma_i \notin \bigcup_{t' \in]\tau_1, \tau_2]} \bigcup_{\mathcal{J}' \in \mathcal{B}(t')} \mathcal{N}(\mathcal{J}', \tau_2) \quad \forall\, i,$$

i.e. $\mathcal{J} \in \Gamma(\tau_1, \tau_2)$ if and only if each front of $\mathcal{J}$ is not originated by a bad interaction occurring in the time interval $]\tau_1, \tau_2]$.

LEMMA 8.9. *For a 2×2 system satisfying the assumption* (**A**), *there exists constants $K_4 > K_3 > K_2$ and $0 < \delta_3 < \delta_2$ (δ_2 being the constant provided by Proposition 3.2) independent of ε, such that the following holds. Let $\{v^\theta\}_{\theta \in]a,b[}$ be a family of ε-solutions that satisfy the assumptions of Proposition 3.5, and such that $v^\theta(0) \in \mathcal{D}^{\varepsilon, \delta_3}$ for all $\theta \in]a, b[$ ($\mathcal{D}^{\varepsilon, \delta_3}$ denoting the domain at (3.33) with $\delta = \delta_3$). Given $0 \leq \tau_1 < \tau_2$, assume that, for every interaction of bad type occurring within the interval $]\tau_1, \tau_2]$, letting $\widehat{t}_0$ be the corresponding ITG1, one of the following two cases occurs. Either $\widehat{t}_0 > \tau_2$, or $\widehat{t}_0$ does not satisfy condition* VI) *of Definition 8.1. Then, letting $\Upsilon_\xi = \Upsilon_\xi(t)$ be the functional defined at (8.4), there holds*
(8.142)

$$\Upsilon_\xi(\tau_2+) \leq \Upsilon_\xi(\tau_1+) - 2 \sum_{\substack{\mathcal{J}_\ell \in \Gamma(\tau_1, t_\ell) \\ \tau_1 < t_\ell \leq \tau_2}} \sum_{i=1}^{2} \Big[\mathcal{F}_i(\mathcal{J}_\ell; t_\ell, t_\ell) + \mathcal{G}_i(\mathcal{J}_\ell; t_\ell, t_\ell) + \mathcal{H}_i(\mathcal{J}_\ell; t_\ell, t_\ell)\Big] +$$

$$- 2 \sum_{\substack{\mathcal{J}_\ell \in \Gamma(\tau_1, t_\ell) \\ \tau_1 < t_\ell \leq \tau_2}} \sum_{i=1}^{2} \Big[\widehat{\mathcal{F}}_{i,\xi}(\mathcal{J}_\ell; t_\ell, t_\ell) + \widehat{\mathcal{G}}_{i,\xi}(\mathcal{J}_\ell; t_\ell, t_\ell) + \widetilde{\mathcal{H}}_{i,\xi}(\mathcal{J}_\ell; t_\ell, t_\ell) + \mathcal{M}_i(\mathcal{J}_\ell; t_\ell, t_\ell)\Big] +$$

$$+ \sum_{i \in NGNL} \sum_{\substack{\mathcal{J}_\ell \in \mathcal{B}^i(t_\ell) \\ \tau_1 < t_\ell \leq \tau_2}} \Xi_i(\mathcal{J}_\ell; \tau_2) \Big[\mathcal{E}_i(\mathcal{J}_\ell; \tau_2) - 2K_2 \widehat{\mathcal{E}}_{i,\xi}(\mathcal{J}_\ell; \tau_2) - 2\mathcal{F}_i(\mathcal{J}_\ell; t_\ell, \tau_2) - 2\mathcal{G}_i(\mathcal{J}_\ell; t_\ell, \tau_2) - 2\mathcal{H}_i(\mathcal{J}_\ell; t_\ell, \tau_2)\Big] +$$

$$+ \sum_{i \in NGNL} \sum_{\substack{\mathcal{J}_\ell \in \mathcal{B}^i(t_\ell) \\ \tau_1 < t_\ell \leq \tau_2}} \Xi_i(\mathcal{J}_\ell; \tau_2) \Big[2K_2 \widehat{\mathcal{E}}_{i,\xi}(\mathcal{J}_\ell; \tau_2) - 2\widehat{\mathcal{F}}_{i,\xi}(\mathcal{J}_\ell; t_\ell, \tau_2) - 2\widehat{\mathcal{G}}_{i,\xi}(\mathcal{J}_\ell; t_\ell, \tau_2) - 2\widetilde{\mathcal{H}}_{i,\xi}(\mathcal{J}_\ell; t_\ell, \tau_2) - 2\mathcal{M}_i(\mathcal{J}_\ell; t_\ell, \tau_2)\Big]$$

(t_ℓ being the interaction times of the fronts in v^θ, and $\mathcal{J}_\ell$ denoting the pair of fronts interacting at t_ℓ).

PROOF. We will establish (8.141) proceeding by induction on the number of interactions occurring in the time interval $]\tau_1, \tau_2]$. Namely, letting $\tau_1 < t_1 < t_2 < \cdots < t_\ell < t_{\ell+1}, \cdots < t_N \leq \tau_2$, be the times when the interactions take place, and denoting with $\mathcal{J}_\ell, 1 \leq \ell \leq N$, the pair of fronts interacting at t_ℓ, we will show by induction on $\ell \geq 1$ that there holds

$(8.143)_\ell$

$$\Upsilon_\xi(t_\ell+) \leq \Upsilon_\xi(\tau_1+) - 2 \sum_{\substack{\mathcal{J}_h \in \Gamma(\tau_1, t_h) \\ \tau_1 < t_h \leq t_\ell}} \sum_{i=1}^{2} \Big[\mathcal{F}_i(\mathcal{J}_h; t_h, t_h) + \mathcal{G}_i(\mathcal{J}_h; t_h, t_h) + \mathcal{H}_i(\mathcal{J}_h; t_h, t_h) \Big] +$$

$$- 2 \sum_{\substack{\mathcal{J}_h \in \Gamma(\tau_1, t_h) \\ \tau_1 < t_h \leq t_\ell}} \sum_{i=1}^{2} \Big[\widehat{\mathcal{F}}_{i,\xi}(\mathcal{J}_h; t_h, t_h) + \widehat{\mathcal{G}}_{i,\xi}(\mathcal{J}_h; t_h, t_h) + \widetilde{\mathcal{H}}_{i,\xi}(\mathcal{J}_h; t_h, t_h) + \mathcal{M}_i(\mathcal{J}_h; t_h, t_h) \Big] +$$

$$+ \sum_{i \in NGNL} \sum_{\substack{\mathcal{J}_h \in \mathcal{B}^i(t_h) \\ \tau_1 < t_h \leq t_\ell}} \Xi_i(\mathcal{J}_h; t_\ell) \Big[\mathcal{E}_i(\mathcal{J}_h; t_\ell) - 2K_2 \widehat{\mathcal{E}}_{i,\xi}(\mathcal{J}_h; t_\ell) + \\ - 2\mathcal{F}_i(\mathcal{J}_h; t_h, t_\ell) - 2\mathcal{G}_i(\mathcal{J}_h; t_h, t_\ell) + \\ - 2\mathcal{H}_i(\mathcal{J}_h; t_h, t_\ell) \Big] +$$

$$+ \sum_{i \in NGNL} \sum_{\substack{\mathcal{J}_h \in \mathcal{B}^i(t_h) \\ \tau_1 < t_h \leq t_\ell}} \Xi_i(\mathcal{J}_h; t_\ell) \Big[2K_2 \widehat{\mathcal{E}}_{i,\xi}(\mathcal{J}_h; t_\ell) - 2\widehat{\mathcal{F}}_{i,\xi}(\mathcal{J}_h; t_h, t_\ell) + \\ - 2\widehat{\mathcal{G}}_{i,\xi}(\mathcal{J}_h; t_h, t_\ell) - 2\widetilde{\mathcal{H}}_{i,\xi}(\mathcal{J}_h; t_h, t_\ell) + \\ - 2\mathcal{M}_i(\mathcal{J}_h; t_h, t_\ell) \Big].$$

Indeed, if t_ℓ is not an interaction time of **1st good type** for some $\mathcal{B}^i(t_h)$, $t_h \in \,]\tau_1, t_{\ell-1}]$, one can directly verify that the inductive hypotheses $(8.143)_{\ell-1}$ and the estimate (8.78), together, imply $(8.143)_\ell$. On the other hand, if t_ℓ is an ITG1 for some $\mathcal{B}^i(t_h)$, $t_h \in \,]\tau_1, t_{\ell-1}]$, we know by assumption that t_ℓ does not satisfy condition vi) of Definition 8.1. Hence, applying Lemmas 8.6-8.7, by a direct computation we derive $(8.143)_\ell$ from $(8.143)_{\ell-1}$. By induction we thus obtain $(8.143)_N$ which proves (8.142). □

REMARK 8.6. We observe that, in the setting of Lemma 8.9, if both the characteristic families are GNL or LD, we can choose $K_2 = K_3$ in (8.1)-(8.2), and the conclusion of the Lemma still holds.

LEMMA 8.10. *In the same setting of Lemma 8.9, given a PFB of v^θ $\mathcal{J}_0 \in \mathcal{B}^i(t_0)$, $i \in \{1, 2\} \cap NGNL$, and letting $\widehat{t}_0$ be the ITG1 for t_0, assume that, for any other interaction of bad type occurring within the interval $]t_0, \widehat{t}_0[$, letting $\widehat{t}'_0$ be the corresponding ITG1, one of the following two cases occurs. Either $\widehat{t}'_0 > \widehat{t}_0$, or $\widehat{t}'_0$ does not satisfy condition* VI) *of Definition 8.1. Then, letting $\Upsilon_\xi = \Upsilon_\xi(t)$ be*

the functional defined at (8.4), the following estimates hold
(8.144)
$$\Upsilon_\xi(\widehat{t}_0+) \leq \Upsilon_\xi(t_0-) - \sum_{\substack{\mathcal{J}_\ell \in \Gamma(t_0,t_\ell) \\ t_0 \leq t_\ell \leq \widehat{t}_0}} \sum_{i=1}^2 \Big[\mathcal{F}_i(\mathcal{J}_\ell;t_\ell,t_\ell) + \mathcal{G}_i(\mathcal{J}_\ell;t_\ell,t_\ell) + \mathcal{H}_i(\mathcal{J}_\ell;t_\ell,t_\ell)\Big] +$$

$$- \sum_{\substack{\mathcal{J}_\ell \in \Gamma(t_0,t_\ell) \\ t_0 \leq t_\ell \leq \widehat{t}_0}} \sum_{i=1}^2 \Big[\widehat{\mathcal{F}}_{i,\xi}(\mathcal{J}_\ell;t_\ell,t_\ell) + \widehat{\mathcal{G}}_{i,\xi}(\mathcal{J}_\ell;t_\ell,t_\ell) + \widetilde{\mathcal{H}}_{i,\xi}(\mathcal{J}_\ell;t_\ell,t_\ell) + \mathcal{M}_i(\mathcal{J}_\ell;t_\ell,t_\ell)\Big] +$$

$$+ \sum_{i \in NGNL} \sum_{\substack{\mathcal{J}_\ell \in \mathcal{B}^i(t_\ell) \\ t_0 < t_\ell \leq \widehat{t}_0}} \Xi_i(\mathcal{J}_\ell;\widehat{t}_0) \Big[\mathcal{E}_i(\mathcal{J}_\ell;\widehat{t}_0) - 2K_2 \widehat{\mathcal{E}}_{i,\xi}(\mathcal{J}_\ell;\widehat{t}_0) - \mathcal{F}_i(\mathcal{J}_\ell;t_\ell,\widehat{t}_0) - \mathcal{G}_i(\mathcal{J}_\ell;t_\ell,\widehat{t}_0) - \mathcal{H}_i(\mathcal{J}_\ell;t_\ell,\widehat{t}_0)\Big] +$$

$$+ \sum_{i \in NGNL} \sum_{\substack{\mathcal{J}_\ell \in \mathcal{B}^i(t_\ell) \\ t_0 < t_\ell \leq \widehat{t}_0}} \Xi_i(\mathcal{J}_\ell;\widehat{t}_0) \Big[2K_2 \widehat{\mathcal{E}}_{i,\xi}(\mathcal{J}_\ell;\widehat{t}_0) - \widehat{\mathcal{F}}_{i,\xi}(\mathcal{J}_\ell;t_\ell,\widehat{t}_0) - \widehat{\mathcal{G}}_{i,\xi}(\mathcal{J}_\ell;t_\ell,\widehat{t}_0) - \widetilde{\mathcal{H}}_{i,\xi}(\mathcal{J}_\ell;t_\ell,\widehat{t}_0) - \mathcal{M}_i(\mathcal{J}_\ell;t_\ell,\widehat{t}_0)\Big],$$

(8.145)
$$\Upsilon_\xi(\widehat{t}_0+) \leq \Upsilon_\xi(t_0+) - \sum_{\substack{\mathcal{J}_\ell \in \Gamma(t_0,t_\ell) \\ t_0 < t_\ell \leq \widehat{t}_0}} \sum_{i=1}^2 \Big[\mathcal{F}_i(\mathcal{J}_\ell;t_\ell,t_\ell) + \mathcal{G}_i(\mathcal{J}_\ell;t_\ell,t_\ell) + \mathcal{H}_i(\mathcal{J}_\ell;t_\ell,t_\ell)\Big] +$$

$$- \sum_{\substack{\mathcal{J}_\ell \in \Gamma(t_0,t_\ell) \\ t_0 < t_\ell \leq \widehat{t}_0}} \sum_{i=1}^2 \Big[\widehat{\mathcal{F}}_{i,\xi}(\mathcal{J}_\ell;t_\ell,t_\ell) + \widehat{\mathcal{G}}_{i,\xi}(\mathcal{J}_\ell;t_\ell,t_\ell) + \widetilde{\mathcal{H}}_{i,\xi}(\mathcal{J}_\ell;t_\ell,t_\ell) + \mathcal{M}_i(\mathcal{J}_\ell;t_\ell,t_\ell)\Big] +$$

$$+ \sum_{i \in NGNL} \sum_{\substack{\mathcal{J}_\ell \in \mathcal{B}^i(t_\ell) \\ t_0 < t_\ell \leq \widehat{t}_0}} \Xi_i(\mathcal{J}_\ell;\widehat{t}_0) \Big[\mathcal{E}_i(\mathcal{J}_\ell;\widehat{t}_0) - 2K_2 \widehat{\mathcal{E}}_{i,\xi}(\mathcal{J}_\ell;\widehat{t}_0) - \mathcal{F}_i(\mathcal{J}_\ell;t_\ell,\widehat{t}_0) - \mathcal{G}_i(\mathcal{J}_\ell;t_\ell,\widehat{t}_0) - \mathcal{H}_i(\mathcal{J}_\ell;t_\ell,\widehat{t}_0)\Big] +$$

$$+ \sum_{i \in NGNL} \sum_{\substack{\mathcal{J}_\ell \in \mathcal{B}^i(t_\ell) \\ t_0 < t_\ell \leq \widehat{t}_0}} \Xi_i(\mathcal{J}_\ell;\widehat{t}_0) \Big[2K_2 \widehat{\mathcal{E}}_{i,\xi}(\mathcal{J}_\ell;\widehat{t}_0) - \widehat{\mathcal{F}}_{i,\xi}(\mathcal{J}_\ell;t_\ell,\widehat{t}_0) - \widehat{\mathcal{G}}_{i,\xi}(\mathcal{J}_\ell;t_\ell,\widehat{t}_0) - \widetilde{\mathcal{H}}_{i,\xi}(\mathcal{J}_\ell;t_\ell,\widehat{t}_0) - \mathcal{M}_i(\mathcal{J}_\ell;t_\ell,\widehat{t}_0)\Big]$$

(t_ℓ being the interaction times of the fronts in v^θ, and $\mathcal{J}_\ell$ denoting the pair of fronts interacting at t_ℓ).

PROOF. Fix $\tau_1 < t_0$ so that no interaction occurs in the interval $[\tau_1, t_0[$, and observe that, in the case $\widehat{t}_0$ does not satisfy condition VI) of Definition 8.1, one immediately obtains the estimates (8.144)-(8.145) applying Lemma 8.9 on $[\tau_1, \widehat{t}_0]$. Next, assume that $\widehat{t}_0$ satisfies condition VI) of Definition 8.1, and fix $\tau_1 \in]t_0, \widehat{t}_0[$, $\tau_2 > \widehat{t}_0$ so that no interaction occurs in the intervals $]t_0, \tau_1]$, $[\widehat{t}_0, \tau_2[$. Observe that, by (8.35), (8.41)-(8.45), and by the choice of the bounds ε'_0, δ_2 on the parameter ε of the approximate solution, and on the total strength of the initial data (cfr. Proposition 3.2), there holds

$$\frac{\Upsilon_\xi(t_0-)}{4} \geq \mathcal{E}_i(\mathcal{J}_0, t_0) \geq \Upsilon_\xi(t_0+) - \Upsilon_\xi(t_0-),$$

(8.146)
$$\frac{\Upsilon_\xi(t_0-)}{8} \geq \sum_{i=1}^{2} \Big[\mathcal{F}_i(\mathcal{J}_0; t_0, t_0) + \mathcal{G}_i(\mathcal{J}_0; t_0, t_0) + \mathcal{H}_i(\mathcal{J}_0; t_0, t_0)\Big] +$$
$$+ \sum_{i=1}^{2} \Big[\widehat{\mathcal{F}}_{i,\xi}(\mathcal{J}_0; t_0, t_0) + \widehat{\mathcal{G}}_{i,\xi}(\mathcal{J}_0; t_0, t_0) + \widetilde{\mathcal{H}}_{i,\xi}(\mathcal{J}_0; t_0, t_0) +$$
$$+ \mathcal{M}_i(\mathcal{J}_0; t_0, t_0)\Big].$$

Thus, using (8.29), (8.146), and applying Lemma 8.9 on $[\tau_1, \tau_2]$, we find

$$\Upsilon_\xi(\widehat{t}_0+) \leq \frac{3\Upsilon_\xi(t_0-)}{4} + \frac{\Upsilon_\xi(\widehat{t}_0+) - \Upsilon_\xi(t_0-)}{2} \leq$$
$$\leq \Upsilon_\xi(t_0-) + \frac{\Upsilon_\xi(\widehat{t}_0+) - \Upsilon_\xi(t_0+)}{2} +$$
$$- \sum_{i=1}^{2} \Big[\mathcal{F}_i(\mathcal{J}_0; t_0, t_0) + \mathcal{G}_i(\mathcal{J}_0; t_0, t_0) + \mathcal{H}_i(\mathcal{J}_0; t_0, t_0)\Big] +$$
$$- \sum_{i=1}^{2} \Big[\widehat{\mathcal{F}}_{i,\xi}(\mathcal{J}_0; t_0, t_0) + \widehat{\mathcal{G}}_{i,\xi}(\mathcal{J}_0; t_0, t_0) + \widetilde{\mathcal{H}}_{i,\xi}(\mathcal{J}_0; t_0, t_0) +$$
$$+ \mathcal{M}_i(\mathcal{J}_0; t_0, t_0)\Big]$$
$$\leq \Upsilon_\xi(t_0-) - \sum_{\substack{\mathcal{J}_\ell \in \Gamma(t_0, t_\ell) \\ t_0 \leq t_\ell \leq \widehat{t}_0}} \sum_{i=1}^{2} \Big[\mathcal{F}_i(\mathcal{J}_\ell; t_\ell, t_\ell) + \mathcal{G}_i(\mathcal{J}_\ell; t_\ell, t_\ell) + \mathcal{H}_i(\mathcal{J}_\ell; t_\ell, t_\ell)\Big] +$$
$$- \sum_{\substack{\mathcal{J}_\ell \in \Gamma(t_0, t_\ell) \\ t_0 \leq t_\ell \leq \widehat{t}_0}} \sum_{i=1}^{2} \Big[\widehat{\mathcal{F}}_{i,\xi}(\mathcal{J}_\ell; t_\ell, t_\ell) + \widehat{\mathcal{G}}_{i,\xi}(\mathcal{J}_\ell; t_\ell, t_\ell) +$$
$$+ \widetilde{\mathcal{H}}_{i,\xi}(\mathcal{J}_\ell; t_\ell, t_\ell) + \mathcal{M}_i(\mathcal{J}_\ell; t_\ell, t_\ell)\Big] +$$
$$+ \sum_{i \in NGNL} \sum_{\substack{\mathcal{J}_\ell \in \mathcal{B}^i(t_\ell) \\ t_0 < t_\ell \leq \widehat{t}_0}} \Xi_i(\mathcal{J}_\ell; \widehat{t}_0) \Big[\mathcal{E}_i(\mathcal{J}_\ell; \widehat{t}_0) - 2K_2 \widehat{\mathcal{E}}_{i,\xi}(\mathcal{J}_\ell; \widehat{t}_0) +$$
$$- \mathcal{F}_i(\mathcal{J}_\ell; t_\ell, \widehat{t}_0) - \mathcal{G}_i(\mathcal{J}_\ell; t_\ell, \widehat{t}_0) +$$
$$- \mathcal{H}_i(\mathcal{J}_\ell; t_\ell, \widehat{t}_0)\Big] +$$
$$+ \sum_{i \in NGNL} \sum_{\substack{\mathcal{J}_\ell \in \mathcal{B}^i(t_\ell) \\ t_0 < t_\ell \leq \widehat{t}_0}} \Xi_i(\mathcal{J}_\ell; \widehat{t}_0) \Big[2K_2 \widehat{\mathcal{E}}_{i,\xi}(\mathcal{J}_\ell; \widehat{t}_0) - \widehat{\mathcal{F}}_{i,\xi}(\mathcal{J}_\ell; t_\ell, \widehat{t}_0) +$$
$$- \widehat{\mathcal{G}}_{i,\xi}(\mathcal{J}_\ell; t_\ell, \widehat{t}_0) - \widetilde{\mathcal{H}}_{i,\xi}(\mathcal{J}_\ell; t_\ell, \widehat{t}_0) +$$

$$- \mathcal{M}_i(\mathcal{J}_\ell; t_\ell, \widehat{t}_0)\Big]$$

which proves (8.144). The proof of (8.145) is entirely similar, thus we omit it. $\square$

LEMMA 8.11. *For a 2×2 system satisfying the assumption* **(A)**, *there exist constants ε_0'', and C_{24}, C_{25} so that, in the same setting of Lemma 8.9, given a family of ε-solutions $\{v^\theta\}_{\theta \in \,]a,b[}$ that satisfy the assumptions of Proposition 3.5, with $\varepsilon \leq \varepsilon_0''$, for any $0 < t \leq T$ the following estimates hold*

(8.147) $$\Upsilon_\xi(t+) \leq C_{24} \cdot \big(\Upsilon_\xi(0) + \Upsilon_\xi(t-) \cdot \varepsilon\big),$$

(8.148) $$\Upsilon_\xi(t) \leq C_{25} \cdot \Upsilon_\xi(0).$$

PROOF. Given $0 < t \leq T$, observe that, proceeding by induction on the number of interactions of bad type occurring in $]0, t]$, and relying on Lemmas 8.9-8.10, we derive
(8.149)
$$\Upsilon_\xi(t+) \leq \Upsilon_\xi(0) + \sum_{i \in NGNL} \sum_{\substack{\mathcal{J}_\ell \in \mathcal{B}^i(t_\ell) \\ 0 < t_\ell \leq t}} \Xi_i(\mathcal{J}_\ell; t) \Big[\mathcal{E}_i(\mathcal{J}_\ell; t) - 2K_2 \widehat{\mathcal{E}}_{i,\xi}(\mathcal{J}_\ell; t) +$$
$$- 2\mathcal{F}_i(\mathcal{J}_\ell; t_\ell, t) - 2\mathcal{G}_i(\mathcal{J}_\ell; t_\ell, t) +$$
$$- 2\mathcal{H}_i(\mathcal{J}_\ell; t_\ell, t)\Big] +$$
$$+ \sum_{i \in NGNL} \sum_{\substack{\mathcal{J}_\ell \in \mathcal{B}^i(t_\ell) \\ 0 < t_\ell \leq t}} \Xi_i(\mathcal{J}_\ell; t) \Big[2K_2 \widehat{\mathcal{E}}_{i,\xi}(\mathcal{J}_\ell; t) - 2\widehat{\mathcal{F}}_{i,\xi}(\mathcal{J}_\ell; t_\ell, t) +$$
$$- 2\widehat{\mathcal{G}}_{i,\xi}(\mathcal{J}_\ell; t_\ell, t) - 2\widetilde{\mathcal{H}}_{i,\xi}(\mathcal{J}_\ell; t_\ell, t) +$$
$$- 2\mathcal{M}_i(\mathcal{J}_\ell; t_\ell, t)\Big].$$

Combining together (8.149) with (8.136), and assuming $\varepsilon \leq \varepsilon_0'' \leq 1/(2C_{23})$, we obtain
$$\Upsilon_\xi(t+) \leq \frac{C_{23}}{1 - C_{23}\,\varepsilon} \cdot \big(\Upsilon_\xi(0) + \Upsilon_\xi(t-) \cdot \varepsilon\big)$$
$$\leq 2\,C_{23} \cdot \big(\Upsilon_\xi(0) + \Upsilon_\xi(t-) \cdot \varepsilon\big)$$

which proves (8.147). Concerning (8.148), let $\tau_0 = 0 < \tau_1 < \cdots < \tau_N \leq t$, be the times when the interactions take place in the interval $[0, t]$. Then, relying on (8.147), assuming $\varepsilon \leq \varepsilon_0'' \leq 1/(2C_{24})$, and proceeding by induction on $\ell \geq 1$, we find

$$\Upsilon_\xi(\tau_\ell+) \leq C_{24}\left((C_{24}\,\varepsilon)^\ell + \sum_{i=0}^{\ell-1}(C_{24}\,\varepsilon)^i\right) \cdot \Upsilon_\xi(0)$$
$$\leq C_{24}\left(1 + e^{C_{24}\,\varepsilon}\right) \cdot \Upsilon_\xi(0)$$

which, in particular, for $\ell = N$, yields (8.148), thus completing the proof of the lemma. $\square$

PROOF OF PROPOSITION 3.5. By (8.3), the estimate (3.47) of Proposition 3.5 follows immediately from (8.148), choosing ε_0'' according with Lemma 8.11. □

PROOF OF PROPOSITION 3.6. Choose δ^* independent of ε such that, for every couple of initial conditions $\bar{v}, \bar{w} \in \mathcal{D}^\varepsilon$, all functions

$$\bar{v}^\theta \doteq \bar{v} \cdot \chi_{]-\infty,\theta]} + \bar{w}\chi_{]\theta,+\infty[}$$

belong to the domain $\mathcal{D}^{\varepsilon,\delta_3}$ at (3.33) with $\delta = \delta_3$. The path $\gamma : \theta \mapsto \bar{v}^\theta$ is clearly a pseudopolygonal path and satisfies

$$C_0^{-1}\|\gamma\|_\varepsilon \leq \|\gamma\|_{\mathbb{L}^1} = \|\bar{v} - \bar{w}\|_{\mathbb{L}^1} \leq C_0\|\gamma\|_\varepsilon$$

because of (3.46). Thus, we may apply Proposition 3.5 and recover the estimate (3.49) from (3.47). □

9. COMPLETION OF THE PROOF

In this section we complete the proof of Theorem 1 stated in the Introduction, by establishing Proposition 3.7. Fix a sequence $\{\varepsilon_n\}_{n\in\mathbb{N}}$ of positive real numbers tending to zero. Then, for any initial data $\bar{v} \in \mathcal{D}$, we consider a sequence $\{\bar{v}_n\}_{n\in\mathbb{N}}$ converging to $\bar{v}$ in $\mathbb{L}^1$, with $\bar{v}_n \in \mathcal{D}^{\varepsilon_n}$, and satisfying

$$\limsup_{n\to+\infty} \left[V^{\varepsilon_n}(\bar{v}_n) + Q^{\varepsilon_n}(\bar{v}_n)\right] \leq \frac{\delta^*}{2}.$$

In order to prove that the limit semigroup S at (3.51) is well defined, we will show that the sequence of ε-approximate solutions $\{S_T^{\varepsilon_n}\bar{v}_n\}_{n\in\mathbb{N}}$ evaluated at time $T > 0$ is a Cauchy sequence in $\mathbb{L}^1$. To estimate the $\mathbb{L}^1$ distance between two approximate solutions, we will use the following lemma established in [**BC1**, Lemma 27].

LEMMA 9.1. *Let $S : [0, +\infty[\times\mathcal{D} \to \mathcal{D}$ be a globally lipschitzean semigroup. Let $\bar{v} \in \mathcal{D}$ and let $v : [0,T] \to \mathcal{D}$ be a continuous map whose values are piecewise constant in the (t, x)-plane, with jumps occurring along finitely many polygonal lines. Calling L the Lipschitz constant of the semigroup, one then has*
(9.1)
$$\|v(T) - S_T\bar{v}\|_{\mathbb{L}^1} \leq L\left\{\|v(0) - \bar{v}\|_{\mathbb{L}^1} + \int_0^T \left(\limsup_{h\to 0+} \frac{\|v(t+h) - S_hv(t)\|_{\mathbb{L}^1}}{h}\right) dt\right\}.$$

PROOF OF PROPOSITION 3.7.

1. We will first show that the sequence $\{S_T^{\varepsilon_n}\bar{v}_n\}_{n\in\mathbb{N}}$ is Cauchy in $\mathbb{L}^1$. Applying Lemma 9.1, the difference between two trajectories can be bounded by

(9.2) $\left\|S_T^{\varepsilon_n}\bar{v}_n - S_T^{\varepsilon_m}\bar{v}_m\right\|_{\mathbb{L}^1} \leq L\|\bar{v}_n - \bar{v}_m\|_{\mathbb{L}^1} +$

$$+ L\int_0^T \left(\limsup_{h\to 0+} \frac{\left\|S_h^{\varepsilon_n}(S_t^{\varepsilon_m}\bar{v}_m) - S_{t+h}^{\varepsilon_n}\bar{v}_m\right\|_{\mathbb{L}^1}}{h}\right) dt.$$

By assumption, as $n \to +\infty$ we have $\|\bar{v}_n - \bar{v}_m\|_{\mathbb{L}^1} \to 0$. Hence we have to prove that

$$\sup_{0 \leq t \leq T} \limsup_{h \to 0+} \frac{\left\|S_h^{\varepsilon_n}\left(S_t^{\varepsilon_m} \bar{v}_m\right) - S_{t+h}^{\varepsilon_n} \bar{v}_m\right\|_{\mathbb{L}^1}}{h} \to 0, \qquad \text{as} \quad n \to +\infty.$$

To fix the ideas, assume $\varepsilon_m > \varepsilon_n$ and fix any time t where no interactions occur involving wave-fronts of $v(t,\cdot) \doteq S_t^{\varepsilon_m} \bar{v}_m$. Call $x_1 < \cdots < x_N$ the points where $v(t,\cdot)$ suffers a discontinuity and v_α^L and v_α^R respectively the left and right states of $v(t,\cdot)$ at x_α. Let σ_α, $\alpha = 1, \ldots, N$, be the size of the wave solving the Riemann problem at x_α with initial data (v_α^L, v_α^R), and denote by $k_\alpha \in \{1, 2\}$ its characteristic family. Hence, by (3.20), one of the following two cases holds:

(9.3)
$$v_\alpha^R = \Psi_{k_\alpha}^{\varepsilon_m}(v_\alpha^L, \sigma_\alpha) \qquad \text{if} \qquad \text{the } k_\alpha\text{-th fam is LD, or GNL, or NGNL1,}$$

(9.4)
$$v_\alpha^L = \Psi_{k_\alpha}^{\varepsilon_m}(v_\alpha^R, \sigma_\alpha) \qquad \text{if} \qquad \text{the } k_\alpha\text{-th fam is NGNL2,}$$

with $\Psi_{k_\alpha}^\varepsilon$ defined at (4.55)-(4.57). If w_α^m and w_α^n are respectively the ε_m- and the ε_n-approximate solution of the Riemann problem with initial data (v_α^L, v_α^R), by the self-similarity of such solutions one has

$$(9.5) \quad \limsup_{h \to 0+} \frac{\left\|S_h^{\varepsilon_n}\left(S_t^{\varepsilon_m} \bar{v}_m\right) - S_{t+h}^{\varepsilon_n} \bar{v}_m\right\|_{\mathbb{L}^1}}{h} = \sum_{\alpha=1}^N \int_{-\infty}^{+\infty} \left| w_\alpha^m(1, x) - w_\alpha^n(1, x) \right| dx.$$

In order to estimate the right-hand side of (9.5), we now consider several cases depending on the size and on the characteristic family of σ_α. For sake of simplicity, we assume that σ_α is a wave of the first characteristic family, and that the first family is either GNL, or LD, or NGNL1 i.e. (9.3) holds with $k_\alpha = 1$. Moreover, we shall assume that, if the first family is NGNL1, the state v_α^L belongs to the set $\mathcal{V}_1^+$ defined in (2.20), i.e. $r_1 \bullet \lambda_1(v_\alpha^L) > 0$. The analysis of the other cases is entirely similar. In the following, we call Λ_1^m the speed of σ_α.

CASE 1: Either the first family is LD and σ_α is any wave-front, or the first family is GNL or NGNL1 and $\sigma_\alpha < 0$.
By Definition 4.2, one has $v_\alpha^R = S_1^{\psi,\varepsilon_m}(v_\alpha^L, \sigma_\alpha)$. On the other hand, by (3.20), the ε_n-solution w_α^n will contain, in general, an intermediate state v_α^M, such that

$$v_\alpha^M = \Psi_1^{\varepsilon_n}(v_\alpha^L, \sigma_\alpha') = S_1^{\psi,\varepsilon_n}(v_\alpha^L, \sigma_\alpha'),$$

$$v_\alpha^R = \Psi_2^{\varepsilon_n}(v_\alpha^M, \sigma_\alpha'') \qquad \text{if} \qquad \text{the 2nd fam is LD, or GNL, or NGNL1,}$$

$$v_\alpha^M = \Psi_2^{\varepsilon_n}(v_\alpha^R, \sigma_\alpha'') \qquad \text{if} \qquad \text{the 2nd fam is NGNL2,}$$

for some σ_α', σ_α'', i.e. w_α^n will consist of a single front of the first family of size σ_α', that travels with speed $\Lambda_1^n \doteq \lambda_1^{\psi,\varepsilon_n}(v_\alpha^L, \sigma_\alpha')$, and of possibly several fronts of the second family of total size σ_α''. Then, by (3.10), and applying Lemma 4.2 or Remark 4.5, one finds

$$\left|\sigma_\alpha - \sigma'_\alpha\right| + \left|\sigma''_\alpha\right| = \mathcal{O}(1) \cdot \left|S_1^{\psi,\varepsilon_m}(v_\alpha^L, \sigma_\alpha) - S_1^{\psi,\varepsilon_n}(v_\alpha^L, \sigma_\alpha)\right| = \mathcal{O}(1) \cdot \left|\sigma_\alpha\right|\varepsilon_m,$$

$$\left|\Lambda_1^m - \Lambda_1^n\right| = \mathcal{O}(1) \cdot \left|\lambda_1^{\psi,\varepsilon_m}(v_\alpha^L, \sigma_\alpha) - \lambda_1^{\psi,\varepsilon_n}(v_\alpha^L, \sigma_\alpha)\right| = \mathcal{O}(1) \cdot \varepsilon_m,$$

which, together, yield

$$(9.6) \qquad \int_{-\infty}^{+\infty} \left|w_\alpha^m(1, x) - w_\alpha^n(1, x)\right| dx = \mathcal{O}(1) \cdot \left|\sigma_\alpha\right|\varepsilon_m.$$

CASE 2: Either the first family is GNL and $\sigma_\alpha \in]0, \varepsilon_m]$, or the first family is NGNL1, $v_\alpha^R \in \mathcal{V}_1^+$, and $\sigma_\alpha \in]0, \varepsilon_m]$.

In this case one has $v_\alpha^R = R_1(v_\alpha^L, \sigma_\alpha)$, while w_α^n consists of a piecewise constant rarefaction fan of the first characteristic family of size σ_α, that contains several wave-fronts traveling with the discrete speed $\widetilde{\Lambda}_1^{n,\ell} \doteq \widetilde{\lambda}_1^\ell(\omega_\ell)$ defined in (3.4), where

$$\omega_\ell = (\ell\varepsilon_n, v_{\alpha,2}^L), \qquad \ell = \lfloor v_\alpha^L\rfloor_1, \ldots, \bar{\ell},$$

$$\bar{\ell} = \begin{cases} \lfloor v_\alpha^R\rfloor_1 & \text{if } \lfloor v_\alpha^R\rfloor_1\varepsilon \neq v_{\alpha,1}^R, \\ \lfloor v_\alpha^R\rfloor_1 - 1 & \text{otherwise.} \end{cases}$$

Then, observing that, for all ℓ, one has $|\Lambda_1^m - \widetilde{\Lambda}_1^{n,\ell}| = \mathcal{O}(1) \cdot |\sigma_\alpha|$, we derive

$$(9.7) \qquad \int_{-\infty}^{+\infty} \left|w_\alpha^m(1, x) - w_\alpha^n(1, x)\right| dx = \mathcal{O}(1) \cdot \left|\sigma_\alpha\right|^2 = \mathcal{O}(1) \cdot \left|\sigma_\alpha\right|\varepsilon_m.$$

CASE 3: The first family is NGNL1, $v_\alpha^R \in \mathcal{V}_1^-$, and $\sigma_\alpha > 2\sqrt{\varepsilon_m}$.

In this case one has $v_\alpha^R = S_1^{\psi,\varepsilon_m}(v_\alpha^L, \sigma_\alpha)$, while, by the same observations in Case 1, we deduce that w_α^n consists of a possibly composed wave of the first family of total size σ'_α, and of possibly many wave-fronts of the second family of total size σ''_α, separated by the intermediate state $v_\alpha^M = \Psi_1^{\varepsilon_n}(v_\alpha^L, \sigma'_\alpha)$. Call $\widehat{\Lambda}_1^n$ the speed of the shock component of σ'_α, and let $\widetilde{\Lambda}_1^{n,\ell}$, $\ell = 1, \ldots \bar{\ell}$, be the speeds of the rarefaction wave-fronts of σ'_α. Then, recalling the expressions (4.46), (4.56) of $T_1^{\varepsilon_n}, \Psi_1^{\varepsilon_n}$, it can be easily seen that, thanks to Lemma 4.2 and to (4.51), there holds

$$(9.8) \qquad \begin{aligned}\left|\sigma_\alpha - \sigma'_\alpha\right| + \left|\sigma''_\alpha\right| &= \mathcal{O}(1) \cdot \left(\left|S_1^{\psi,\varepsilon_m}(v_\alpha^L, \sigma_\alpha) - S_1^{\psi,\varepsilon_n}(v_\alpha^L, \sigma_\alpha)\right| + \right.\\ &\qquad \left. + \left|\nu_1^{\varepsilon_m}(v_\alpha^L) - \nu_1^{\varepsilon_n}(v_\alpha^L)\right|\right)\\ &= \mathcal{O}(1) \cdot \varepsilon_m = \mathcal{O}(1) \cdot \left|\sigma_\alpha\right|\sqrt{\varepsilon_m},\end{aligned}$$

and

$$(9.9) \qquad \begin{aligned}\left|\Lambda_1^m - \widetilde{\Lambda}_1^{n,\ell}\right| + \left|\Lambda_1^m - \widehat{\Lambda}_1^n\right| &= \mathcal{O}(1) \cdot \left(\left|\lambda_1^{\psi,\varepsilon_m}(v_\alpha^L, \sigma_\alpha) - \lambda_1^{\psi,\varepsilon_n}(v_\alpha^L, \sigma_\alpha)\right| + \right.\\ &\qquad \left. + \left|\nu_1^{\varepsilon_m}(v_\alpha^L) - \nu_1^{\varepsilon_n}(v_\alpha^L)\right|\right)\\ &= \mathcal{O}(1) \cdot \varepsilon_m.\end{aligned}$$

Hence, relying on (9.8)-(9.9), we get

$$\text{(9.10)} \qquad \int_{-\infty}^{+\infty} \left| w_\alpha^m(1, x) - w_\alpha^n(1, x) \right| dx = \mathcal{O}(1) \cdot |\sigma_\alpha| \sqrt{\varepsilon_m}.$$

CASE 4: The first family is NGNL1, $v_\alpha^R \in \mathcal{V}_1^-$, and $\sigma_\alpha \in]0, 2\sqrt{\varepsilon_m}]$.
In this case, by (3.2) one has $v_\alpha^R = R_1(v_\alpha^L, \sigma_\alpha)$, while the ε_n-solution w_α^n consists of a possibly composed wave of the first family of total size σ_α', and of possibly many wave-fronts of the second family of total size σ_α''. Then, recalling the expressions (4.46), (4.56) of $T_1^{\varepsilon_n}, \Psi_1^{\varepsilon_n}$, and using Lemma 4.2 (and the Taylor expansion of S_1: cfr. [**BC1, B5**]), we get

$$\text{(9.11)} \qquad \begin{aligned} |\sigma_\alpha - \sigma_\alpha'| + |\sigma_\alpha''| &= \mathcal{O}(1) \cdot \left| S_1^{\psi, \varepsilon_n}(v_\alpha^L, \sigma_\alpha) - R_1(v_\alpha^L, \sigma_\alpha) \right| \\ &= \mathcal{O}(1) \cdot |\sigma_\alpha|^3 = \mathcal{O}(1) \cdot |\sigma_\alpha| \varepsilon_m, \end{aligned}$$

while (9.9) still holds (adopting the same notations of Case 3). Hence, relying on (9.9), (9.11), we deduce again (9.6).

Using (9.6), (9.7), (9.10), and the a priori bound on the total variation provided by Proposition 3.1, from (9.5) we obtain

$$\text{(9.12)} \qquad \limsup_{h \to 0+} \frac{\left\| S_h^{\varepsilon_n}(S_t^{\varepsilon_m} \bar{v}_m) - S_{t+h}^{\varepsilon_n} \bar{v}_m \right\|_{\mathbb{L}^1}}{h} = \mathcal{O}(1) \cdot \sqrt{\varepsilon_m}.$$

This establishes that the sequence of approximate solutions is a Cauchy sequence in $\mathbb{L}^1$ and hence it does converge. It follows that the map $S : [0, +\infty[\times \mathcal{D} \to \mathcal{D}$ at (3.51) is well defined.

2. Now it remains to prove that the properties *(i)-(v)* stated in Theorem 1 actually hold. Observe that *(i)-(iv)* are easy consequences of the properties of approximate semigroups. Indeed:

(i) holds due to the uniform estimates on the approximate solutions with initial data in the domains $\mathcal{D}^\varepsilon$ and $\mathcal{D}^{\varepsilon, \delta_3}$, provided by Proposition 3.4;

(ii) holds since each S^ε satisfies the semigroup-like property $S_{t+s}^\varepsilon \bar{v} = S_t^\varepsilon S_s^\varepsilon \bar{v}$;

(iii) holds since the maps S^ε are uniformly lipschitzean (i.e. their Lipschitz constants do not depend on ε) w.r.t. the initial data; moreover they are uniformly lipschitzean w.r.t. the t variable, since the domains $\mathcal{D}^\varepsilon$, $\varepsilon > 0$, contain initial data with uniformly bounded total variation.

(iv) holds since, for any Riemann problem, as $\varepsilon \to 0$ the ε-approximate solution constructed in Sections 3-4 approaches the corresponding unique, admissible (in the sense of Lax), self-similar exact solution (see [**AM2**]).

Concerning *(v)*, we must prove that each trajectory $t \mapsto u(t, \cdot) = S_t \bar{u}$ is a weak solution of the Cauchy problem (1.1)-(1.2), i.e. that
$$\text{(9.13)}$$
$$\int_{-\infty}^{+\infty} \phi(0, x) \bar{u}(x) \, dx + \int_0^{+\infty} \int_{-\infty}^{+\infty} \left[\phi_t(t, x) u(t, x) + \phi_x(t, x) F(u(t, x)) \right] dx dt = 0$$

for any smooth function ϕ having compact support in $\mathbb{R}^2$. Let $u_n(t, \cdot) \doteq S_t^{\varepsilon_n} \bar{u}_n$ be a sequence of approximate solutions, with $\| u_n(t, \cdot) - u(t, \cdot) \|_1 \to 0$ uniformly on compact subsets of $[0, +\infty[$ as $\varepsilon_n \to 0+$. By the regularity assumptions on F, it suffices to prove that

$$(9.14) \quad \lim_{n \to +\infty} \left\{ \int_{-\infty}^{+\infty} \phi(0, x) \overline{u}_n(x) \, dx + \right.$$
$$\left. + \int_0^{+\infty} \int_{-\infty}^{+\infty} \left[\phi_t(t, x) u_n(t, x) + \phi_x(t, x) F(u_n(t, x)) \right] dx \, dt \right\} = 0.$$

Fix $T > 0$ so that $\phi(t, x) = 0$ whenever $t > T$. At any time $t \in [0, T]$, call $x_{i,1}(t) \ldots x_{i,N_i}(t)$ the points where $u_n(t, \cdot)$ suffers a discontinuity of the i-th characteristic family and set

$$\Delta u_n(t, x_{i,\alpha}) \doteq u_n(t, x_{i,\alpha}+) - u_n(t, x_{i,\alpha}-),$$
$$\Delta F(u_n(t, x_{i,\alpha})) \doteq F(u_n(t, x_{i,\alpha}+)) - F(u_n(t, x_{i,\alpha}-)), \quad i = 1, 2.$$

Moreover, in connection with a k-th NGNL family, letting $F^{k,\varepsilon}$ be the approximate flux functions constructed at (4.5)-(4.6), set

$$\Delta F^{k,\varepsilon_n}(u_n(t, x_{k,\alpha})) \doteq F^{k,\varepsilon_n}(u_n(t, x_{k,\alpha}+)) - F^{k,\varepsilon_n}(u_n(t, x_{k,\alpha}-)).$$

Then, using the divergence theorem, we can write the argument of the limit at (9.14) as
(9.15)
$$\int_0^T \sum_{i \notin NGNL} \sum_{\alpha=1}^{N_i} \left[\dot{x}_{i,\alpha}(t) \cdot \Delta u_n(t, x_{i,\alpha}) - \Delta F(u_n(t, x_{i,\alpha})) \right] \phi(t, x_{i,\alpha}(t)) dt +$$
$$+ \int_0^T \sum_{i \in NGNL} \sum_{\alpha=1}^{N_i} \left[\dot{x}_{i,\alpha}(t) \cdot \Delta u_n(t, x_{i,\alpha}) - \Delta F^{i,\varepsilon_n}(u_n(t, x_{i,\alpha})) \right] \phi(t, x_{i,\alpha}(t)) dt +$$
$$+ \int_0^T \sum_{i \in NGNL} \sum_{\alpha=1}^{N_i} \left[\Delta F^{i,\varepsilon_n}(u_n(t, x_{i,\alpha})) - \Delta F(u_n(t, x_{i,\alpha})) \right] \phi(t, x_{i,\alpha}(t)) dt.$$

Call $\sigma_{i,\alpha}$ the size of the wave located at $x_{i,\alpha}$. Observe that, if the i-th family is GNL or LD, by Definition 4.2 the i-th discontinuities of $u_n(t)$ coincide with the waves of the corresponding approximate solution provided by the algorithm of [**BC1**] and hence, with the same arguments in [**BC1**, Section 10], we deduce
(9.16)
$$\left| \dot{x}_{i,\alpha}(t) \cdot \Delta u_n(t, x_{i,\alpha}) - \Delta F(u_n(t, x_{i,\alpha})) \right| = \mathcal{O}(1) |\sigma_{i,\alpha}| \sqrt{\varepsilon_n} \quad \forall \alpha, \quad \text{if} \quad i \notin NGNL.$$

On the other hand, if the i-th family is NGNL, relying on the ε Rankine-Hugoniot equations $(4.20)_i$ satisfied by the approximate shock curve S_i^ε, and applying the estimates stated in Lemma 4.2 and Remark 4.4, we obtain
(9.17)
$$\left| \dot{x}_{i,\alpha}(t) \cdot \Delta u_n(t, x_{i,\alpha}) - \Delta F^{i,\varepsilon_n}(u_n(t, x_{i,\alpha})) \right| = \mathcal{O}(1) |\sigma_{i,\alpha}| \varepsilon_n,$$
$$\forall \alpha, \quad i \in NGNL.$$
$$\Delta F^{i,\varepsilon_n}(u_n(t, x_{i,\alpha})) - \Delta F(u_n(t, x_{i,\alpha})) = \mathcal{O}(1) |\sigma_{i,\alpha}| \varepsilon_n,$$

By the uniform (w.r.t. ε) a priori bound (3.32) on the total variation of the approximate solutions $u_n(t, \cdot)$, the estimates (9.15)-(9.17), together, yield (9.14). This completes the proof of Proposition 3.7 and hence of Theorem 1. □

10. CONCLUSION

The problem of the $\mathbb{L}^1$ stability of solutions for a strictly hyperbolic system of conservation laws has been extensively studied in the last years. The proof of the existence of a Lipschitz continuous semigroup of solutions generated by (1.1) produced in the present paper is based on the same type of homotopy and linearization technique developed by A. Bressan, R.M. Colombo in [**BC1**], for systems of two equations, and by A. Bressan, G. Crasta, B. Piccoli in [**BCP**] for $n \times n$ systems with GNL or LD characteristic fields. In order to estimate the distance between two solutions u and v, one constructs a one parameter path $\gamma_t : \theta \mapsto u^\theta(t)$ of solutions joining u with v, and then study how the $\mathbb{L}^1$ length of γ_t varies in time. By studying a linearized evolution equation for a generalized tangent vector, one derives an a-priori estimate on a suitably defined weighted norm for such tangent vectors. This, in turn, provides a bound on the length of γ_t and hence on the distance $\|u(t) - v(t)\|_{\mathbb{L}^1}$. Unfortunately, due to the lack of regularity of the solutions, the rigorous proof of this result, both in the GNL or LD case pursued in [**BC1**, **BCP**], and in the NGNL case established here, is obtained at the price of heavy technicalities and the implementation of this approach, in the general case of $n \times n$ NGNL systems, appears to be an almost hopeless task.

An alternative proof of the Lipschitz continuity of the semigroup of solutions generated by a strictly hyperbolic system of conservation laws (1.1) is provided in the recent paper [**BB**] of S. Bianchini and A. Bressan by studying the evolution of the first order (infinitesimal) perturbation $z = z(t, x)$ to a solution of the viscous parabolic approximation

$$(10.1) \qquad u_t + F(u)_x = \varepsilon\, u_{xx}$$

of (1.1). The advantage of dealing with the parabolic system (10.1) instead of the hyperbolic system (1.1) is that all solutions of the former are smooth. Therefore, as soon as one establishes an a-priori bound on the $\mathbb{L}^1$ norm of the first order perturbation $z(t, \cdot)$, uniformly valid for all times, and independent of the viscous coefficient ε, the same type of homotopy argument mentioned above immediately yields the (uniform in ε) $\mathbb{L}^1$ stability of solutions of (10.1), which, in turn, implies the Lipschitz continuity of the flow of (1.1) w.r.t. the initial data.

An entirely different approach to establish the Lipschitz stability estimates for solutions of (1.1) was introduced by T.P. liu and T. Yang in [**LY2**, **LY3**] defining explicitly a functional $\Gamma = \Gamma(u, v)$ which is equivalent to the $\mathbb{L}^1$ distance

$$(10.1) \qquad \frac{1}{C} \cdot \|u - v\|_{\mathbb{L}^1} \leq \Gamma(u, v) \leq C \cdot \|u - v\|_{\mathbb{L}^1} \qquad \forall\, u, v,$$

and which is decreasing in time along any pair of solutions of (1.1):

$$(10.2) \qquad \Gamma\big(u(t), v(t)\big) \leq \Gamma\big(u(s), v(s)\big) \qquad \forall\, t > s \geq 0\,.$$

A functional of this type was first introduced in the case of systems with coinciding Hugoniot and rarefaction curves [**LY2**]. The construction of an appropriate functional was then carried out in [**LY3**] for systems of two equations without this coincidence property, and in [**BLY, LY4**] in the general case of $n \times n$ systems with LD or GNL characteristic fields. The functional $\Gamma(t) \doteq \Gamma(u(t,\cdot), v(t,\cdot))$ constructed in all of these works is defined explicitly in terms of the wave patterns in $u(t,\cdot)$ and $v(t,\cdot)$. It consists of three parts: a linear term $\mathcal{L}_\Gamma(t)$ measuring the $\mathbb{L}^1$ distance between $u(t,\cdot)$ and $v(t,\cdot)$, a quadratic term $\mathcal{Q}_\Gamma(t)$ measuring the nonlinear coupling of waves (in both solutions) and distance (between the two solutions) of different characteristic families, which makes use of the strict hyperbolicity of the system, and a generalized "entropy functional" $\mathcal{E}_\Gamma(t)$ that describes the nonlinear coupling of waves and distance of the same characteristic family due to the genuine nonlinearity of the corresponding characteristic field.

The main step towards the construction of a robust functional $\Gamma(t)$ for strictly hyperbolic NGNL systems of conservation laws, will be to provide a quadratic functional $\mathcal{E}_\Gamma(t)$ that captures the nonlinearity of scalar nonconvex conservation laws and plays the same role of the entropy functional introduced by T.P. Liu and T. Yang [**LY1**] for convex conservation laws. We conjecture that the basic idea here should be to construct a functional $\mathcal{E}_\Gamma = \mathcal{E}_\Gamma(u, v)$ defined in terms of a suitable wave decomposition of the distance between u and v, which, in a sense, reflects the different types of nonlinearities of a nonconvex flux $f(u)$. Roughly speaking, in the case of a flux $f(u)$ with a single inflection point, for any given pair of states u_0, v_0, the value of the functional $\mathcal{E}_\Gamma(u_0, v_0)$ should depend on three types of parameters measuring the (possibly null) size of a wave contained in (u_0, v_0) related with the region of increasing characteristic speed, with the region of decreasing characteristic speed, and with the region of maximum (or minimum) characteristic speed around the inflection point. To understand the basic idea behind this conjecture consider the following

EXAMPLE 10.1. For the nonconvex scalar equation

$$u_t + \left[-\frac{u^3}{3}\right]_x = 0, \tag{10.3}$$

consider the one-parameter family of pairs of initial conditions

$$\overline{u}(x) \doteq -\chi_{]0,3[} - 2 \cdot \chi_{]3,4[}, \tag{10.4}$$

$$\overline{v}^\varepsilon(x) \doteq \varepsilon \cdot \chi_{]1,2[}, \qquad \varepsilon \geq 0, \tag{10.5}_\varepsilon$$

where χ_I denotes the characteristic function of the interval I. Call $u(t,x)$ and $v^\varepsilon(t,x)$ the solution of (10.3) with initial condition, respectively, (10.4) and (10.5)$_\varepsilon$. A direct calculation yields

$$u(t,x) = \begin{cases} 0 & \text{if } x < -t/3 \text{ or } x > 4, \\ -1 & \text{if } -t/3 < x < 3 - (7/3)\,t, \\ -3 & \text{if } 3 - (7/3)\,t < x < 4 - 4\,t, \\ -\sqrt{(4-x)/t} & \text{if } 4 - 4\,t < x < 4, \end{cases} \tag{10.6}$$

$$\text{(10.7)} \quad v^\varepsilon(t,x) = \begin{cases} 0 & \text{if} \quad x < 1-(\varepsilon^2/3)\,t \quad \text{or} \quad x > 2, \\ \varepsilon & \text{if} \quad 1-(\varepsilon^2/3)\,t < x < 2-\varepsilon^2\,t, \\ \sqrt{(2-x)/t} & \text{if} \quad 2-\varepsilon^2\,t < x < 2, \end{cases}$$

for any $t < 1/2$, and for all $\varepsilon \leq 4$. The entropy functional introduced in [**LY1**] for a convex conservation law can be expressed in the form

(10.8)
$$\mathcal{E}[\varepsilon](t) = \int_{-\infty}^{+\infty} \lfloor v^\varepsilon(t,x) - u(t,x) \rfloor_+ \Big\{ \text{T.V.}\{u(t,\cdot);\,(x,+\infty)\} +$$
$$+ \text{T.V.}\{v^\varepsilon(t,\cdot);\,(-\infty,x)\} \Big\} dx +$$
$$+ \int_{-\infty}^{+\infty} \lfloor v_\varepsilon(t,x) - u(t,x) \rfloor_- \Big\{ \text{T.V.}\{u(t,\cdot);\,(-\infty,x)\} +$$
$$+ \text{T.V.}\{v^\varepsilon(t,\cdot);\,(x,+\infty)\} \Big\} dx$$
$$\doteq \mathcal{E}_+[\varepsilon](t) + \mathcal{E}_-[\varepsilon](t),$$

where $\lfloor a \rfloor_+ \doteq \max\{a,0\}$ and $\lfloor a \rfloor_- \doteq \max\{-a,0\}$ denote the positive and the negative part of a real number a, while $\text{T.V.}\{u(t,\cdot);\,I\}$ represents the total variation of the function $u(t,\cdot)$ over the interval I. Then, one has

$$\text{(10.9)} \quad \mathcal{E}[\varepsilon](t) = (11 + 8\varepsilon + \varepsilon^2) + \frac{(\varepsilon^4 + 4\varepsilon^3 + 16\varepsilon - 8)}{6} t \qquad \forall\, t \leq 1/12.$$

Observe that there exists some $\bar\varepsilon \in\,]0,1[$ such that, as long as $\varepsilon < \bar\varepsilon$, the map $t \mapsto \mathcal{E}[\varepsilon](t)$ remains decreasing over the interval $[0, 1/12]$. Indeed, for such ε and for $t \leq 1/12$, both solutions $u(t,\cdot)$ and $v^\varepsilon(t,\cdot)$ take values in a sufficiently small neighborhood of $(-\infty, 0]$, i.e. of the region of convexity of the flux function $f(u) = -u^3/3$, so that the functional $\mathcal{E}[\varepsilon]$ may preserve also in this case his basic property of controlling the nonlinear coupling of the wave patterns of the two solutions.

On the other hand, one can easily check that, for $\varepsilon \geq 1$, the functional $\mathcal{E}[\varepsilon]$ is increasing in time. Notice that, for sake of simplicity we have chosen the initial data $\bar u,\, \bar v^\varepsilon$ as in (10.4)-(10.5)$_\varepsilon$ taking values in the interval $[-2, \varepsilon]$, but the same type of computations remains valid for initial conditions with values in an arbitrary small neighborhood of the inflection point $u = 0$. In fact, if we consider the pairs of rescaled initial conditions

$$\text{(10.10)} \qquad \bar u_\rho(x) \doteq \rho \cdot \bar u(x), \qquad \bar v^\varepsilon_\rho(x) \doteq \rho \cdot \bar v^\varepsilon, \qquad \rho > 0,$$

one can verify as above that the corresponding entropy functional (10.8) remains decreasing as long as $\varepsilon < \bar\varepsilon_\rho$, for some $\bar\varepsilon_\rho \in\,]0, \rho[$, while it results to be increasing for $\varepsilon \geq \rho$. This is the consequence of the fact that the definition (10.8) of the functional $\mathcal{E}[\varepsilon]$ in terms of the $\mathbb{L}^1$ distance between the two solutions is not sufficiently robust to deal, at the same time, with the different types of nonlinearities related with the region of convexity and with the region of concavity of the flux function $f(u)$.

To achieve this property, we conjecture that one should define a linear functional $\mathcal{L}' \doteq \mathcal{L}'(u,v)$ and a quadratic functional $\mathcal{E}' \doteq \mathcal{E}'(u,v)$, which depend on a

wave decomposition of the distance between u and v that reflects the three different types of nonlinearities determined by a nonconvex flux function with a single inflection point. In the case of the flux $f(u) = -u^3/3$ of the previous example, for any pair of functions $u(x) \leq 0$ and $v(x)$, this corresponds to consider the following three scalar quantities (see Figure (10.1)):
(10.11)
$$q(x) \doteq q\big(u(x), v(x)\big) = \begin{cases} \lfloor v(x) - u(x) \rfloor_+ & \text{if} \quad v(x) \leq 0, \\ \lfloor -v(x) - u(x) \rfloor_+ & \text{otherwise}, \end{cases}$$

$$p(x) \doteq p\big(u(x), v(x)\big) = \begin{cases} \min\{2v(x), -2u(x)\} & \text{if} \quad u(x) < 0, \ v(x) > 0, \\ 0 & \text{otherwise}, \end{cases}$$

$$m(x) \doteq m\big(u(x), v(x)\big) = v(x) - u(x) - q(x) - p(x).$$

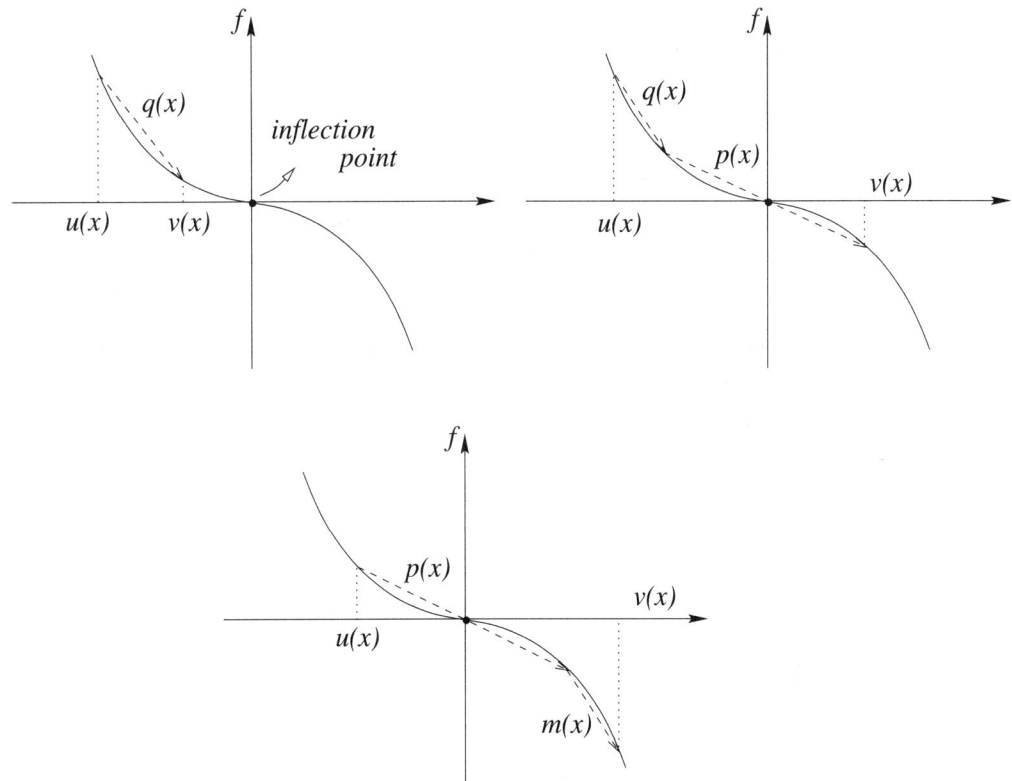

FIGURE 10.1

Here the parameter $q(x)$, $p(x)$ and $m(x)$ represent, respectively, the size of a rarefaction shock, of a fast undercompressive shock, and of a Lax shock, contained in a suitably defined non Laxian solution of the Riemann problem with initial data $(u(x), v(x))$. We then define the functionals

(10.12)
$$\mathcal{L}'(u,v) \doteq \int_{-\infty}^{+\infty} \Big\{ |q(x)| + c_p \cdot |p(x)| + c_m \cdot |m(x)| \Big\} dx,$$

(10.13)
$$\mathcal{E}'(u,v) \doteq \int_{-\infty}^{+\infty} |q(x)| \Big\{ \text{T.V.}\{u;\, (x,+\infty)\} + \text{T.V.}\{v;\, (-\infty,x)\} \Big\} dx +$$
$$+ \int_{-\infty}^{+\infty} |p(x)| \Big\{ \text{T.V.}\{u;\, (x,+\infty)\} + \text{T.V.}\{v;\, (x,+\infty)\} \Big\} dx$$
$$+ \int_{-\infty}^{+\infty} |m(x)| \Big\{ \text{T.V.}\{u;\, (-\infty,x)\} + \text{T.V.}\{v;\, (x,+\infty)\} \Big\} dx$$
$$\doteq \mathcal{E}'_q(u,v) + \mathcal{E}'_p(u,v) + \mathcal{E}'_m(u,v),$$

where c_p, c_m ($c_m > c_p > 1$) are suitable positive constants. Roughly speaking, for any given pair of states $u(x)$, $v(x)$, one may regard the jump $(u(x), v(x))$ either as consisting of a single wave (whenever only one parameter among $q(x)$, $m(x)$, $p(x)$, is non zero) or as the superposition of two waves (in the case where two parameters among $q(x)$, $m(x)$, $p(x)$ are non zero, while the other vanishes). These waves evolve with different behaviours determined by the region of increasing, decreasing, or maximum characteristic speed to which they are related.

Notice that, in the particular case where two solutions $u(t,x)$, $v(t,x)$ take all values on the same side w.r.t. the inflection point, say on its left, according with the definition (10.11) one has $q(t,x) = \lfloor v(t,x) - u(t,x) \rfloor_+$, $p(t,x) = 0$, and $m(t,x) = \lfloor v(t,x) - u(t,x) \rfloor_-$. Therefore, our new functional $\mathcal{E}'(u(t,\cdot), v(t,\cdot))$ reduces to the functional $\mathcal{E}(u(t,\cdot), v(t,\cdot))$ defined as in (10.8), and hence we recover precisely the same functional introduced in [**LY1**] for a convex conservation law.

Computing the functionals (10.12)-(10.13) for the pair of solutions $u(t,x)$, $v^\varepsilon(t,x)$ at (10.6)-(10.7), one finds, for $\varepsilon \geq 1$:

$$\mathcal{L}'\big(u(t,\cdot), v^\varepsilon(t,\cdot)\big) = (4 + \varepsilon + 2c_p - c_m) + \frac{(1-\varepsilon^2)(1 - 2c_p + c_m)}{3} t,$$

$$\mathcal{E}'\big(u(t,\cdot), v^\varepsilon(t,\cdot)\big) = (15 + 8\varepsilon + \varepsilon^2) - \frac{(48 - 4\varepsilon - 4\varepsilon^2 - 4\varepsilon^3 + \varepsilon^4)}{6} t.$$

A direct calculation shows that, if one takes $c_p = 5$, $c_m = 15$, the map

$$t \mapsto \mathcal{L}'\big(u(t,\cdot), v^\varepsilon(t,\cdot)\big) + \mathcal{E}'\big(u(t,\cdot), v^\varepsilon(t,\cdot)\big)$$

is decreasing for any $\varepsilon \leq 4$ as desired. Again, observe that the same type of result holds true if we evaluate the functional $\mathcal{L}' + \mathcal{E}'$ along the pairs of rescaled initial data $\overline{u}_\rho$, $\overline{v}_\rho^\varepsilon$ in (10.10), taking values within the interval $[-2\rho, \rho \cdot \varepsilon]$.

The general construction for NGNL systems of a Lypaunov type functional $\Gamma(t)$ enjoying properties (10.1)-(10.2) is currently pursued by the authors according with the above conjecture. The possible accomplishment of this construction would provide a new, true hyperbolic, proof of the existence of a Lipschitz continuous flow generated by a strictly hyperbolic system of conservation laws with NGNL characteristic fields.

Bibliography

[AM1] F. Ancona, A. Marson, *A note on the Riemann problem for general $n \times n$ conservation laws*, J. Math. Anal. Appl. **260** (2001), 279–293.

[AM2] F. Ancona, A. Marson, *Basic estimates on a front tracking algorithm for general 2×2 conservation laws*, Math. Models Meth. Appl. Sci.-M3AS **12**, No. 2 (2002), 155-182.

[AM3] F. Ancona, A. Marson, *Well-posedness for non genuinely nonlinear conservation laws*, Hyperbolic problems: theory, numerics, applications, Vol. I, II (Magdeburg, 2000), Birkhäuser, Basel, 2001, pp. 29-38.

[AM4] F. Ancona, A. Marson, *A wave-front tracking algorithm for $N \times N$ non genuinely nonlinear conservation laws*, J. Differential Equations **177** (2001), 454–493.

[BJ] P. Baiti, H.K. Jenssen, *On the front-tracking algorithm*, J. Math. Anal. Appl. **217** (1998), 395–404.

[BLP] P. Baiti, P. LeFloch, B. Piccoli, *Uniqueness of classical and nonclassical solutions for nonlinear hyperbolic systems*, J. Differential Equations **172** (2001), 59–82.

[BB] S. Bianchini, A. Bressan, *Vanishing viscosity solutions of nonlinear hyperbolic systems*, preprint SISSA 86/2001/M (2001).

[B1] A. Bressan, *Contractive metrics for nonlinear hyperbolic systems*, Indiana Univ. J. Math. **37** (1988), 409–421.

[B2] A. Bressan, *Global Solutions to systems of conservation laws by wave-front Tracking*, J. Math. Anal. Appl. **170** (1992), 414–432.

[B3] A. Bressan, *A locally contractive metric for systems of conservation laws*, Ann. Scuola Norm. Sup. Pisa **IV-22** (1995), 109–135.

[B4] A. Bressan, *The unique limit of the Glimm scheme*, Arch. Rational Mech. Anal. **130** (1995), 205–230.

[B5] A. Bressan, *Hyperbolic systems of conservation laws. The one-dimensional Cauchy problem*, Oxford University Press, Oxford, 2000.

[BC1] A. Bressan, R.M. Colombo, *The semigroup generated by 2×2 conservation laws*, Arch. Rational Mech. Anal. **133** (1995), 1–75.

[BC2] A. Bressan, R.M. Colombo, *Unique solutions of 2×2 conservation laws with large data*, Indiana Univ. Math. J. **44** (1995), 677–725.

[BCP] A. Bressan, G. Crasta, B. Piccoli, *Well-posedness of the Cauchy problem for $n \times n$ systems of conservation laws*, Mem. Amer. Math. Soc. **146** (2000), no. 694, viii+134.

[BG] A. Bressan, P. Goatin, *Oleinik type estimates and uniqueness for $n \times n$ conservation laws*, J. Differential Equations **156** (1999), 26–49.

[BLF1] A. Bressan, P. LeFloch, *Uniqueness of weak solutions to systems of conservation laws*, Arch. Rational Mech. Anal. **140** (1997), 301–317.

[BLF2] A. Bressan, P. LeFloch, *Structural stability and regularity of entropy solutions to hyperbolic systems of conservation laws*, Indiana Univ. Math. J. **48** (1999), 43–84.

[BL] A. Bressan, M. Lewicka, *A uniqueness condition for hyperbolic systems of conservation laws*, Discrete Contin. Dynam. Systems **6** (2000), 673–682.

[BLY] A. Bressan, T.P. Liu, T. Yang, $\mathbb{L}^1$ *stability estimates for $n \times n$ conservation laws*, Arch. Rational Mech. Anal. **149** (1999), 1–22.

[BM1] A. Bressan, A. Marson, *A variational calculus for discontinuous solutions of systems of conservation laws*, Comm. Partial Differential Equations **20** (1995), 1491–1552.

[BM2] A. Bressan, A. Marson, *Error bounds for a deterministic version of the Glimm scheme*, Arch. Rational Mech. Anal. **142** (1998), 155–176.

[D] C. Dafermos, *Hyperbolic conservation laws in continuum physics*, Springer-Verlag, Berlin Heidelberg, 2000.

[DP1] R. DiPerna, *Global existence of solutions to nonlinear hyperbolic systems of conservation laws*, J. Differential Equations **20** (1976), 187–212.

[DP2] R. DiPerna, *Uniqueness of solutions to hyperbolic conservation laws*, Indiana Univ. Math. J. **28** (1979), 137–188.

[DP3] R. DiPerna, *Convergence of approximate solutions to conservation laws*, Arch. Rational Mech. Anal. **82** (1983), 27–70.

[G] J. Glimm, *Solutions in the large for nonlinear hyperbolic systems of equations*, Comm. Pure Appl. Math. **18** (1965), 697–715.

[GL] J. Glimm, P.D. Lax, *Decay of solutions of systems of nonlinear hyperbolic conservation laws*, Mem. Amer. Math. Soc. No.101 (1970), xvii+112.

[K] S.N. Kruzkov, *First Order Quasilinear Equations in Several Independent Variables*, Math. USSR Sbornik **10** (1970), 217–243.

[La] P.D. Lax, *Hyperbolic Systems of Conservation Laws* II, Comm. on Pure and Applied Math. **10** (1957), 537–566.

[Li1] T.P. Liu, *The Riemann problem for general 2×2 conservation laws*, Trans. Amer. Math. Soc. **199** (1974), 89–112.

[Li2] T.P. Liu, *The Riemann problem for general systems of conservation laws*, J. Differential Equations **18** (1975), 218–234.

[Li3] T.P. Liu, *Admissible solutions of hyperbolic conservation laws*, Memoir Amer. Math. Soc. No.240 **30** (1981), iv+78.

[LY1] T.P. Liu, T. Yang, *A new entropy functional for a scalar conservation law*, Comm. Pure Appl. Math. **52** (1999), 1472–1442.

[LY2] T.P. Liu, T. Yang, $\mathbb{L}^1$ *stability of conservation laws with coinciding Hugoniot and characteristic curves*, Indiana Univ. math. J. **48** (1999), 237–247.

[LY3] T.P. Liu, T. Yang, $\mathbb{L}^1$ *stability of weak solutions for 2×2 systems of hyperbolic conservation laws*, J. Amer. Math. Soc. **12** (1999), 729–774.

[LY4] T.P. Liu, T. Yang, *Well-posedness theory for hyperbolic conservation laws*, Comm. Pure Appl. Math. **52** (1999), 1553–1586.

[LY5] T.P. Liu, T. Yang, *Weak solutions of general systems of hyperbolic conservation laws*, Comm. Math. Phys. **230** (2002), 289-327.

[Ri] N.H.Risebro, *A Front-Tracking Alternative to the Random Choice Method*, Proc. Amer. Math. Soc. **117** (1993), 1125–1139.

[RMS1] T.Ruggeri, A.Muracchini, L.Seccia, *Continuum Approach to Phonon Gas and Shape Canges of Second Sound via Shock Waves Theory*, Nuovo Cimento D **16** (1994), 15–44.

[RMS2] T.Ruggeri, A.Muracchini, L.Seccia, *Second sound and characteristic temperature in solids*, Physical Review B **54** (1996), 332–339.

[Sm] J. Smoller, *Shock Waves and Reaction-Diffusion Equations*, Springer-Verlag, New York, 1984.

Editorial Information

To be published in the *Memoirs*, a paper must be correct, new, nontrivial, and significant. Further, it must be well written and of interest to a substantial number of mathematicians. Piecemeal results, such as an inconclusive step toward an unproved major theorem or a minor variation on a known result, are in general not acceptable for publication. Papers appearing in *Memoirs* are generally longer than those appearing in *Transactions*, which shares the same editorial committee.

As of February 1, 2004, the backlog for this journal was approximately 4 volumes. This estimate is the result of dividing the number of manuscripts for this journal in the Providence office that have not yet gone to the printer on the above date by the average number of monographs per volume over the previous twelve months, reduced by the number of volumes published in four months (the time necessary for preparing a volume for the printer). (There are 6 volumes per year, each containing at least 4 numbers.)

A Consent to Publish and Copyright Agreement is required before a paper will be published in the *Memoirs*. After a paper is accepted for publication, the Providence office will send a Consent to Publish and Copyright Agreement to all authors of the paper. By submitting a paper to the *Memoirs*, authors certify that the results have not been submitted to nor are they under consideration for publication by another journal, conference proceedings, or similar publication.

Information for Authors

Memoirs are printed from camera copy fully prepared by the author. This means that the finished book will look exactly like the copy submitted.

The paper must contain a *descriptive title* and an *abstract* that summarizes the article in language suitable for workers in the general field (algebra, analysis, etc.). The *descriptive title* should be short, but informative; useless or vague phrases such as "some remarks about" or "concerning" should be avoided. The *abstract* should be at least one complete sentence, and at most 300 words. Included with the footnotes to the paper should be the 2000 *Mathematics Subject Classification* representing the primary and secondary subjects of the article. The classifications are accessible from www.ams.org/msc/. The list of classifications is also available in print starting with the 1999 annual index of *Mathematical Reviews*. The Mathematics Subject Classification footnote may be followed by a list of *key words and phrases* describing the subject matter of the article and taken from it. Journal abbreviations used in bibliographies are listed in the latest *Mathematical Reviews* annual index. The series abbreviations are also accessible from www.ams.org/publications/. To help in preparing and verifying references, the AMS offers MR Lookup, a Reference Tool for Linking, at www.ams.org/mrlookup/. When the manuscript is submitted, authors should supply the editor with electronic addresses if available. These will be printed after the postal address at the end of the article.

Electronically prepared manuscripts. The AMS encourages electronically prepared manuscripts, with a strong preference for $\mathcal{AMS}$-LaTeX. To this end, the Society has prepared $\mathcal{AMS}$-LaTeX author packages for each AMS publication. Author packages include instructions for preparing electronic manuscripts, the *AMS Author Handbook*, samples, and a style file that generates the particular design specifications of that publication series. Though $\mathcal{AMS}$-LaTeX is the highly preferred format of TeX, author packages are also available in $\mathcal{AMS}$-TeX.

Authors may retrieve an author package from e-MATH starting from `www.ams.org/tex/` or via FTP to `ftp.ams.org` (login as `anonymous`, enter username as password, and type `cd pub/author-info`). The *AMS Author Handbook* and the *Instruction Manual* are available in PDF format following the author packages link from `www.ams.org/tex/`. The author package can be obtained free of charge by sending email to `pub@ams.org` (Internet) or from the Publication Division, American Mathematical Society, 201 Charles St., Providence, RI 02904, USA. When requesting an author package, please specify $\mathcal{AMS}$-LaTeX or $\mathcal{AMS}$-TeX, Macintosh or IBM (3.5) format, and the publication in which your paper will appear. Please be sure to include your complete mailing address.

Sending electronic files. After acceptance, the source file(s) should be sent to the Providence office (this includes any TeX source file, any graphics files, and the DVI or PostScript file).

Before sending the source file, be sure you have proofread your paper carefully. The files you send must be the EXACT files used to generate the proof copy that was accepted for publication. For all publications, authors are required to send a printed copy of their paper, which exactly matches the copy approved for publication, along with any graphics that will appear in the paper.

TeX files may be submitted by email, FTP, or on diskette. The DVI file(s) and PostScript files should be submitted only by FTP or on diskette unless they are encoded properly to submit through email. (DVI files are binary and PostScript files tend to be very large.)

Electronically prepared manuscripts can be sent via email to `pub-submit@ams.org` (Internet). The subject line of the message should include the publication code to identify it as a Memoir. TeX source files, DVI files, and PostScript files can be transferred over the Internet by FTP to the Internet node `e-math.ams.org` (130.44.1.100).

Electronic graphics. Comprehensive instructions on preparing graphics are available at `www.ams.org/jourhtml/graphics.html`. A few of the major requirements are given here.

Submit files for graphics as EPS (Encapsulated PostScript) files. This includes graphics originated via a graphics application as well as scanned photographs or other computer-generated images. If this is not possible, TIFF files are acceptable as long as they can be opened in Adobe Photoshop or Illustrator. No matter what method was used to produce the graphic, it is necessary to provide a paper copy to the AMS.

Authors using graphics packages for the creation of electronic art should also avoid the use of any lines thinner than 0.5 points in width. Many graphics packages allow the user to specify a "hairline" for a very thin line. Hairlines often look acceptable when proofed on a typical laser printer. However, when produced on a high-resolution laser imagesetter, hairlines become nearly invisible and will be lost entirely in the final printing process.

Screens should be set to values between 15% and 85%. Screens which fall outside of this range are too light or too dark to print correctly. Variations of screens within a graphic should be no less than 10%.

Inquiries. Any inquiries concerning a paper that has been accepted for publication should be sent directly to the Electronic Prepress Department, American Mathematical Society, 201 Charles St., Providence, RI 02904, USA.

Editors

This journal is designed particularly for long research papers, normally at least 80 pages in length, and groups of cognate papers in pure and applied mathematics. Papers intended for publication in the *Memoirs* should be addressed to one of the following editors. In principle the Memoirs welcomes electronic submissions, and some of the editors, those whose names appear below with an asterisk (*), have indicated that they prefer them. However, editors reserve the right to request hard copies after papers have been submitted electronically. Authors are advised to make preliminary email inquiries to editors about whether they are likely to be able to handle submissions in a particular electronic form.

*Algebra to ROBERT GURALNICK, Department of Mathematics, University of Southern California, Los Angeles, CA 90089-1113; email: guralnic@math.usc.edu

Algebraic geometry to DAN ABRAMOVICH, Department of Mathematics, Boston University, 111 Cummington St., Boston, MA 02215; email: abramovic@bu.edu

*Algebraic number theory to V. KUMAR MURTY, Department of Mathematics, University of Toronto, 100 St. George Street, Toronto, ON M5S 1A1, Canada; email: murty@math.toronto.edu

Algebraic topology and cohomology of groups to STEWART PRIDDY, Department of Mathematics, Northwestern University, 2033 Sheridan Road, Evanston, IL 60208-2730; email: priddy@math.nwu.edu

Combinatorics and Lie theory to SERGEY FOMIN, Department of Mathematics, University of Michigan, Ann Arbor, Michigan 48109-1109; email: fomin@umich.edu

Complex analysis and complex geometry to DUONG H. PHONG, Department of Mathematics, Columbia University, 2990 Broadway, New York, NY 10027-0029; email: phong@math.columbia.edu

*Differential geometry and global analysis to LISA C. JEFFREY, Department of Mathematics, University of Toronto, 100 St. George St., Toronto, ON Canada M5S 3G3; email: jeffrey@math.toronto.edu

Dynamical systems and ergodic theory to ROBERT F. WILLIAMS, Department of Mathematics, University of Texas, Austin, Texas 78712-1082; email: bob@math.utexas.edu

*Functional analysis and operator algebras to MARIUS DADARLAT, Department of Mathematics, Purdue University, 150 N. University St., West Lafayette, IN 47907-2067; email: mdd@math.purdue.edu

*Geometric analysis to TOBIAS COLDING, Courant Institute, New York University, 251 Mercer St., New York, NY 10012; email: colding@cims.nyu.edu

*Geometric analysis to MLADEN BESTVINA, Department of Mathematics, University of Utah, 155 South 1400 East, JWB 233, Salt Lake City, Utah 84112-0090; email: bestvina@math.utah.edu

Harmonic analysis to ALEXANDER NAGEL, Department of Mathematics, University of Wisconsin, 480 Lincoln Drive, Madison, WI 53706-1313; email: nagel@math.wisc.edu

Harmonic analysis, representation theory, and Lie theory to ROBERT J. STANTON, Department of Mathematics, The Ohio State University, 231 West 18th Avenue, Columbus, OH 43210-1174; email: stanton@math.ohio-state.edu

*Logic to STEFFEN LEMPP, Department of Mathematics, University of Wisconsin, 480 Lincoln Drive, Madison, Wisconsin 53706-1388; email: lempp@math.wisc.edu

Number theory to HAROLD G. DIAMOND, Department of Mathematics, University of Illinois, 1409 W. Green St., Urbana, IL 61801-2917; email: diamond@math.uiuc.edu

*Ordinary differential equations, and applied mathematics to PETER W. BATES, Department of Mathematics, Michigan State University, East Lansing, MI 48824-1027; email: peter@math.msu.edu

*Partial differential equations to PATRICIA E. BAUMAN, Department of Mathematics, Purdue University, West Lafayette, IN 47907-1395; email: bauman@math.purdue.edu

*Probability and statistics to KRZYSZTOF BURDZY, Department of Mathematics, University of Washington, Box 354350, Seattle, Washington 98195-4350; email: burdzy@math.washington.edu

*Real analysis and partial differential equations to DANIEL TATARU, Department of Mathematics, University of California, Berkeley, Berkeley, CA 94720; email: tataru@ math.berkeley.edu

All other communications to the editors should be addressed to the Managing Editor, WILLIAM BECKNER, Department of Mathematics, University of Texas, Austin, TX 78712-1082; email: beckner@math.utexas.edu.

Titles in This Series

803 **Michael Field and Matthew Nicol,** Ergodic theory of equivariant diffeomorphisms: Markov partitions and stable ergodicity, 2004

802 **Martin W. Liebeck and Gary M. Seitz,** The maximal subgroups of positive dimension in exceptional algebraic groups, 2004

801 **Fabio Ancona and Andrea Marson,** Well-posedness for general 2×2 systems of conservation laws, 2004

800 **V. Poénaru and C. Tanasi,** Equivariant, almost-arborescent representations of open simply-connected 3-manifolds; A finiteness result, 2004

799 **Barry Mazur and Karl Rubin,** Kolyvagin systems, 2004

798 **Benoît Mselati,** Classification and probabilistic representation of the positive solutions of a semilinear elliptic equation, 2004

797 **Ola Bratteli, Palle E. T. Jorgensen, and Vasyl' Ostrovs'kyĭ,** Representation theory and numerical AF-invariants, 2004

796 **Marc A. Rieffel,** Gromov-Hausdorff distance for quantum metric spaces/Matrix algebras converge to the sphere for quantum Gromov-Hausdorff distance, 2004

795 **Adam Nyman,** Points on quantum projectivizations, 2004

794 **Kevin K. Ferland and L. Gaunce Lewis, Jr.,** The $RO(G)$-graded equivariant ordinary homology of G-cell complexes with even-dimensional cells for $G = \mathbb{Z}/p$, 2004

793 **Jindřich Zapletal,** Descriptive set theory and definable forcing, 2004

792 **Inmaculada Baldomá and Ernest Fontich,** Exponentially small splitting of invariant manifolds of parabolic points, 2004

791 **Eva A. Gallardo-Gutiérrez and Alfonso Montes-Rodríguez,** The role of the spectrum in the cyclic behavior of composition operators, 2004

790 **Thierry Lévy,** Yang-Mills measure on compact surfaces, 2003

789 **Helge Glöckner,** Positive definite functions on infinite-dimensional convex cones, 2003

788 **Robert Denk, Matthias Hieber, and Jan Prüss,** $\mathcal{R}$-boundedness, Fourier multipliers and problems of elliptic and parabolic type, 2003

787 **Michael Cwikel, Per G. Nilsson, and Gideon Schechtman,** Interpolation of weighted Banach lattices/A characterization of relatively decomposable Banach lattices, 2003

786 **Arnd Scheel,** Radially symmetric patterns of reaction-diffusion systems, 2003

785 **R. R. Bruner and J. P. C. Greenlees,** The connective K-theory of finite groups, 2003

784 **Desmond Sheiham,** Invariants of boundary link cobordism, 2003

783 **Ethan Akin, Mike Hurley, and Judy A. Kennedy,** Dynamics of topologically generic homeomorphisms, 2003

782 **Masaaki Furusawa and Joseph A. Shalika,** On central critical values of the degree four L-functions for GSp(4): The Fundamental Lemma, 2003

781 **Marcin Bownik,** Anisotropic Hardy spaces and wavelets, 2003

780 **S. Marmi and D. Sauzin,** Quasianalytic monogenic solutions of a cohomological equation, 2003

779 **Hansjörg Geiges,** h-principles and flexibility in geometry, 2003

778 **David B. Massey,** Numerical control over complex analytic singularities, 2003

777 **Robert Lauter,** Pseudodifferential analysis on conformally compact spaces, 2003

For a complete list of titles in this series, visit the AMS Bookstore at **www.ams.org/bookstore/**.